T0186634

INTELLIGENT SYSTEMS

Technology and Applications

VOLUME IV

Database and Learning Systems

Edited by
Cornelius T. Leondes

INTELLIGENT SYSTEMS

Technology and Applications

VOLUME IV

Database and Learning Systems

CRC PRESS

Boca Raton   London   New York   Washington, D.C.

Library of Congress Cataloging-in-Publication Data

Intelligent systems : technology and applications / edited by Cornelius T. Leondes.
     p. cm.
     Includes bibliographical references and index.
     Contents: v. 1. Implementation techniques -- v. 2. Fuzzy systems, neural networks, and
expert systems -- v. 3. Signal, image, and speech processing -- v. 4. Database and
learning systems -- v. 5. Manufacturing, industrial, and management systems -- v. 6.
Control and electric power systems.
     ISBN 0-8493-1121-7 (alk. paper)
     1. Intelligent control systems. I. Leondes, Cornelius T.

TJ217.5 .I5448 2002
629.8--dc21
                                            2002017473

Visit the CRC Press Web site at www.crcpress.com

Foreword

Intelligent Systems: Technology and Applications is a significant contribution to the artificial intelligence (AI) field. Edited by Professor Cornelius Leondes, a leading contributor to intelligent systems, this set of six well-integrated volumes on the subject of intelligent systems techniques and applications provides a valuable reference for researchers and practitioners. This landmark work features contributions from more than 60 of the world's foremost AI authorities in industry, government, and academia.

Perhaps the most valuable feature of this work is the breadth of material covered. Volume I looks at the steps in implementing intelligent systems. Here the reader learns from some of the leading individuals in the field how to develop an intelligent system. Volume II covers the most important technologies in the field, including fuzzy systems, neural networks, and expert systems. In this volume the reader sees the steps taken to effectively develop each type of system, and also sees how these technologies have been successfully applied to practical real-world problems, such as intelligent signal processing, robotic control, and the operation of telecommunications systems. The final four volumes provide insight into developing and deploying intelligent systems in a wide range of application areas. For instance, Volume III discusses applications of signal, image, and speech processing; Volume IV looks at intelligent database management and learning systems; Volume V covers manufacturing, industrial, and business applications; and Volume VI considers applications in control and power systems. Collectively this material provides a tremendous resource for developing an intelligent system across a wide range of technologies and application areas.

Let us consider this work in the context of the history of artificial intelligence. AI has come a long way in a relatively short time. The early days were spent in somewhat of a probing fashion, where researchers looked for ways to develop a machine that captured human intelligence. After considerable struggle, they fortunately met with success. Armed with an understanding of how to design an intelligent system, they went on to develop useful applications to solve real-world problems. At this point AI took on a very meaningful role in the area of information technology.

Along the way there were a few individuals who saw the importance of publishing the accomplishments of AI and providing guidance to advance the field. Among this small group I believe that Dr. Leondes has made the largest contribution to this effort. He has edited numerous books on intelligent systems that provide a wealth of information to individuals in the field. I believe his latest work discussed here is his most valuable contribution to date and should be in the possession of all individuals involved in the field of intelligent systems.

Jack Durkin

Preface

For most of our history the wealth of a nation was limited by the size and stamina of the work force. Today, national wealth is measured in intellectual capital. Nations possessing skillful people in such diverse areas as science, medicine, business, and engineering, produce innovations that drive the nations to a higher quality of life. To better utilize these valuable resources, intelligent systems technology has evolved at a rapid and significantly expanding rate to accomplish this purpose. Intelligent systems technology can be utilized by nations to improve their medical care, advance their engineering technology, and increase their manufacturing productivity, as well as play a significant role in a very wide variety of other areas of activity of substantive significance.

Intelligent systems technology almost defines itself as the replication to some effective degree of human intelligence by the utilization of computers, sensor systems, effective algorithms, software technology, and other technologies in the performance of useful or significant tasks. Widely publicized earlier examples include the defeat of Garry Kasparov, arguably the greatest chess champion in history, by IBM's intelligent system known as "Big Blue." Separately, the greatest stock market crash in history, which took place on Monday, October 19, 1987, occurred because of a poorly designed intelligent system known as computerized program trading. As was reported, the Wall Street stockbrokers watched in a state of shock as the computerized program trading system took complete control of the events of the day. Alternatively, a significant example where no intelligent system was in place and which could have, indeed no doubt would have, prevented a disaster is the Chernobyl disaster which occurred at 1:15 A.M. on April 26, 1987. In this case the system operators were no doubt in a rather tired state and an effectively designed class of intelligent system known as "backward chaining" EXPERT System would, in all likelihood, have averted this disaster.

The techniques which are utilized to implement Intelligent Systems Technology include, among others:

Knowledge-Based Systems Techniques
EXPERT Systems Techniques
Fuzzy Theory Systems
Neural Network Systems
Case-Based Reasoning Methods
Induction Methods
Frame-Based Techniques
Cognition System Techniques

These techniques and others may be utilized individually or in combination with others.

The breadth of the major application areas of intelligent systems technology is remarkable and very impressive. These include:

Agriculture	Law
Business	Manufacturing
Chemistry	Mathematics
Communications	Medicine
Computer Systems	Meteorology
Education	Military
Electronics	Mining
Engineering	Power Systems
Environment	Science
Geology	Space Technology
Image Processing	Transportation
Information Management	

It is difficult now to find an area that has not been touched by Intelligent Systems Technology. Indeed, a perusal of the tables of contents of these six volumes, *Intelligent Systems: Technology and Applications*, reveals that there are substantively significant treatments of applications in many of these areas.

Needless to say, the great breadth and expanding significance of this field on the international scene requires a multi-volume set for an adequately substantive treatment of the subject of intelligent systems technology. This set of volumes consists of six distinctly titled and well-integrated volumes. It is appropriate to mention that each of the six volumes can be utilized individually. In any event, the six volume titles are:

1. Implementation Techniques
2. Fuzzy Systems, Neural Networks, and Expert Systems
3. Signal, Image, and Speech Processing
4. Database and Learning Systems
5. Manufacturing, Industrial, and Management Systems
6. Control and Electric Power Systems

The contributors to these volumes clearly reveal the effectiveness and great significance of the techniques available and, with further development, the essential role that they will play in the future. I hope that practitioners, research workers, students, computer scientists, and others on the international scene will find this set of volumes to be a unique and significant reference source for years to come.

Cornelius T. Leondes
Editor

About the Editor

Cornelius T. Leondes, B.S., M.S., Ph.D., Emeritus Professor, School of Engineering and Applied Science, University of California, Los Angeles, has served as a member or consultant on numerous national technical and scientific advisory boards. Dr. Leondes served as a consultant for numerous Fortune 500 companies and international corporations. He has published over 200 technical journal articles and has edited and/or co-authored more than 120 books. Dr. Leondes is a Guggenheim Fellow, Fulbright Research Scholar, and IEEE Fellow, as well as a recipient of the IEEE Baker Prize award and the Barry Carlton Award of the IEEE.

Contributors

Ho Tu Bao
Japan Advanced Institute of
  Science and Technology
Ishikawa, Japan

Nick Bassiliades
Aristotle University of Thessaloniki
Thessaloniki, Greece

Shyi-Ming Chen
National Taiwan University of
  Science and Technology
Taipei, Taiwan, R.O.C.

Su-Shing Chen
University of Missouri
Columbia, Missouri

Eliseo Clementini
University of L'Aquila
L'Aquila, Italy

Chao Deng
Columbia Genome Center
Columbia University
New York

Shing-Hwang Doong
Shu-Te University
Kaohsiung County, Taiwan, R.O.C.

Nguyen Trong Dung
Japan Advanced Institute of Science
  and Technology
Ishikawa, Japan

Frank Höppner
University of Applied Sciences
Wolfenbüttel, Germany

Annette Keller
German Aerospace Center
Braunschweig, Germany

Frank Klawonn
University of Applied
  Sciences
Wolfenbüttel, Germany

Fotis Kokkoras
Aristotle University of Thessaloniki
Thessaloniki, Greece

Chih-Chin Lai
Shu-Te University
Kaohsiung County, Taiwan, R.O.C.

Yun-Shyang Lin
National Taiwan University of
  Science and Technology
Taipei, Taiwan, R.O.C.

Chai Quek
Nanyang Technological University
Singapore

Othoniel Rodriguez
University of Missouri
Columbia, Missouri

Dimitrios Sampson
Informatics and Telematics Institute
Thessaloniki, Greece

Yi Shang
University of Missouri
Columbia, Missouri

Hongchi Shi
University of Missouri
Columbia, Missouri

Ioannis Vlahavas
Aristotle University of Thessaloniki
Thessaloniki, Greece

L.H. Wong
Institute of Systems Science
Singapore

Chih-Hung Wu
Shu-Te University
Kaohsiung County, Taiwan, R.O.C.

Jae Dong Yang
Chonbuk National University
Chonju, South Korea

Ming-Shiou Yeh
National Chiao Tung University
Hsinchu, Taiwan, R.O.C.

Contents

Contents

Volume I: Implementation Techniques

Contents

Volume II: Fuzzy Systems, Neural Networks, and Expert Systems

Contents

Contents

Contents

1

Model Selection in Knowledge Discovery and Data Mining

Ho Tu Bao
Japan Advanced Institute of Science and Technology

Nguyen Trong Dung
Japan Advanced Institute of Science and Technology

Abstract.   The process of knowledge discovery in databases inherently consists of several steps that are necessarily iterative and interactive. In each application, to go through this process, the user has to exploit different algorithms and their settings that usually yield different discovered models. The selection of appropriate discovered models or algorithms to achieve such models, referred to as model selection—which requires meta-knowledge on algorithms/models and model performance metrics—is generally a difficult task for the user. Taking this difficulty into account, we consider that the ease of model selection is crucial in the success of real-life knowledge discovery activities. Different from most related work that aims at automatic model selection, in our view, model selection should be a semiautomatic work requiring an effective collaboration between the user and the discovery system. For such collaboration, our solution is to give the user the ability to try various alternatives easily and to compare competing models quantitatively by performance metrics and qualitatively by effective visualization. This chapter presents our research on model selection in the development of a knowledge discovery system called D2MS. The chapter first addresses the motivation of model selection in knowledge discovery and related work, an overview of D2MS, and its solution to model selection and visualization. It then presents the usefulness of D2MS model selection in two case studies of discovering medical knowledge from meningitis and stomach cancer data using three methods, i.e., decision trees, conceptual clustering, and rule induction.

1.1   Introduction

Knowledge discovery in databases (KDD)—the rapidly growing interdisciplinary field of computing that evolves from its roots in database management, statistics, and machine learning—aims at finding useful knowledge from large databases. The process of knowledge discovery is complicated and should be seen inherently as a process containing several steps. The first step is to understand the application domain, to formulate the problem, and to collect data. The second step is to preprocess the data. The third step is that of data mining with the aim of extracting useful knowledge as patterns or models hidden in data. The fourth step is to post-process discovered knowledge, and the fifth step is to put discovered knowledge in practical use (Mannila, 1997). These steps are inherently iterative and interactive, i.e., one cannot expect to extract useful knowledge by just pushing a large amount of data into a black box once without the user's participation (Figure IV.1.1).

In this work, we adopt the view of Hand (1998) regarding a pattern as a local structure and a model as a global representation of a structure. While a pattern is related to just a handful of variables and a few cases such as a rule, a model summarizes the systematic component underlying the data or describes how the data may have arisen, e.g., a decision tree or a regression hyperplane. In a broader sense, we will refer to such a model or the set of patterns discovered in an application as model.

Two KDD primary goals of prediction and description are concerned with different tasks in the data mining step, such as those for characterization, discrimination, association, classification, and clustering (Han and Kamber, 2001). Also, there are different tasks of data cleaning, integration, transformation, and reduction in the preprocessing step, and those of interpretation, evaluation, exportation, and visualization of results in the post-processing step. Moreover, each of these tasks can be done with different methods and algorithms. To solve a given discovery problem, the user usually has to go through these steps several times, each time corresponding to an exploitation of a series of algorithms.

It is known that there is no inherent superior method/model in terms of generalization performance. The "no free lunch" theorem (Wolpert, 1995) states that in the absence of prior information about the problem, there are no reasons to prefer one learning algorithm or classifier model over another. The problem of model selection—choosing appropriate discovered models or algorithms and their settings for obtaining such models in a given application—is difficult and non-trivial because it requires empirical comparative evaluation of discovered models and meta-knowledge on models/algorithms. The user often has to go through a trial-and-error process to select the most suitable models from competing ones. Clearly, trying all possible options is impractical, and an informed search process is needed to attain expected models. Informed search requires both performance metrics and model characteristics that are often not available for the user. Moreover, the user's interest in discovered

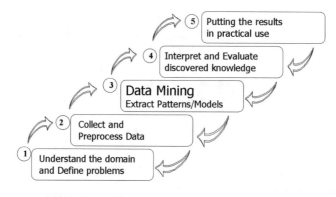

FIGURE IV.1.1   Main steps of the KDD process.

models is a subjective matter that depends much on his/her domain knowledge and sometimes is very independent of performance metrics provided by the system. Current data mining provides multiple algorithms within a single system, but the selection and combination of these algorithms are external to the system and specified by the user. This makes the KDD process difficult and possibly less efficient in practice.

Unlike the major research tendency that aims to provide the user with meta-knowledge for an automatic model selection, as described in the next section, in our view, model selection should be semiautomatic and requires an effective collaboration between the user and the discovery system. In such a collaboration, visualization plays an indispensable role because it can give a deep understanding of complicated models that the user cannot have if using only performance metrics. The research on visualization integrated with model selection is significant because there is currently very limited visualization support for the process of building and selecting models in knowledge discovery (Crapo et al., 2000).

The goal of this work is to develop a research system for knowledge discovery with support for model selection. The system called D2MS (data mining with model selection) first provides the user with the ability of trying various alternatives of algorithm combinations and their settings. Each combination of algorithms is registered in a list called a plan. After executing registered plans, the user is supported in evaluating competing models quantitatively and qualitatively before making his/her final selection. The quantitative evaluation can be obtained by performance metrics provided by the system, while the qualitative evaluation can be obtained by effective visualization of the discovered models.

Within D2MS, several data mining methods have been implemented, including the decision tree induction method CABRO (Nguyen and Ho, 1999), CABROrule (Nguyen et al., 2001), the conceptual clustering method OSHAM (Ho, 1997) and its variants (Ho, 2000), and the rule induction method LUPC (Ho et al., 2001). These programs have been integrated with several advanced visualization techniques in D2MS such as tightly coupled views (Kumar et al., 1997), fish-eye views (Furnas, 1981), and, particularly, the proposed technique T2.5D (Tree 2.5 Dimensions) for large hierarchical structures (Nguyen et al., 2000).

This chapter presents model selection in knowledge discovery system D2MS. Section 1.2 discusses related work, and Section 1.3 gives an overview of the system. Section 1.4 introduces D2MS's solution to model selection integrated with visualization. Section 1.5 briefly summarizes three data mining methods—decision tree induction, conceptual clustering, and rule induction—that were implemented in D2MS and presents two case studies of discovering knowledge in meningitis and stomach cancer data with these methods. Section 1.6 summarizes the chapter and outlines further research.

1.2 Related Work

Model selection is a general statistical notion, and a probability model is generally a mathematical expression representing some underlying phenomenon (Hand, 1997). The model selection has usually been viewed in the statistics community in terms of estimation of expected discrepancy $E\Delta$ (f, g_θ) between the underlying model f (that is unknown in practice) and approximating models g_θ, where θ is the parameter vector (Zucchini, 2000). The standard methods of model selection in statistics include classical hypothesis testing, maximum likelihood, Bayes method, minimum description length, cross-validation, bootstrap, and Akaike's information criteria (see, for example, Forster, 2000; Zucchini, 2000).

The model selection problem has attracted researchers in machine learning and, recently, in knowledge discovery by its significant importance. Most research on model selection in this community has focused on two issues, the preselection of models/algorithms, which is typically concerned with finding measures of relation between the data set characteristics and the models'/algorithms' performance (referred to as model characteristics), and the post-selection, concerned with the evaluation of models/algorithms using multi-criteria of performance (referred to as performance metrics).

Brodley (1995) introduced an approach that uses knowledge about the representational biases of a set of learning algorithms coded in form of rules to model/algorithm preselection. Brazdil and Soares (2000) introduced three ranking methods of classification algorithms using average ranks, success rate ratios, and significant wins. These methods generate rankings by results obtained by algorithms on the training datasets, and the rankings are evaluated by their distance to the ideal ranking built in advance. Hilario and Kalousis (2000) proposed using learning algorithm profiles to address the model selection problem. The algorithm profiles consist of meta-level feature-value vectors that describe learning algorithms in terms of their representation, functionality, efficiency, robustness, and practicality. NOEMON (Kalousis and Theoharis, 1999) is a prototype system in the context of the classification task that suggests the most appropriate classifier, decided on the basis of morphological similarity between the new data set and the existing collection.

Most research on performance metrics focuses on multi-criteria for doing post-selection of algorithms. Nakhaeizadeh and Schnabl (1997) proposed an evaluation multi-criteria metric that aims to take into account both positive and negative properties of data mining algorithms and provide a fast and comprehensive evaluation of various algorithms. Kohavi (1995) strengthened the practical role of stratified cross-validation in model selection for many common application domains by performing a large number of careful experiments.

Most of the mentioned related work aims to find meta-knowledge towards an automatic selection of models/algorithms and does not take into account the participation of the user during model selection. Also, researchers often limit themselves to the classification problem with supervised data, where performance metrics can be computed and served for the prediction goal. The situation is somewhat different in the case of the clustering problem, where the main goal is description, and several usual performance metrics cannot be computed from unsupervised data, e.g., accuracy. Besides model characteristics and performance metrics for a quantitative evaluation, the major distinguishing feature of the D2MS system is its tight integration of visualization with model selection that supports a qualitative evaluation so that the user can make a better final selection based on his/her insight into discovered models.

Many KDD systems provide visualization of data and knowledge, and different visualization techniques have been developed or adapted to the KDD process. System MineSet (Brunk et al., 1997) provides several 3D visualizers, in particular a 3D tree visualizer. Ankerst et al. (2000) developed an interactive visualization in decision tree construction for supporting an effective cooperation of the user and the computer in classification. Han and Cercone (2000) integrated tight visualization with five steps of the KDD process.

Our D2MS shares common features with the above systems in human–machine interaction and process visualization and contributes two additional features. First, it integrates visualization into model selection by providing visual information of competing models for a qualitative evaluation of the user. Second, it provides the effective view T2.5D for large hierarchical structures that can be seen as an alternative to powerful techniques for representing large hierarchical structures such as cone trees (Robertson et al., 1991) or hyperbolic trees (Lamping and Rao, 1997). For large trees, T2.5D can give a more effective view than that of 3D visualization (Nguyen et al., 2000).

D2MS is a KDD research system that shares several common points with some KDD research systems. MLC++ (Kohavi et al., 1997) is a powerful library of classification algorithms with a guide to compare and select appropriate algorithms for the classification task. It also provides visualization and interfaces between programs for both end users and software developers. Sharing the class library approach, while MLC++ is only suitable for well-trained programmers, D2MS orients to the user without much computer knowledge. DBMiner (Han and Kamber, 2001) is a KDD system that provides a wide spectrum of data mining functions. As interactive data mining environments, DBMiner focuses on a tight integration of data mining methods with online analytical processing, while D2MS focuses on a tight integration of data mining methods with model selection and visualization. FlexiMine (Domshlak et al., 1998) is a flexible platform for KDD research and application development. It is designed as a testbed for data mining research and a generic knowledge discovery tool. Sharing with FlexiMine the integration of most KDD functions and the flexibility, D2MS is a more user-centered system.

1.3 Overview of the System

Figure IV.1.2 presents the conceptual architecture of D2MS. The system consists of eight modules: graphical user interface, data interface, data preprocessing, data mining, post-processing and evaluation, plan management, visualization, and application. Additionally, it has a plan base and a model base to store registered plans and discovered models, respectively. Importantly, these modules follow the cooperative mechanism: the modules are constructed in such a way that each of them can cooperate with the others to achieve the final goal.

D2MS is able to deal with various aspects on knowledge discovery: large data sets and high dimensionality, the degree of supervision, different types of data in different forms, the changing nature of data and knowledge, and interaction and visualization (Ho, 2000). This makes the choice of algorithms a strong influence on discovered results.

Data interface—The native input format of the system is a flat form of data files such as that of C4.5 (Quinlan, 1993) data files format (filestem.data and filestem.names). The system supports other common database formats by using standard interfaces (e.g., ODBC).

Graphical user interface—A graphical user interface is designed to facilitate the complicated interaction between the user and the system in the knowledge discovery process. It allows the user to register different combinations of algorithms in the KDD process and their parameter settings via plans. Also, it provides the user with facilities to participate in the KDD process. More importantly, the graphical user interface supports the user in evaluating competing models in order to select the appropriate ones.

Data preprocessing—This module is to provide different programs for four main tasks of data cleaning, integration, transformation, and reduction in the preprocessing step. The current version of D2MS has algorithms to solve the most common problems in preprocessing: discretization of continuous attributes, filling up missing values, and instance and feature selection. Discretization

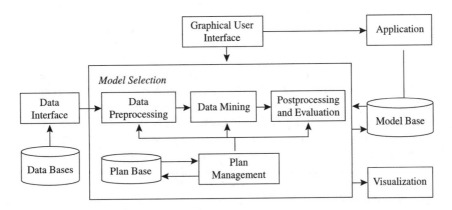

FIGURE IV.1.2 Conceptual architecture of D2MS.

methods include rough-set based and entropy-based programs for supervised data, and k-means based discretization program for unsupervised data (Dougherty et al., 1995). Missing value techniques include three programs for filling up missing values (Fujikawa, 2001): natural-cluster based mean-and-mode (NCBMM), attribute-rank cluster-based mean-and-mode (ARBMM), and k-means clustering-based mean-and-mode (kMCMM). Sampling programs and feature selection method SFG (Liu and Motoda, 1998) are available in D2MS.

Data mining—The system provides several algorithms for prediction and description. In the current version of D2MS, classification algorithms consist of k-nearest neighbors, Bayesian classification, decision tree induction method CABRO (Nguyen and Ho, 1999), CABROrule (Nguyen et al., 2001), and LUPC (Ho et al., 2001) where CABRO and CABROrule have their variants with R-measure, gain ratio (Quinlan, 1993), gini-index (Breiman et al., 1984), and chi2 (Minger, 1989a), and each of them can produce trees or rule sets by either error-based pruning (Quinlan, 1993) or the post-pruning technique (Furnkranz, 1999). Clustering algorithms include k-means and its variants for mixed numerical and categorical attributes using an efficient mixed similarity measure MSM (Nguyen and Ho, 2000), and the conceptual clustering algorithm OSHAM (Ho, 1997; Ho, 2000). These programs can read data in the common input form and represent discovered models in a form readable by other related modules (post-processing, visualization, and plan management).

Post-processing and evaluation—This module provides tools for post-processing and evaluation of discovered models. Its available tools include those for k-fold stratified cross-validation integrated with data mining programs, for post-pruning of decision trees, for automatically generating tables containing synthesized information of discovered models, for exporting competing models into other easy-to-understand forms for the user, e.g., the spreadsheet format (Ho, 2000), and for generating forms readable for visualization programs, e.g., T2.5D for visualizing the hierarchical structures (Nguyen et al., 2000).

Plan management—The main function of this module is to manage a plan base containing plans declared by the user. It also coordinates modules on data mining, evaluation, and visualization in model selection.

Data and knowledge visualization—This module consists of three visualizers for data, rules, and hierarchical structures. The algorithms in other modules can frequently invoke this module during their running in order to help the user take part in it effectively.

Application—This module contains utilities that help the user use discovered models. A model can be used to match an unknown instance or to generate an interactive dialogue where the system conducts a series of questions/answers in order to predict the outcome.

1.4 Model Selection in D2MS

1.4.1 Model Selection

The interesting nature of discovered patterns/models is commonly characterized by several criteria: evidence indicates the significance of a finding measured by a statistical criterion; redundancy amounts to the similarity of a finding with respect to other findings and measures to what degree one finding follows from another; usefulness relates a finding to the goal of the users; novelty includes the deviation from prior knowledge of the user or system; simplicity refers to the syntactical complexity of the presentation of a finding; and generality is determined by the fraction of the population a finding refers to. The interestingness can be seen as a function of the above criteria (Fayyad et al., 1996) and strongly depends on the user as well as his/her domain knowledge.

Working with medical experts for more than two years, we eventually came to realize that their evaluations are not always concerned with performance metrics provided by discovery systems. In our case studies on meningitis and stomach cancer domains presented in Section 1.5, domain experts were often interested in models that do not necessarily have the highest predictive accuracy but are related to their background knowledge or novel knowledge. It turns out that model selection in a KDD system should be more a user-centered process in which the participation of the user is crucial.

The key idea of our solution to model selection in D2MS is to support an effective participation of the user in this process. Concretely, D2MS first supports the user in doing trials on combinations of algorithms and their parameter settings in order to produce competing models and then supports the user to evaluate them quantitatively and qualitatively by providing both performance criteria as well as visualization of these models.

The model selection in D2MS mainly involves three steps of data processing, data mining, and post-processing, as shown in Figure IV.1.1. There are three phases of doing model selection in D2MS, and all are managed by the plan management module: (1) registering plans of selected algorithms and their settings; (2) executing the plans to discover models; (3) selecting appropriate models by a comparative evaluation of competing models.

The first phase is to register plans, and this is commonly done before the data preprocessing step. A plan is an ordered list of algorithms associated with their parameter settings that can yield a model or an intermediate result when being executed. The plans are represented in a tree form called a plan tree whose nodes are selected algorithms associated with their settings (the top-left window in Figure IV.1.3). The nodes on a path of the plan tree must follow the order of preprocessing, data mining, and post-processing. A plan may contain several algorithms (nodes) of the preprocessing and post-processing steps, for example, filling missing values by natural-cluster based mean-and-mode then discretizing continuous attributes using an entropy-based algorithm in preprocessing.

The plan management module provides the user with a friendly user interface to register, modify, or delete plans on the plan tree. D2MS allows the user to register plans fully or gradually. A plan can be fully registered by immediately determining all of its component algorithms and their settings so that it can yield a model. However, a plan can be registered gradually by determining algorithms and their settings step-by-step. This mode of registration is useful in mining large data sets. For example, to do a task in the KDD process, the user can try several algorithms/settings, then evaluate their results, and select only promising ones for doing further tasks.

The plan management module maintains profiles of algorithms available in preprocessing, data mining, and post-processing of D2MS. An algorithm profile contains information about the algorithm functions, its requirements, types, and effect of its parameters. The registration of an algorithm is done via a dialog box when it is added into a plan. For example, to generate a decision tree by CABRO, the user needs to choose the minimum number of instances at leaf nodes, the lowest accepted error rate, which attributes to be grouped, encoding missing values, etc. The top-left window in Figure IV.1.3 shows a plan tree created for mining the meningitis data (Section 1.5) on which the plans consist of NCBMM for inputting missing values, a discretization based on entropy and rough sets, and CABRO with different selection measures or LUPC with several parameter settings.

The second phase is to execute registered plans. While gradually registering a plan, the user can run an algorithm just after adding it to the plan, then evaluate its results before deciding whether to continue this plan from its current stage with other algorithms or to backtrack and try the plan with another algorithm or setting. The user also can run a plan after fully registering it, or even register a number of plans then run them altogether. The intermediate results, the discovered models, and their summaries and exported forms will be automatically created and stored in the model base.

The third phase is to select appropriate models by the user. D2MS provides a summary table presenting performance metrics of discovered models according to executed plans (the bottom-left window in Figure IV.1.3). However, the user can evaluate each model in depth by visualizing it, browsing its structure, checking its relationship with the data set, etc. (the top-right and bottom-right

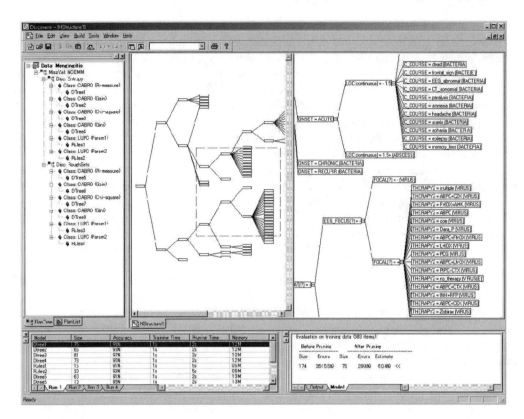

FIGURE IV.1.3   A screen shot of D2MS in selecting models learned from meningitis data: the top-left window shows the plan tree; the bottom-left window shows the summary table of performance metrics of discovered models; the bottom-right window shows detailed information of the model being activated (highlighted); and the top-right window shows tightly coupled views of the model where the detail view in the right part corresponds to the field-of-view specified in the left part.

windows in Figure IV.1.3). The user can also visualize several models simultaneously for comparing them (see Figure IV.1.5). By getting a real insight into the competing models, the user certainly can make a better selection of models.

The performance metrics on discovered models are reported according to the types of algorithms. D2MS reports a multidimensional measure for classification of supervised algorithms consisting of the following submeasures: (1) accuracy, (2) size (number of leaf nodes or rules), (3) training time, (4) pruning time, and (5) resource demand (memory) as well as that for unsupervised clustering algorithms with the following submeasures: (1) size (number of clusters), (2) number of hierarchy levels, (3) training time, and (4) resource demand (memory).

Section 1.4.2 addresses the visualization in model selection, and Section 1.5 will illustrate the D2MS's model selection with more detail in medicine applications.

1.4.2   Visualization Support for Model Selection

In D2MS, visualization helps the user interpret data and models as well their connections in order to understand and evaluate models better. Through visualization, the user can also better see the effect of algorithm settings on resultant models so that he/she can adjust settings to reach adequate models.

After a description of available visualizers in D2MS, this subsection focuses on two highlighted characteristics of its visualization: visualization of hierarchical structures and a tight integration of visualization with model selection.

1.4.2.1 Visualizers

There are three visualizers in D2MS: the data visualizer, the rule visualizer, and the hierarchy visualizer. The data visualizer provides the user with graphical views on the statistics of the input data and relations between attributes. These include multiple forms of viewing data such as tables, cross-tabulations, pie graphs, charts, and cubes. The data visualizer supports the user in the pre-processing step and helps to select algorithms when registering plans. The rule visualizer allows the user to view rules generated by CABROrule or LUPC. The visualizer provides graphical views on statistics of conditions and conclusions of a rule, correctly and wrongly matched cases in both training and testing data, and links between rules and data. The hierarchy visualizer visualizes hierarchical structures generated by CABRO and OSHAM. The visualizer provides different views that may be suitable for different types and sizes of hierarchies. The user can view an overall structure of a hierarchy together with the detailed information about each node. D2MS provides several visualization techniques that allow the user to effectively visualize large hierarchical structures.

The tightly coupled views (Kumar et al., 1997) simultaneously display a hierarchy in normal size and tiny size that allows the user to quickly determine the field-of-view and to pan to the region of interest (Figure IV.1.3). The fisheye view (Furnas, 1981) distorts the magnified image so that the center of interest is displayed at a high magnification, and the rest of the image is progressively compressed. Also, the new technique T2.5D (Nguyen et al., 2000) is implemented in D2MS for visualizing very large hierarchical structures.

1.4.2.2 Tree 2.5 Dimensions

The user might find it difficult to navigate a very large hierarchy, even with tightly coupled and fish-eye views. To overcome this difficulty, we have been developing a new technique called T2.5D (tree 2.5 dimensions).

T2.5D is inspired by the work of Reingold and Tilford (1991) that draws tidy trees in a reasonable time and storage. Different from tightly coupled and fisheye views that are considered location-based views, T2.5D can be seen as a relation-based view. The starting point of T2.5D is the observation that a large tree consists of many subtrees that are not usually or necessarily viewed simultaneously.

The key idea of T2.5D is to represent a large tree in a virtual 3D space (subtrees are overlapped to reduce occupied space), while each subtree of interest is displayed in a 2D space. To this end, T2.5D determines the fixed position of each subtree (its root node) in two axes X and Y, and, in addition, it dynamically computes a Z-order for this subtree in an imaginary axis Z. A subtree with a given Z-order is displayed "above" its siblings that have higher Z-orders.

When visualizing and navigating a tree, at each moment, the Z-order of all nodes on the path from the root to a node in focus in the tree is set to zero by T2.5D. The active wide path to a node in focus, which contains all nodes on the path from the root to this node in focus and their siblings, is displayed in the front of the screen with highlighted colors to give the user a clear view. Other parts of the tree remain in the background to provide an image of the overall structure. With Z-order, T2.5D can give the user an impression that trees are drawn in a 3D space. The user can easily change the active wide path by choosing another node in focus.

We have experimented T2.5D with various real and artificial data sets, and it was shown that T2.5D can handle trees with nearly 20,000 nodes (from U.S. census data), and more than 1000 nodes can be displayed together on the screen. Figure IV.1.4 illustrates a pruned tree of 1795 nodes learned from stomach cancer data and drawn by T2.5D (note that the original screen with colors gives a better view than this black and white screen).

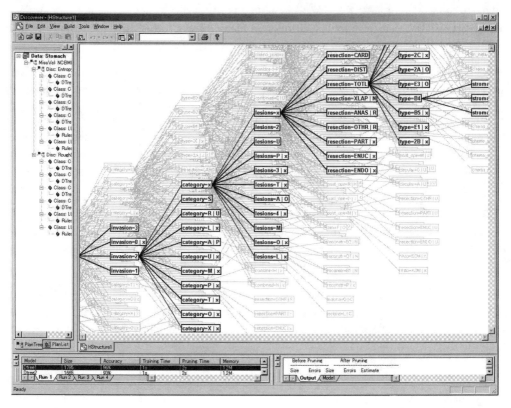

FIGURE IV.1.4   T2.5D provides views in between 2D and 3D. The pruned tree learned from the stomach cancer data by CABRO has 1795 nodes where the active wide path to a node in focus displays all of its direct relatives in a 2D view.

1.4.2.3   Visualization in Model Selection

In D2MS, visualization is integrated with the steps of the KDD process and closely associated with the plan management module in support for model selection. The user can have either views of executing a plan or comparative views of discovered models. If the user is interested in following the execution of one plan, he/she can view, for example, the input data, the derived data after preprocessing, the generated models with chosen settings, or the exported results. Thus, the user can follow and verify the process of discovery by each plan, and change settings to reach alternative results.

If the user is interested in comparative evaluation of competing models generated by different plans, he/she can have multiple views on these models. The user can compare performance metrics of all activated plans that are always available in the summary table. Whenever the user highlights a row in the table, he/she will see the associated model. Several windows can be opened simultaneously to display competing models in forms of trees, concept hierarchies, or rule sets. Figure IV.1.5 illustrates a view of two different concept hierarchies generated by OSHAM from the meningitis data with different sizes and structures.

1.5   Case Studies

This section illustrates the utility of model selection and visualization of D2MS in two knowledge discovery case studies. It starts with a brief introduction to three data mining methods, CABRO,

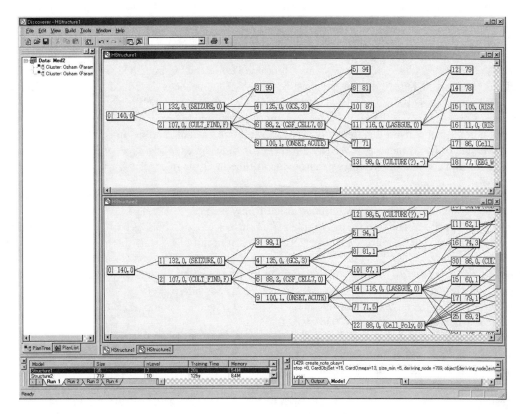

FIGURE IV.1.5   Comparative view of competing models obtained by different OSHAM parameter settings in mining useful concepts from the meningitis data. Two concept hierarchies show different possible descriptions in terms of size and structure.

LUPC, and OSHAM, implemented in D2MS and two medical data sets on meningitis and stomach cancer. In these two case studies, D2MS shows many advantages in comparison with other systems.

1.5.1   CABRO, LUPC, and OSHAM

1.5.1.1   Decision Tree Induction Method CABRO

Decision tree induction (DTI) is one of the most widely used knowledge discovery methods for supervised data. From a given set of labelled instances, DTI induces a model (classifier) in the form of a decision tree that predicts classes of unknown instances. Typical decision tree methods include C4.5 (Quinlan, 1993) and CART (Breiman et al., 1984).

The two main problems in DTI are attribute selection and pruning. Attribute selection means to decide which attribute will be chosen to branch a decision node (Minger, 1989a). Pruning means to avoid over-fitting by reducing the complexity of the tree (Minger, 1989b). In our current system, decision tree program CABRO (construction automatique d'une base de règles à partir d'observations), or its variant CABROrule (Nguyen and Ho, 1999; Nguyen et al., 2001), allows the user to generate different tree models or rule sets by combining one of four attribute selection measures including gain-ratio (Quinlan, 1993), the gini-index (Breiman et al., 1984), Chi2 (Minger, 1998a), and R-measure (Nguyen and Ho, 1999) with one of pruning algorithms including error-complexity, reduced-error, and pessimistic error (Breiman et al., 1984; Quinlan, 1993).

The techniques resulting from our research on DTI have been implemented in the system: efficient visualization of hierarchical structures including T2.5D views and rule post-pruning of large trees in the post-processing step (Nguyen et al., 2000; 2001). They offer advantages in data mining where data sets are often large and, consequently, produce large decision trees. One example is the case of the U.S. census bureau database consisting of 199,523 instances described by 40 numeric and symbolic attributes (103 Mbytes). On that database, C4.5 (Quinlan, 1993) produces a pruned tree of about 18,500 decision and leaf nodes that is almost impossible to be comprehended in its text form. Also, it could not convert this tree into a rule set after two days of running on a Sun workstation of 128 Mbytes of RAM memory. CABRO, however, provides an excellent view of its discovered tree by T2.5D and converts it into rules in a near-linear time of O(nlogn).

1.5.1.2   Rule Induction Method LUPC

Another approach to learning descriptive rules from supervised data is LUPC. LUPC (learning unbalanced positive class) considers the rule induction task as a two-classes problem. LUPC permits two learning modes: learning one target class or learning all classes. If the user is interested in learning one target class, this class will be considered as the positive class C+ and all other classes as the negative class C−. With a suitable voting procedure and by sequentially taking each class as the target class, LUPC can learn all classes. It has been shown to be very effective when learning a minority class as the target class in an unbalanced data set, and it is comparable to other well-known methods when learning all classes (Ho et al., 2001).

LUPC yields different rule sets depending on the settings of its four parameters: (1) minimum accuracy of a rule α; (2) minimum coverage of a rule β; (3) number of candidate conditions η; and (4) number of candidate rules γ. There are three main features that distinguish LUPC from related methods and allow it to learn effectively in unbalanced data sets. First, it efficiently combines separate-and-conquer rule induction (Furnkranz, 1999) with association rule mining by finding $\alpha\beta$-strong rules biased on accuracy and cover ratio with adaptive thresholds. Second, it exploits a proven property of imbalance[1] to make an effective search with a large beam search parameter and to use data in negative class to speed up the mining process on a minority positive class. Third, it integrates prepruning and post-pruning in a way that can avoid over-pruning.

1.5.1.3   Conceptual Clustering Method OSHAM

Conceptual clustering is a typical knowledge discovery method for unsupervised data. Its basic task is to simultaneously find, from a given set of unlabelled instances, a hierarchical model that determines useful object subsets and their intensional definitions. These two problems relate to another crucial problem of interpreting concept hierarchies discovered by conceptual clustering methods.

The method called OSHAM (making automatically hierarchies of structured objects, in reverse) employs a hybrid representation of concepts that combines advantages of classical, prototype, and exemplar views on concepts (Ho, 1997). Depending on parameter settings, OSHAM can extract non-disjoint or disjoint hierarchical models of concepts. OSHAM is a non-incremental divisive algorithm that works recursively, and, at each step, it seeks an acceptable solution according to a quality function defined on its hybrid representation.

The form and the size of OSHAM's hierarchical models depend on plans and settings of the following parameters: (1) the discovered hierarchy is disjoint or non-disjoint; (2) the minimum size of each node; (3) the threshold about the concept dispersion; and (4) the number of competitors for beam search. OSHAM is associated with an interpretation procedure to use discovered models.

---

[1]Given α, a rule R is not $\alpha\beta$-strong for any arbitrary β if $\text{cov} - (R) \geq ((1 - \alpha)/\alpha) \times \text{cov} + (R)$, where $\text{cov} + (R)$ and $\text{cov} - (R)$ denote the positive and negative coverage of the rule R.

There are a couple of variants of OSHAM: incremental I-OSHAM that can learn when databases are incrementally updated, and approximate A-OSHAM that can learn approximate concepts when data are uncertain and imprecise (Ho, 2000).

Like other unsupervised methods, OSHAM yields descriptive models with description of nodes in the hierarchy, but there is not any evaluation of confidence of discovered models as in those available for models obtained by supervised mining methods, e.g., CABRO and LUPC. In such cases, to obtain an appropriate model, the user needs to interact with the system and try different plans with various parameters. The plan management module facilitates this task by its support for comparing models, data, and hierarchy visualization.

1.5.2   The Meningitis and Stomach Cancer Data Sets

1.5.2.1   Meningitis Data

The meningitis data set was collected at the Medical Research Institute, Tokyo Medical and Dental University, from 1979 to 1993. It contains data of patients who suffered from meningitis and who were admitted to the department of emergency and neurology in several hospitals. There are 38 numeric and categorical attributes presenting patients' history, physical examination, laboratory examination, diagnosis, therapy, clinical course, final status, and risk factors. The objectives of medical experts are to extract factors important for (1) diagnosis of meningitis, (2) detection of bacteria or virus, and (3) predicting prognosis (Tsumoto, 2000).

1.5.2.2   Stomach Cancer Data

Every year, about 50,000 people in Japan die of stomach cancer. The stomach cancer data set collected at the National Cancer Center in Tokyo during the period 1962–1991 is a very precious source for research. It contains data of 7520 patients described originally by 83 numeric and categorical attributes. These include information on patients, symptoms, type of cancer, longitudinal and circular location, serosal invasion, metastasis, pre-operative complication, post-operative complication, etc. One problem is the use of attributes containing patient information before operation to predict the patient status after the operation. The domain experts are particularly interested in finding predictive and descriptive rules for the class of patients who died within 90 days after operation amidst a total of 5 classes.

1.5.3   Knowledge Extraction from the Meningitis Data

The main feature of the problem of knowledge extraction from meningitis data is its multi-tasks. Each of the tasks (1), (2), and (3) can be done by two alternative target attributes. For example, the task of mining diagnosis knowledge can be done by using either attribute DIAG2 that groups diagnosis results into two classes, VIRUS and BACTERIA, or attribute DIAG that groups diagnosis results into six classes (ABSCESS, BACTERIA, BACTE(E), TB(E), VIRUS(E), and VIRUS). Similarly, two tasks of mining predictive and descriptive knowledge on the detection of bacteria/virus and prognosis can be carried out with pairs of attributes CULTFIND and CULTURE (2 and 13 classes) and COURSE and CCOURSE (2 and 12 classes), respectively. There are a total of six target attributes in this data set. To find interesting knowledge in such a complicated problem, the interaction and iteration with various parameter settings of data mining methods are indispensable in order to generate and select models. In our experiments, it was very difficult to run iteratively with different parameter settings and to evaluate comratively other data mining systems such as C4.5 (Quinlan, 1993), its commercial version See5.0, or CBA (Liu et al., 1998).

Different plans have been created in D2MS for doing these tasks. Each of them is typically a sequence of methods for discretization of numerical attributes, data mining, and visualization of extracted knowledge or its exportation to the Excel format if applicable. Two discretization methods in D2MS yield two different derived data sets from the meningitis data. The derived data set based on entropy-based methods ignores many attributes and often has few discretized values on the others. The other set, based on rough set-based methods, discretizes continuous attributes into more values and does not ignore any attributes. When applying data mining methods CABRO (CABROrule), LUPC, and OSHAM with their different settings to these two derived data sets, different possible models can be generated.

The plans associated with CABRO produce eight competing decision trees for each target attribute by alternatively using four attribute selection measures (R-measure, gain ratio, gini-index, and χ^2) on two derived data sets. Thus, only with the default parameter setting, 48 models can be generated for 6 target attributes by CABRO. Figure IV.1.3 illustrates such generated decision tree models. The bottom-left window reports the performance metrics on each discovered model. Each of them, when highlighted, will be visualized on the top-right window, and detailed information of the model can be seen in the bottom-right window. The generated trees can be visualized and navigated by fisheyes and T2.5D visualizers. The utility of D2MS is supported by the fact that it is relatively easy to induce decision trees from a data set, but it is much more difficult to select the best among them. The convenient ways of evaluating and comparing generated decision trees in D2MS allow the user to do this work efficiently.

The plans associated with LUPC (also CABROrule) produce competing rule sets. LUPC runs in two modes: (1) learning all classes, and (2) learning one target class of the target attribute (one model generated for one class). D2MS allows us to easily try LUPC with various parameter settings and to evaluate the obtained rule sets by the performance metrics and visualization of rules and data. With the easy evaluation of multiple models, we have determined a good parameter setting for this data set: minimum accuracy of a rule $\alpha = 95\%$, minimum coverage of a rule $\beta = 2$, number of candidate attribute–value pairs $\eta = 100$, number of candidate rules $\gamma = 30$. With these parameters, we obtained 73 models for all target attributes from two derived data sets by two learning modes (some classes with very few training instances cannot be learned for target attributes DIAG, CULTURE, and CCOURSE). Such a number of models can be changed when parameter settings change. Rules found for identifying factors important for diagnosis, detection of bacteria or virus, and predicting prognosis were highly accepted by domain experts (Ho et al., 2001).

Each concept hierarchy presents a hierarchical model of concepts with multiple links (Ho, 1997). Each concept is a cluster of related objects characterized by a conjunction of common attribute–value pairs (the most significant attribute–value pairs for the cluster). Below is an example of a concept found by OSHAM from the meningitis data with parameters (1, 7, 0.3, 3) that can be seen when navigating the concept hierarchy. The concept covers 88 cases, of which 2 are considered as exceptional (they will not be classified further in the next levels).

```
CONCEPT 59
Level = 4, SuperConcepts = {25}, SubConcepts = {60 61 62 63 64}
Concept_dispersion = 0.297297
Local_instance_dispersion = 0.297297
Concept probability = 0.628571
Concept local conditional probability = 1.000000
Features = {(WBC,0),(SEIZURE,1),(EEG_FOCUS(?),-),(CSF_CELL,0)}
Local_instances/Covered_instances = 2/88
Local_instance_dispersion = 0.297297
Local_instance_conditional_probability = 1.000000
```

Different running modes of OSHAM and the visualization of competing concept hierarchies (hierarchy structure and information of concept) allows us to answer a number of important questions in doing model selection, for example, which attribute–value pairs are the most significant in discovered clusters? Also, what is the relationship between the model size (simplicity) and the quality of

TABLE IV.1.1 Number and Quality of Rules Discovered by Various Data Mining Methods

Methods	# rules			Method LUPC	# rules
	cover ≥ 1 acc ≥ 50%	cover ≥ 10 acc ≥ 50%	cover ≥ 7 acc ≥ 80%	cover ≥ 1, acc ≥ 50%	225
				cover ≥ 7, acc = 100%	4
				cover ≥ 7, acc ≥ 80%	22
See 5.0 (CF = 25%)	16	1	0	cover ≥ 10, acc ≥ 80%	0
See 5.0 (CF = 99%)	178	1	2	cover ≥ 10, acc ≥ 70%	6
CAB-CB	106	0	1	cover ≥ 10, acc ≥ 60%	23
CAB-CAR	1,496	45	1	cover ≥ 10, acc ≥ 50%	45
Rosetta	19,564	0	5		

discovered concepts (e.g., its dispersion, occurrence probability, etc.)? Figure IV.1.5 illustrates that, thanks to the plan management module and visualization tools, competing models can be discovered and compared. In this example, the navigation of two concept hierarchies consisting of 75 and 114 nodes—generated by two plans containing OSHAM with parameter settings (1, 7, 0.3, 3) and (1, 4, 0.3, 4), respectively—supports the domain experts in comparatively examining two different possible descriptions of concepts from the meningitis data that have different sizes and structures.

1.5.4 Knowledge Extraction from Stomach Cancer Data

The data mining methods C4.5 (Quinlan, 1993) and its successor See5.0, CBA (Liu et al., 1998), and Rosetta (Ohrn, 1999) have been applied to extract knowledge from stomach cancer data, in particular to discover prediction and description rules for the class of patients who died within 90 days after the operations. However, the obtained results were far from expected: they have low support and confidence, and usually relate to only a small percentage of patients of the target class.

Different plans have been created in D2MS for finding rules of this minority class. They typically combine the feature selection algorithm SFG (Liu and Motoda, 1998), rule induction method LUPC, and visualization and exportation of discovered rules. SFG orders attributes by information gain, and the user can choose different subsets of attributes in decreasing order of information gain. This method has also been used for preprocessing when applying See5, CBA, and Rosetta, but only with sequences of combined methods in D2MS could we evaluate and choose the best subset of attributes from various possible subsets. The plans also depend strongly on LUPC parameter settings.

We used the default values for number of candidate attribute–value pairs $\eta = 100$ and number of candidate rules $\gamma = 20$, while varying two parameters, i.e., minimum accuracy of a rule α and minimum coverage of a rule β. Given α and β, there are two modes of varying α and β with the search biases on accuracy or coverage. In the former, LUPC finds rules for the target class with accuracy as high as possible, while coverage remains equal to or greater than β. In the latter, rules are found with coverage as large as possible, while accuracy satisfies α.

The left part of Table IV.1.1 shows the number of rules discovered by See5.0, CBA (two modes CBA-CB and CBA-CAR), Rosetta, and the right part of the table shows the number of rules discovered by LUPC (in the right part), according to the required minimum of coverage (cover) and accuracy (acc) of rules shown in the table, respectively. Thanks to its support for model selection and visualization, LUPC allows us to find more and higher quality rules than the other systems.

For example, when the required minimum coverage and accuracy are low (1 and 50%, respectively) See5.0, CBA, and Rosetta can find many rules, but when these thresholds are set high (10 and 50%, 7 and 80%, respectively) to find higher quality rules, these methods can discover only a few rules. However, with required minimum coverage and accuracy, LUPC can find more rules, as shown in the right part of Table IV.1.1.

TABLE IV.1.2   Some Rules Found by LUPC and Evaluation of Domain Experts

Discovered Rules	Acceptability	Novelty	Utility
liver_metastasis = 3 ∧ hepatic = 1 → dead < 90 days (conf = 52.4%) (10/21)	1.2.3.4.<u>5</u>	<u>1</u>.2.3.4.5	1.2.3.4.5
liver_metastasis = 3 ∧ dcancer = x ∧ age ≤ 73 ∧ type = B1 → dead < 90 days (conf = 75.0%) (1/4)	1.2.<u>3</u>.4.5	1.2.3.<u>4</u>.5	1.2.3.<u>4</u>.5

As analyzed in more detail by Ho et al. (2001), See5.0 induces rules with an average error of 30.5% on testing data (taken randomly of 30% of the stomach cancer data), but with a very high false positive rate of 98.9% (i.e., rules found for this class recognize wrongly 98.9% cases from other classes). Similarly, CBA and Rosetta also give poor results on the class "died within 90 days" even when they produce a large number of rules with small thresholds.

The domain experts have evaluated discovered rules in each model by three criteria of acceptability, novelty, and utility with points from 1 (lowest) to 5 (highest). Table IV.1.2 illustrates two discovered rules and the evaluation of the domain experts. The confidence of these rules (59.1% and 75.0%) were estimated on training data, and numbers in parentheses, e.g., (10/21), show that on testing data, 21 cases matched the rules, from which 10 cases belong to the target class. The underlined numbers are evaluation of domain experts who are very interested in rules with a high degree of novelty.

1.6   Conclusion

We have presented the knowledge discovery system D2MS with support for model selection integrated with visualization. We have emphasized the crucial role of the user's participation in the model selection process and designed D2MS to support such participation. The basic idea is to provide the user with the ability to visually try various alternatives of algorithm combinations and their settings and to provide the user with performance metrics as well as effective visualization so that the user can gain insight into the discovered models before making his/her final selection. D2MS, with its model selection support, has been used and shown advantages in extracting knowledge from two real-world applications on meningitis and stomach cancer data.

The following issues are under investigation in our current and future work: (1) to continue to fully build the system with described properties; (2) to validate and improve the effectiveness of the system through the use of the system in real applications, in particular, medical data with the participation of domain experts; and (3) to enrich the system by adding other techniques for preprocessing, postprocessing, and data mining, or by adding meta-knowledge in algorithm profiles and integrating online rules to the phase of plan registration.

References

1. Ankerst, M., Ester, M., and Kriegel, H.P., Towards an effective cooperation of the user and the computer for classification, *Sixth International Conference on Knowledge Discovery and Data Mining KDD '00*, AAAI Press, Menlo Park, CA, 2000, 197.
2. Brazdil, P.B. and Soares, C., A comparison of ranking methods for classification algorithm selection, *Eleventh European Conference on Machine Learning ICML 2000*, 2000, 63.
3. Breiman, L. et al., *Classification and Regression Trees*, Wadsworth, Belmont, CA, 1984.

4. Brodley, C.E., Recursive automatic bias selection for classifier construction, *Machine Learning*, 20, 63, 1995.

5. Brunk, C., Kelly, J., and Kohavi, R., MineSet: an integrated system for data mining, *Third International Conference on Knowledge Discovery and Data Mining KDD '97*, AAAI Press, Menlo Park, CA, 1997, 135.

6. Crapo, A.W. et al., Visualization and the process of modeling: a cognitive-theoretic view, *Sixth International Conference on Knowledge Discovery and Data Mining KDD '00*, AAAI Press, Menlo Park, CA, 2000, 218.

7. Domshlak, C. et al., FlexiMine—a flexible platform for KDD research and application construction, *Fourth International Conference on Knowledge Discovery and Data Mining KDD '98*, AAAI Press, Menlo Park, CA, 1998, 184.

8. Fayyad, U.M. et al., From data mining to knowledge discovery: an overview, in *Advances in Knowledge Discovery and Data Mining*, Fayyad, U.M. et al., Eds., AAAI Press/MIT Press, Menlo Park, CA, 1996, 1.

9. Fujikawa, Y. and Ho, T.B., Cluster-based algorithms for filling in missing values, *Sixth Pacific-Asia Conference on Knowledge Discovery and Data Mining, Lecture Notes in Artificial Intelligence 2336*, Springer-Verlag, Heidelberg, 2001, 549.

10. Furnas, G.W., The FISHEYE view: a new look at structured files, Bell Laboratories Technical Memorandum #81-11221-9, 1981.

11. Furnkranz, J., Separate-and-conquer rule learning, *J. Artif. Intelligence Rev.*, 13, 3, 1999.

12. Han, J. and Kamber, M., *Data Mining: Concepts and Techniques*, Morgan Kaufmann, Palo Alto, CA, 2001.

13. Han, J. and Cercone, N., RuleViz: a model for visualizing knowledge discovery process, *Sixth International Conference on Knowledge Discovery and Data Mining*, AAAI Press, Menlo Park, CA, 2000, 244.

14. Hand, D.J., Data mining: statistics or more?, *Am. Statistician*, 52(2), 112, 1998.

15. Hilario, M. and Kalousis, A., Building algorithm profiles for prior model selection in knowledge discovery systems, *Eng. Intelligent Syst.*, 8(2), 77, 2000.

16. Ho, T.B., Discovering and using knowledge from unsupervised data, *Decision Support Syst.*, 21(1), 27, 1997.

17. Ho, T.B., Knowledge discovery from unsupervised data in support of decision making, in *Knowledge Based Systems: Techniques and Applications*, Leondes, C.T., Ed., Academic Press, New York, 2000, 435.

18. Ho, T.B., Kawasaki, S., and Nguyen, D.D., Extracting predictive knowledge from meningitis data by integration of rule induction and association mining, *International Workshop Challenge KDD, Lecture Notes in Artificial Intelligence*, Springer-Verlag, Heidelberg, 2001.

19. Kalousis, A. and Theoharis, T., NOEMON: design, implementation and performance results for an intelligent assistant for classifier selection, *Intelligent Data Anal. J.*, 3(5), 319, 1999.

20. Kohavi, R., A study of cross-validation and bootstrap for accuracy estimation and model selection, *International Joint Conference on Artificial Intelligence IJCAI '95*, Morgan Kaufman, Palo Alto, CA, 1995, 1137.

21. Kohavi, R., Sommerfield D., and Dougherty J., Data mining using MLC++, a machine learning library in C++, *Int. J. Artif. Intelligence Tools*, 6(4), 537, 1997.

22. Kumar, H.P., Plaisant, C., and Shneiderman, B., Browsing hierarchical data with multi-level dynamic queries and pruning, *Int. J. Human Comput. Studies*, 46(1), 103, 1997.

23. Lamping, J. and Rao, R., The hyperbolic browser: a focus plus context techniques for visualizing large hierarchies, *J. Visual Languages Comput.*, 7(1), 33, 1997.

24. Liu, B., Hsu, W., and Ma, Y., Integrating classification and association rule mining, *Fourth International Conference on Knowledge Discovery and Data Mining KDD '98*, AAAI Press, Menlo Park, CA, 1998, 80.

25. Liu, H. and Motoda, H., *Feature Selection for Knowledge Discovery and Data Mining*, Kluwer Academic Publishers, Dordrecht, 1998.
26. Mannila, H., Methods and problems in data mining, *International Conference on Database Theory*, Springer-Verlag, Heidelberg, 1997, 41.
27. Mingers, J., An empirical comparison of selection measures for decision tree induction, *Machine Learning*, 3, 319, 1989a.
28. Mingers, J., An empirical comparison of pruning methods for decision tree induction, *Machine Learning*, 4, 227, 1989b.
29. Nakhaeizadeh, G. and Schnabl, A., Development of multi-criteria metrics for evaluation of data mining algorithms, *Third International Conference on Knowledge Discovery and Data Mining KDD '97*, AAAI Press, Menlo Park, CA, 1997, 37.
30. Nguyen, T.D. and Ho, T.B., An interactive graphic system for decision tree induction, *J. Japanese Soc. Artif. Intelligence*, 14(1), 131, 1999.
31. Nguyen, T.D., Ho, T.B., and Shimodaira, H., A visualization tool for interactive learning of large decision trees, *Twelfth IEEE International Conference on Tools with Artificial Intelligence ICTAI 2000*, IEEE Computer Society Press, New York, 2000, 28.
32. Nguyen, T.D., Ho, T.B., and Shimodaira, H., A scalable algorithm for rule post-pruning of large decision trees, *Fifth Pacific-Asia Conference on Knowledge Discovery and Data Mining PAKDD'01*, LNAI 2035, Springer-Verlag, Heidelberg, 2001, 467.
33. Nguyen, N.B. and Ho, T.B., A mixed similarity measure in near-linear computational complexity for distance-based methods, *4th European Conference on Principles of Data Mining and Knowledge Discovery PKDD 2000*, LNAI 1910, Springer, 2000, 211.
34. Ohrn, A., *Rosetta Technical Reference Manual*, Norwegian University of Science and Technology, 1999.
35. Quinlan, J.R., *C4.5: Programs for Machine Learning*, Morgan Kaufmann, Palo Alto, CA, 1993.
36. Reingold, E.M. and Tilford, J.S., Tidier drawings of trees, *IEEE Trans. Software Eng.*, SE-7(2), 223, 1991.
37. Robertson, G.G., Mackinlay, J.D., and Card, S.K., Cone trees: animated 3D visualization of hierarchical information, *ACM Conference on Human Factors in Computing Systems*, 1991, 189.
38. Tsumoto, S., Comparison and evaluation of knowledge obtained by KDD methods, *J. Japanese Soc. Artif. Intelligence*, 15(5), 790, 2000.
39. Wolpert, D.H., The relationship between PAC, the statistical physics framework, the Bayesian framework, and the VC framework, in *The Mathematics of Generalization*, Wolpert, D.H., Ed., Addison-Wesley, Reading, MA, 1995, 117.
40. Zucchini, W., An introduction to model selection, *J. Math. Psychol.*, 44, 41, 2000.

2

Mining Chaotic Patterns in Relational Databases and Rule Extraction by Neural Networks

Chao Deng
Columbia University

Abstract. Knowledge discovery in databases (KDD) is one of the rapidly emerging and very important research areas in the context of computer science. Presently, the problems solved by KDD not only come from customers' databases and transaction databases, but also expand to multimedia databases, spatial databases, temporal databases, and the information widely distributed onto the Internet. As a result, the advanced technologies and techniques are joined, intercrossed, and mixed together at this point. This chapter systematically studies neural network methods in KDD and their applications.

As we know, there exist a lot of chaotic patterns in natural and social phenomena. Thus, databases storing this kind of data also contain chaotic patterns. Generally, chaotic patterns are characteristics of great random fluctuation, although they often appear between deterministic and stochastic patterns of knowledge discovery in a database. Therefore, chaotic patterns are always treated as random fluctuation

0-8493-1121-7/03/$0.00+$1.50

distributions and ignored. This viewpoint can be illustrated by Zytkow's Forty-Niner,[5,6] a pattern discovery platform based on statistics. A new network approach to discover and predict chaotic patterns in databases is proposed. This approach, together with the Forty-Niner, can not only discover the chaotic pattern but also predict it efficiently. By the way, this approach is very suitable to deal with large databases and has extensive applicable prospects in the vivid research fields of KDD.

In addition, a rule extraction method is proposed on the basis of neural networks' two-time convergence. Knowledge acquisition is fulfilled by the first convergence of the primary network, while rule extraction is realized by the second convergence of the mapping network. Through the dynamic mechanism of neural networks, this method greatly increases the efficiency of searching for rules from databases and has a good performance of solving for problems with large-scale data sets. Finally, this method is successfully applied to the extraction of the fertility ranking rules from real soil databases and appraised with a high score by the domain experts.

Key words: Knowledge discovery in databases (KDD), neural networks, chaotic pattern, discovery and predication of patterns in databases, chaotic pattern discovery network, rule extraction, two-time convergent network.

2.1 Introduction

Recently, knowledge discovery in databases (KDD) has become one of the rapidly emerging and very important research areas in the context of computer science and artificial intelligence. The problems solved by KDD not only come from customers' databases and transaction databases, but also expand to multimedia databases, spatial databases, temporal databases, and the information widely distributed onto the Internet. As a result, the advanced technologies and techniques are joined, intercrossed, and mixed together at this point. This chapter systematically studies neural network methods in KDD and their applications.

There exist many large databases that are constructed with the kind of data recording various natural and social phenomena. Relational data within one attribute or between different attributes of these databases usually lay out to be very randomly fluctuating and disordered. Researchers showed that many natural and social recordings (such as the growing and dying of epidemic disease, the rise and fall of stock price, and the change of the weather) are all disordered and mixed.[1] However, they can be described by chaos. The seemingly disordered and mixed relations can be easily confused with random relations. In fact, in these random relations, there might exist chaotic patterns that result from simple nonlinear systems.

Concept, digital logic pattern, elementary pattern, statistical pattern, and regularity are the types of relations to represent the knowledge in KDD.[2] These relations are all deterministic patterns, whether they are expressed in a deterministic or fuzzy manner. Besides, other relations are all treated as completely irrelevant ones that are called random fluctuation relations. Yet, in the random fluctuations, there exist a baffling range of relations that are different from pure noisy relations. Are there any useful patterns in these relations? Theories for chaos reveal that between deterministic patterns and random relations, there exists a middle pattern relation,[3] which is referred to as chaotic patterns in this paper. A so-called chaotic pattern is a time sequence or an attribute relation that can be rebuilt by a low-order nonlinear dynamics system. It can also be viewed as an ordered relation in a disordered system or a dynamics system misgoverned by decision-making theory. Real random data behave as a mass of cobwebs, but chaos is "disorder in order," created by a simple dynamics system. It is gradually clear that many irregular attribute relations that were viewed as stochastic relations can actually be ruled by some deterministic regularities. Since it is distinct from deterministic patterns and pure random relations, a chaotic pattern that is a generalization of the existing deterministic patterns in KDD is an important class of patterns in nature.

The establishment and development of the chaotic and fractal theories have provided a solid basis and sufficient tools for finding and understanding the disordered law (i.e., chaotic relation). If a disordered relation is judged a chaotic relation, it can also be predicted in a short time with high accuracy by nonlinear techniques. The evolution of laws hidden in the chaotic pattern can be dug out and reconstructed by discovering a chaotic pattern. Moreover, discovery of chaotic patterns can find laws according to the character of chaos and recast the existing data into some expressible frame. Thus, it provides a completely new approach and train of thought for the research in KDD.

Since the available methods have difficulties treating chaotic patterns, they are all treated as random relations. It is well known that the statistical technique is one of the important methods that estimate the attribute relations. Statistics in KDD has developed many methods for pattern searching and evaluating which are efficient in analyzing various relations.[4] Nevertheless it cannot efficiently depict the chaotic attribute relations. Rather, it usually disposes of chaotic patterns as random relations. 49er (Forty-Niner), proposed by Zytkow,[5,6] is one of the methods to find useful patterns in databases based on statistics. In this paper, 49er is introduced briefly and is used in estimating chaotic patterns. Analyses and experiments show that 49er is inefficient to find and depict the chaotic patterns and treat them as random relations.

A chaotic pattern discovery net (CPDN) was proposed. Based on the related work,[7–10] CPDN has many distinct advantages. Combined with 49er, it can not only discover chaotic patterns that can not be found by statistical approaches, but it can also predict the trend of the pattern. The CPDN has the ability to both handle large amounts of data and decrease computation complexity. The network resource of CPDN can be allocated online according to the incoming novel information of input. The strategy to prune hidden nodes can efficiently and dynamically wipe out redundant resources of the net. The weights of CPDN can adaptively and sequentially adjust in the learning process so as to find the relations online. The effectiveness of CPDN has been proven by the extensive experiments.[7]

In addition, extracting rules from a database is one of the principal parts of a knowledge discovery process. In particular, the learning rule is an important research direction of machine learning and KDD; machine learning impresses on the creation of computer programs and can learn knowledge, but KDD puts more emphasis on how to discover knowledge from data. To date, the main rule extraction methods in KDD include decision tree inference, neural networks, and induced logic planning (ILP). Among rule extraction methods in KDD, the decision tree method cannot perform multivariable search.[8,9] Because each node of the tree contains one feature only when it builds, it therefore belongs to a single variable algorithm and reduces the correlation between features. Although it links many features into one tree, the relationship among them is a loose format. In addition, the decision tree inference method is pretty sensitive to the noise in data. The main drawback of the ILP method is it does not deal with the problem of uncertain data. Although the hidden knowledge representation in connection with neural networks increases its difficulty to extract rules from data, the neural network's unique advantages, such as nonlinear mapping capability, high error redundancy, and robustness to noise, make it a great prospect in data mining. The neural network method has features of high learning precision and multiple variables search such that it not only extracts classification rules from data but also extracts other types of rules such as association rules, good robustness to noise, and suitable for data uncertainty.

The research to extract rules from data by the use of neural networks is a hot research area in recent neural network technology.[10–12] From the viewpoint of transparency of the rule extraction process, the methods of neural network rule extraction, generally, are catalogued as the decomposition method, pedagogical method, eclectic method, and compositional method.[10,11] The criteria to evaluate these methods are accuracy, confidence, consistency, understandability, and algorithm complexity. Among these criteria, different researchers stress different aspects, but they seldom concern the algorithm complexity.

Very recently, the neural network approach has become one of the most important approaches in KDD,[13–17] and it has great potential. As a new research field, the processed data in KDD has its own features such as huge size, noise, indetermination, and sparsity. Therefore, besides the aforementioned criteria for evaluating mined rules, the computational complexity of an algorithm is a non-neglectable index. Lu, Setiono, and Liu[17] discussed using the neural networks for data mining first. They proposed a neural network method for rule extraction, regarding the classification problem in data mining. Extensive analysis and experiments in their work illustrated the validity of the neural network method. But, in the case of a high dimensional search or big diversity in the hidden layer, this algorithm has a low searching efficiency for rules. In particular, it is impossible to enumerate all cases when the input space or output space is very large. For example, for n dimensional space, 2^n step calculations are needed for enumeration. This kind of exponentially growing computational amount cannot be undertaken for a huge input space in practice. According to the idea of network convergence, this author proposed a rule extraction approach of two-time convergence of the main network and mapping network on the basis of neural networks. This method can extract rules without the enumeration of input and output spaces. Compared to the method proposed by LU, Setiono, and Liu,[17] this method greatly shortened the time of rule extraction and increased the computational efficiency.

2.2   Concept and Analysis of Chaotic Patterns

A chaotic pattern is a special kind of relation that behaves between deterministic and random relations. The chaotic pattern can be found both within an attribute and between different attributes. The discovery of chaotic patterns not only expands the bounds of knowledge but also can increase the credibility of decisions.

The time series is a data type where one attribute represents different moments of time, and the other attributes store information about co-occurring properties of objects. Relational, object-oriented, or special time-series databases can be used to store time series. This paper mainly deals with time-series databases.

It has been proven that the approaches of statistic correlation analysis, logic reasoning, and fuzzy decision are efficient in discovery of non-chaotic patterns in KDD. 49er[5,6] is one of the approaches mainly based on statistics whose main idea of pattern discovery approach is introduced as an example in this chapter. Chaotic patterns always appear to be disordered and are shown to be similar to random patterns in statistic research. As a result, chaotic patterns are usually classified and treated as random fluctuations. Thus, other approaches need to be invented to discover and forecast chaotic patterns. Nonlinearity is the fundamental reason to generate chaotic patterns. Though they are seemingly disordered, chaotic patterns obey certain laws and behave as some correlation in the time delayed state space that gives some memory ability in intra- and inter-attributes. However, this correlation is difficult to express by relation logic, symbolic processing, and analytical methods. Neural networks just possess the ability of such a method of information processing. A chaotic pattern discovery network (CPDN) is proposed in this chapter which can distinguish and discover chaotic patterns from pure random patterns and forecast the pattern trend to a certain degree.

2.2.1   Statistical Approach in Forty-Niner

Some elementary knowledge and analysis approaches of statistics are adopted in 49er which mainly deal with knowledge discovery in relation databases. The knowledge discovered by 49er is expressed in regularities which take the form of PATTERN in RANGE, where RANGE defines a subset of the whole possible events space W, while PATTERN, which can be expressed in either prediction or contingency tables, describes a pattern that holds for the records available within the range. The contingency table with which 49er primarily describes a pattern is a mapping from W to the set of natural numbers, associating each possible event with the number of occurrences of the corresponding record in the database.

49er systematically searches for regularities in the set of all data, then partitions the data into subsets, and searches for regularities in the subsets. Within each range of data, 49er tries to detect 2D regularities for all combinations of attribute pairs admissible by the user. Initially, 49er examines the contingency table for each pair of attributes. If the system finds that the contingency table follows a specific pattern, a subtler discovery mechanism can be invoked, such as a search in the space of equations.

49er applies statistical tests of significance and strength to all hypotheses, qualifying them as regularities if test results exceed acceptance thresholds. Significance is the probability Q that a given pattern has been generated randomly for variables that are really independent. Values close to 0 mean that such a situation is very unlikely, while values close to 1 make it very likely. In other words, the smaller the value of Q, the bigger the probability in which there exists a given pattern relationship between attributes. The value of Q can be calculated according to χ^2 statistics and the freedom degree of contingency table. χ^2 measures the distance between actual and expected events.

$$\chi^2 = \sum_{i,j} \frac{(A_{ij} - E_{ij})^2}{E_{ij}} \tag{IV.2.1}$$

where A_{ij} is the actual distribution of the contingency table and E_{ij} is the expected distribution of the contingency table and can be expressed as

$$E_{ij} = \frac{n_{xi} \cdot n_{yj}}{N} \tag{IV.2.2}$$

Here, N is the number of records, and n_{xi} and n_{yj} respectively represent the number of records when attribute $x = x_i$ and attribute $y = y_j$.

Cramer's coefficient V is adopted by 49er to measure the predictive power of the contingency table. For a given $M_{row} \times M_{col}$ contingency table, Cramer's coefficient V can be defined as

$$V = \sqrt{\frac{\chi^2}{N \min (M_{row} - 1, M_{col} - 1)}} \tag{IV.2.3}$$

If the value of V is larger, it is revealed that the predictive power is stronger and the correlation between attributes is correspondingly tighter.

According to V and Q, 49er can estimate the significance and predictive power of each hypothetical pattern in the process of research. Generally, the default threshold of Q is set as $Q_T = 10^{-5}$ and that of V as $V_T = 0.5$. Q and V should be considered simultaneously so as to achieve correct recognition of the hypothetical pattern. If $Q < Q_T$ and $V > V_T$, the existence of the pattern is likely; otherwise, it is unlikely and the pattern will be discarded.

49er is indicated to have a good recognition capability for 2D patterns. One hundred random fluctuation patterns were tested by 49er, and only one was mistaken as a deterministic pattern. There might, however, exist some chaotic patterns, in the 99 remaining patterns, which cannot be distinguished by 49er.

2.3   Neural Network Approach to Chaotic Pattern Discovery

2.3.1   An Improved Resource Allocation Network

It is well known that neural networks are able to realize arbitrary mapping from a bounded subset A in n-dimensional Euclidean space to a bounded subset $f(A)$ in m-dimensional Euclidean space, $\varphi: A \subset \Re^n \rightarrow \Re^m$.[18] In feedforward neural networks, MLP and RBF are the most commonly used networks to conduct function estimation. Compared to other feedforward neural networks, the RBF network has an optimal approximation property[19] and global approximation performance.[21–25] Therefore, until now, researchers have developed many learning algorithms for RBF neural networks for function estimation and prediction[23–25] such as single-layer network gradient decent method, Newton's method, minimal energy method, project pursuing method, genetic competing method, etc. All of these methods have the drawback that network scale is hard to determine in advance. It is well known that if the network size is small, it is very difficult to estimate the unknown function precisely. In contrast, if the network size is too large, it has to fit each training sample in high precision, which deteriorates the general performance of the network. As a result, many algorithms presently adopt asymptotic approximate learning methods to train and optimize the network. However, this kind of method falls into an off-line learning algorithm and needs to learn all of the training samples recursively. In the case of the dynamic environmental data, this kind of method requires a large amount of memory and computation resources. Hence, a dynamic learning algorithm is necessary in this case. For this purpose, some researchers develop several sequential learning algorithms for networks to efficiently learn under a dynamically changing environment.

Platt[18] first proposed a resource allocation network (RAN) for function approximation. This was a resource allocation radial basis function neural network. Kadirkamanathan[19] elaborated and explained RAN from the viewpoint of function space mapping theoretically and proposed a resource allocation network with extended Kalman filtering by improving RAN.[19,20] This network not only increases the approximation and prediction accuracy of the function but also efficiently utilizes the network resources (basis functions). The two networks have to adjust the network scale from a perspective of resource allocation. Once there is a new feature pattern, the network will allocate a basis function for it, i.e., add a hidden node. The size of the network will increase with the growth of input feature patterns. Although it is very necessary for the case of no redundant hidden nodes in the network, these methods only consider the feature of the current sample and ignore the function of allocated hidden nodes in the network, which results in an imbalance of whole network and saturation over-learning for some problems. Therefore, it is necessary to insert a hidden node pruning mechanism in the learning process so that an optimal allocation can be realized. In contrast, the pruning mechanism of hidden nodes is also needed to avoid over-learning and saturation.[7,20,21]

This author proposed an efficient online learning method for this neural network called P-RAN, which is a resource dynamically allocating network with a pruning mechanism based on the approximation ability and related theory of radial basis function networks.[22–27]

2.3.1.1   Network Architecture

P-RAN is a two-layer feedforward neural network, as shown in Figure IV.2.1, where $\mathbf{x}$ is an input vector. The second layer is a hidden layer that accomplishes the response against the input. Therefore, the output of the RBF network can be expressed as

$$f(x) = \alpha_0 + \sum_{k=1}^{K} \alpha_k \phi_k (r) \tag{IV.2.4}$$

where α_k represents the connection weight from the kth hidden node to the output, and K is the number of hidden nodes. Function ϕ is referred to as a basis function that has multiple forms,[28,29] for example, linear function $\phi (r) = r$, cubic function $\phi (r) = r^3$, Gaussian function $\phi (r) = \exp(-r^2)$,

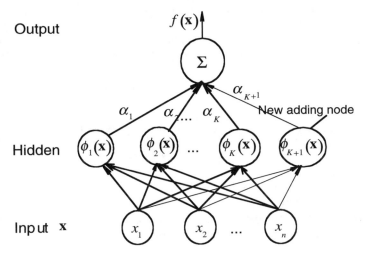

FIGURE IV.2.1   Network structure of P-RAN.

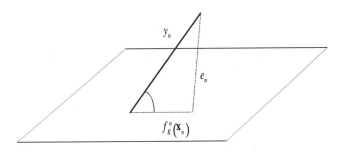

FIGURE IV.2.2   Schematic plot of the projection of function mapping.

thin-plate spline function $\phi(r) = r^2 \log r$, multi-quadrate function $\phi(r) = \sqrt{r^2 + 1}$, and inverse multi-quadrate function $\phi(r) = 1/\sqrt{r^2 + 1}$, where r is the form of

$$r = \frac{\|\mathbf{x} - \mathbf{u}_k\|}{\sigma_k} \tag{IV.2.5}$$

where $\mathbf{u}_k \in \Re^n$ indicates the center of the kth basis function. $\sigma_k \in \Re$ is the width of kth basis function that represents the resolution of the network to the input data. $\|\cdot\|$ stands for the Euclidean norm in $\Re^n$. Due to the local response feature of the Gaussian function, we choose the basis function of $\phi(\mathbf{x})$ as the Gaussian function.

2.3.1.2 Principle of P-RAN

Assume there are K hidden nodes in the network and express the new arriving sample as $(\mathbf{x}_n, y_n)$, as shown in Figure IV.2.2. If the network can estimate the new arriving sample accurately and mark the projection of y_n in K-dimensional Hilbert subspace H_K as $f_K^n(\mathbf{x}_n)$, then the difference between this projection and y_n:

$$e_n = y_n - f_K^n(\mathbf{x}_n) \tag{IV.2.6}$$

must be zero or within a range. In contrast, if e_n exceeds a preset threshold, it means the network is not able to estimate the new arriving sample. In this case, we must add a new hidden node perpendicular to H_K space to express this innovation, i.e., the correct estimation of y_n.

Suppose the P-RAN network is initialized by a null hidden node. When a new observation sample comes, the network has to check whether sufficient innovation is included in the new sample. If yes, allow the network to add one hidden node. Otherwise, adjust the parameters of the network to reduce the prediction error. In such a way, the network is always able to express all of the observed information so far.

In order to discriminate different innovations efficiently, we consider three kinds of measurement parameters and their thresholds as follows:

1. Output error of the network overpasses minimal error threshold

$$e_n = y_n - f(x_n) > e_{\min} \tag{IV.2.7}$$

   where e_{min} is the minimal error threshold.

   Since we know that big output error means there are more innovations in the input sample, the network needs to add a new hidden node to accommodate this innovation.
2. The minimal distance between the input pattern and the stored pattern in a network over a threshold is

$$\|\mathbf{x}_n - \mathbf{u}_{nr}\| > \epsilon_n \tag{IV.2.8}$$

   where $\mathbf{u}_{nr}$ is a stored pattern (basis function) in the input space with the smallest distance to $\mathbf{x}_n$, ϵ_n is the threshold, $\epsilon_n = \max\{\epsilon_{max}\gamma^n, \epsilon_{min}\}$, and γ is a constant between 0 and 1.
3. Output error of the network is greater than correlated mixed error (CME):

$$e_n = y_n - f(x_n) > e_{CME} \tag{IV.2.9}$$

   where

$$e_{CME} = \lambda \cdot e_{n-1} + (1 - \lambda) e_n \qquad (0 \leq \lambda \leq 1) \tag{IV.2.10}$$

   This judgement is in harmony with the pruning process of the hidden node in our proposed algorithm and prevents the learning process from oscillation.

When the three above criteria are satisfied simultaneously, the network is allowed to add a new hidden node and let $K \Leftarrow K+1$. The parameters of the newly added hidden node are able to set up as

$$\mathbf{a}_{K+1} = e_n$$
$$\mathbf{u}_{K+1} = \mathbf{x}_n \tag{IV.2.11}$$
$$\sigma_{K+1} = \kappa \|\mathbf{x}_n - \mathbf{u}_{nr}\|$$

where κ is a covering factor that determines the responding covering degree of the newly added hidden node in the input space.

If the above three criteria are not satisfied simultaneously, the network is not allowed to increase a hidden node but needs to adjust the network parameters to reduce the observation error. Here, we adopt the famous LMS algorithm to adjust the network's parameters due to its properties of simple, fast, and small memory.

2.3.1.3 Adaptive LMS Algorithm

Given the network's parameter vector, which can be expressed as

$$\mathbf{w} = [\alpha_1, u_1^T, \sigma_1, \dots, \alpha_K, u_K^T, \sigma_K]^T \tag{IV.2.12}$$

the parameter update formula is

$$\mathbf{w}^{(n)} = \mathbf{w}^{(n-1)} + \eta e_n \mathbf{a}_n \tag{IV.2.13}$$

where $\mathbf{a}_n = \nabla_w f(\mathbf{x}_n)$ is a gradient vector of function $f(\mathbf{x})$ with respect to parameter vector $\mathbf{w}$ and η is the adaptive step. Here,

$$\mathbf{a}_n = \left[\phi_1(\mathbf{x}_n), \phi_1(\mathbf{x}_n) \frac{2\alpha_1}{\sigma_1^2}(\mathbf{x}_n - \mathbf{u}_1), \phi_1(\mathbf{x}_n) \frac{2\alpha_1}{\sigma_1^3} \|(\mathbf{x}_n - \mathbf{u}_1)\|^2, \ldots, \right.$$

$$\left. \phi_K(\mathbf{x}_n), \phi_K(\mathbf{x}_n) \frac{2\alpha_K}{\sigma_K^2}(\mathbf{x}_n - \mathbf{u}_K), \phi_K(\mathbf{x}_n) \frac{2\alpha_K}{\sigma_K^3} \|(\mathbf{x}_n - \mathbf{u}_K)\|^2 \right]^T \tag{IV.2.14}$$

On the other hand, when a weight from a hidden node to the output is small, which also means little contribution to the output of the network, we enforce a penalty factor ζ ($0 < \zeta < 1$). Once it is smaller than a preset threshold, we can delete it from the network. Specifically, when a new observation sample comes, if $\alpha_j < \alpha_{pu}, j = 1, \ldots, K$, then $\alpha_j = \zeta \alpha_j$. Here, α_{pu} is the penalty threshold. If $\alpha_j < \alpha_{pr}, j = 1, \ldots, K$, then delete the jth hidden node from the network and let $K \Leftarrow K - 1$. Here, α_{pr} is referred to as the pruning threshold of the hidden node.

2.3.2 Neural Approach for Chaotic Pattern Discovery and Prediction

Due to its strong learning and extraction abilities of patterns, the network approach is one kind of technique that has gradually been developed. In recent years, the network approach received extensive attention because it not only could describe the relations difficult to express in formulas but also had the strong ability of relation prediction. The typical network approaches include the single-layer network, multilayer network, feedback network, dynamic network, ART network, etc. Among them, the RBF network is commonly used for pattern recognition and function approximation due to its simple topology structure and concise learning procedure. This chapter proposes an RBF-based network approach for chaotic pattern discovery[7] and applies it to discover the chaotic patterns in databases. This kind of chaotic pattern occurs in both intra-attributes and inter-attributes. CPDN not only finds and learns the chaotic relationship between attributes but also gives very good predictive performance.

The conventional RBF network approach adopts a predetermined network structure (e.g., the number of the basis functions) according to the given problems. This approach is not able to deal with the problem of discovering useful patterns in databases because there are two basic questions in KDD: Is there a pattern? Moreover, what is the pattern in database? Also, this approach is not suitable for large-scale databases.

The proposed CPDN is a two-layer feedforward neural network. Assume $\mathbf{x}$ is an input vector and the second layer is a hidden layer that represents the response to the input. The network output is the linearly weighted sum of all of the output of the hidden nodes. Therefore, the output of this kind of radical basis function neural network can be given in Equation IV.2.4, where $\phi_k(\mathbf{x})$ represents the response of the kth hidden node to the input, which is commonly chosen as the Gaussian function, i.e.,

$$\phi_k(\mathbf{x}) = \exp\left\{ -\frac{1}{\sigma_k^2} \|\mathbf{x} - \mathbf{u}_k\|^2 \right\} \tag{IV.2.15}$$

where $\mathbf{u}_k \in \Re^n$ denotes the center of the kth basis function, and $\sigma_k \in \Re$ denotes the width of the kth basis function for representing the resolution of the network to the data of the input space. $\|\cdot\|$ means the Euclidean norm in $\Re^n$ space.

The main features of the proposed CPDN algorithm can be given as follows:

1. The hidden node of the CPDN is dynamically allocated during the learning process.
2. The CPDN is for the pruning mechanism of the basis function, which is very useful to determine whether there exists a pattern or not.
3. The CPDN has a strong prediction ability due to its good generalization performance.

Specifically, the proposed CPDN algorithm can be summarized as

> *Initialize:* $\epsilon = \epsilon_{\min}$ *and* $\alpha_0 = y_0$
>
> *for* $(n = 1; n < Sample\ Number; n++)$
>
> $\{$　　　computes $f(\mathbf{x}_n)$, $\phi_k(\mathbf{x}_n)$, e_n
>
> 　　　*find the nearest stored pattern* $\mathbf{u}_{nr}$
>
> 　　　*let* $\epsilon_n = \max\{\epsilon_{\max}\gamma^n, \epsilon_{\min}\}$
>
> 　　　*if*　　$e_n > e_{\min}$ & $\|\mathbf{x}_n - \mathbf{u}_{nr}\| > \epsilon_n$ & $e_n > e_{CME}$
>
> 　　　$\{$　　*allocate new hidden unit with*
>
> 　　　　　$\mathbf{a}_{K+1} = e_n$
>
> 　　　　　$\mathbf{u}_{K+1} = \mathbf{x}_n$
>
> 　　　　　$\sigma_{K+1} = \kappa\|\mathbf{x}_n - \mathbf{u}_{nr}\|$
>
> 　　　$\}$ *else* $\{$ *adjusting network parameters:* $\alpha, \mathbf{u}, \sigma$ $\}$
>
> 　　　*pruning the hidden unit according to the pruning strategy*
>
> 　　　$\}$

where $(\mathbf{x}_0, y_0)$ is the first observation sample for initializing the network, e_n denotes the network error, $\mathbf{u}_{nr}$ is the nearest stored pattern of $\mathbf{x}_n$ in the input space, $e_{\min}$ and e_{CME} are two kinds of error thresholds, and ε_n is a threshold of the minimum distance among hidden nodes, which is within the range between $\varepsilon_{\max}$ and $\varepsilon_{\min}$.

2.4　Simulation Results and Discussion

2.4.1　Pattern Discovery

49er can detect 2D patterns efficiently. For the following experiments, two pairs of attributes A_1, A_2 and A_1, A_3 are generated. A_1 is generated randomly in uniform distribution. A_2 is generated similar to A_1 but independently. A_3 has a functional relation with A_1 as $A_3 = a + bA_1^2$, where a and b are real and constant numbers, respectively. The total number of recorders generated is 2000. In the first step of the experiment, the statistical parameters, which are χ^2, V, and Q, are calculated by the 49er (the concrete calculating step is discussed in Section 2.3.2). The values of χ^2, V, and Q are listed in Table IV.2.1.

It can be seen that the Q value between A_1 and A_2 is far greater than the threshold $Q_T = 10^{-5}$, and the value of V is smaller than the threshold V_T. This fact signifies the random relationship between A_1 and A_2. In contrast, the deterministic relationship is revealed between A_1 and A_3 by the fact that its Q value is smaller than the threshold, and its V value is larger than the threshold. The above experimental results agree with the decision rule of 49er, which verifies the effectiveness of 49er.

TABLE IV.2.1   Statistic Parameters of Attribute Pairs A_1 and A_2 and A_1 and A_3

Attributes Relation	χ^2	V	Q
A_1, A_2 (noise)	91.1677	0.1006	0.7245
A_1, A_3 (function)	4761.54	0.72735	2.3e-057

TABLE IV.2.2   Actual Distribution

attri-1	attri-2									
	0.445−	0.528−	0.611−	0.694−	0.777−	0.859−	0.942−	1.025−	1.108−	1.191−
0.473−	8	36	42	38	4	0	0	0	0	0
0.553−	18	12	0	3	31	14	0	0	0	0
0.634−	25	0	13	44	32	27	21	0	0	0
0.714−	25	0	25	41	13	35	50	28	3	0
0.794−	35	0	24	12	0	0	16	33	29	2
0.874−	17	27	19	17	0	4	27	27	13	28
0.954−	0	3	39	52	35	41	89	44	20	31
1.034−	0	0	0	13	36	33	38	80	75	32
1.114−	0	0	0	0	0	25	113	75	22	33
1.194−	0	0	0	0	0	0	0	15	109	29

TABLE IV.2.3   Expected Distribution

attri-1	attri-2									
	0.445−	0.528−	0.611−	0.694−	0.777−	0.859−	0.942−	1.025−	1.108−	1.191−
0.473−	8.192	4.992	10.36	14.08	9.664	11.45	22.65	19.32	17.34	9.920
0.553−	4.992	3.042	6.318	8.580	5.889	6.981	13.80	11.77	10.56	6.045
0.634−	10.36	6.318	13.12	17.82	12.23	14.49	28.67	24.46	21.95	12.55
0.714−	14.08	8.580	17.82	24.20	16.61	19.69	38.94	33.22	29.81	17.05
0.794−	9.664	5.889	12.23	16.61	11.40	13.51	26.72	22.80	20.46	11.70
0.874−	11.45	6.981	14.49	19.69	13.51	16.02	31.68	27.02	24.25	13.87
0.954−	22.65	13.80	28.67	38.94	26.72	31.68	62.65	53.45	47.96	27.43
1.034−	19.64	11.97	24.86	33.77	23.17	27.47	54.33	46.35	41.59	23.79
1.114−	17.15	10.45	21.70	29.48	20.23	23.98	47.43	40.46	36.31	20.77
1.194−	9.792	5.967	12.39	16.83	11.55	13.69	27.08	23.10	20.73	11.85

2.4.2   Discovery of Chaotic Patterns

49er fails to discover chaotic patterns. In detail, a manmade database which contains a Mackey–Glass chaotic pattern is generated, in which there are 5000 tuples with 5 attributes indicated by attri-1, attri-2,..., attri-5, respectively. As an example, by the relationship between attri-1 and attri-2, the following steps are carried out. To begin with, statistic parameters of 49er are calculated for the search of patterns to verify 49er's chaotic pattern discovery capability. Then, the proposed CPDN method is adopted for chaotic pattern discovery.

As described above, χ^2 depends on the number of records so that Q varies with the record number. Therefore, in order to display the change correctly, Q and V must be computed in different numbers

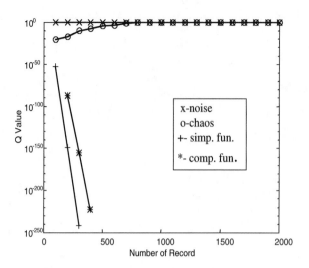

FIGURE IV.2.3   Curves of Q vs. number of records.

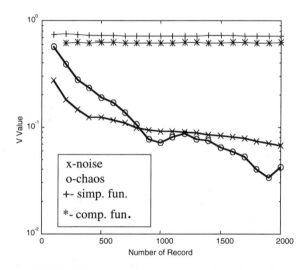

FIGURE IV.2.4   Curves of V vs. number of records.

of records. Tables IV.2.2 and IV.2.3 are, respectively, contingent tables of actual distribution and expected distribution for computing χ^2 when the record number is 2000.

It is discovered by the simulation results that the Q and V of chaotic patterns vary with different record numbers. For convenient comparison, we also calculate the Q and V of the following three patterns, which are illustrated in Figures IV.2.3 and IV.2.4:

1. A_1, A_2 is in a random relationship (generated in the manner of Section 2.3.1).
2. A_1, A_3 is in a simple functional relationship (generated in the manner of Section 2.3.1).
3. A_1, A_4 is in a complex functional relationship that is in the form of $A_4 = a + bA_1^{3*}\exp(A_1^2)$.

In Figures IV.2.3 and IV.2.4, curves with symbol x are values of Q and V for random relationships between A_1 and A_2. Curves with + symbols are for the simple functional relationship of A_1 and A_3, and curves with * are for the complex functional relationship of A_1 and A_4. The Q and V of the chaotic pattern are plotted with the symbol o.

With regard to the random relationship, the Q value is always much larger than the threshold of 10^{-5}, and the V value is always near zero. For deterministic patterns, Q is below the threshold and V

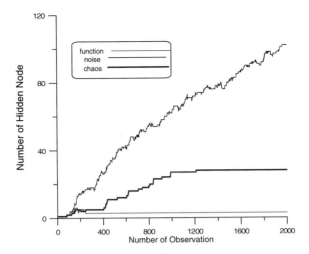

FIGURE IV.2.5 Curves of hidden nodes vs. number of observations.

is close to 1 all the time. It can also be seen from Figures IV.2.3 and IV.2.4 that the chaotic pattern is located between the deterministic pattern and random fluctuation. Nevertheless, 49er cannot discover the chaotic pattern and classify it into a random relationship.

In order to extract chaotic patterns from random fluctuations, one must search for other approaches. In this chapter, a chaotic pattern discovery network (CPDN) is proposed and used for discovering chaotic patterns in databases. CPDN can efficiently distinguish chaotic patterns from random relationships. CPDN can reasonably allocate its basis functions in a strategy of "reward good and punish bad" during the learning process. Since its basis functions are allocated according to the novel input and are also pruned in the learning process, CPDN can not only learn new knowledge but also can overcome the problem of over-learning. The learned network can sufficiently express the environmental information to find both chaotic and deterministic patterns. Next, we apply CPDN to discover the three kinds of patterns (such as the deterministic pattern of A_1 and A_3 and the chaotic pattern and random relationship of A_1 and A_2).

In the learning process, each pair of attribute values are used as input and output of the net, respectively, and samples are sequentially presented in the input-end. The change of hidden nodes during the learning process is plotted in Figure IV.2.5.

In Figure IV.2.5, three lines reflect the change of hidden nodes when CPDN learns the three kinds of patterns. The dashed line denotes the change of hidden nodes of CPDN in learning of the deterministic pattern of A_1 and A_3, while the fine real line represents the random relationship of A_1 and A_2. The hidden node change of chaotic patterns attri-1 and attri-2 is plotted by a thick line. It is shown in Figure IV.2.5 that for the random relationship of A_1 and A_2, the hidden nodes increase all the time and the curve changes acutely and irregularly. Because there is no correlation between data in a random relationship, each of which is novel to the system, the CPDN must allocate new hidden nodes for them. For a deterministic relationship, the number of hidden nodes changes steadily and gradually reaches a stable state. As there exists correlation and a deterministic pattern between A_1 and A_3, the pattern character must be limited. Therefore, CPDN can express the pattern efficiently by a finite number of hidden nodes, which implies that the CPDN is able to filter out patterns from random fluctuation. For a chaotic pattern, the hidden nodes of the CPDN become unchanged at 1300 observation samples during the learning process, which means it is enough to express the chaotic pattern by 24 hidden nodes in this simulation.

2.4.3 Description and Prediction of Chaotic Patterns

In many actual problems, it is required that an efficient method is not only able to discover various useful patterns in large databases but is also able to predict future patterns. The proposed CPDN is qualified for the above requirement. Correspondingly, the CPDN can not only discover and learn useful patterns but also has strong prediction capability that is the foreseen ability of the future data for a pattern. The prediction capability affects the confident degree of pattern decision to the coming patterns.

Furthermore, the former 3000 tuples of the DB are used as samples to train the CPDN, and it produces a proper network of CPDN. Then, the prediction is processed on the next 2000 tuples using the trained CPDN, whose results are shown in Figures IV.2.6 and IV.2.7. In particular, Figure IV.2.6 shows the 2D relationship between attri-1 and attri-2, where the dashed line denotes the relations drawn by data in the database, and the real line represents the prognostic relation by CPDN. A 3D relation graph among attr-1, attr-2, and attr-3 is shown in Figure IV.2.7. Good predictive probability of the CPDN for chaotic patterns can be clearly seen from the two figures.

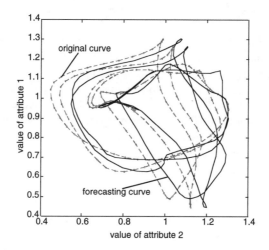

FIGURE IV.2.6   Curves of original and predicted 2D chaotic relations.

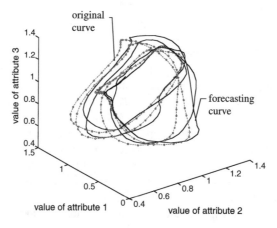

FIGURE IV.2.7   Curves of original and predicted 3D chaotic relations.

2.5   Rule Extraction—United Framework of Data Mining

Classification knowledge and association knowledge are the main parts and common knowledge of data mining targets. A person can utilize different forms of rules to express this knowledge sufficiently.[12] Therefore, the problems of data mining can be united together through rules for synthetic analysis.

2.5.1   United Framework

The problem of data mining can be seen as a problem of discovering rules from a database. Although it cannot include all problems that appear in the data mining community, it contains the majority of catalogs of data mining, such as classification, association, etc.

Definition 1: Assume T is an object set, and $D(m)$ and $R(m)$ are field and range of a method m, respectively. Let M be a method set, for m methods in M, and its field is T or $R(m).t.m$ represents the result of applying method m to object t.

Definition 2: Assume l is the combination of the methods in M. Define expression $p(t.l)$ as a predicate of some methods m in M.

For example, for the predicate senior $(t.age)$, when the age of object t is greater than 65, senior $(t.age)$ is true, where age is a method in M.

Definition 3: The common rule form extracted by the convergent neural network is defined as

$$F(T) \Rightarrow G(T)$$

where the premise F is a join of some formula. Generally, it has two forms:

Form 1 : $\text{atom}_1 \wedge \text{atom}_2 \ldots \wedge \text{atom}_n$
Form 2 : $A_1 \wedge A_2 \ldots \wedge A_n$

where atom_i $(i = 1, \ldots, n)$ in Form 1 is the ith simple atom whose form is attribute = value or $\text{value}_1 \leq \text{attribute} \leq \text{value}_2$. $A_1, \ldots, A_n$ in Form 2 are n attributes, the value of $A_1 \in V_1$, and the value of $A_n \in V_{n^\circ}$.

In the rule's conclusion, G is a formula whose form is of $B_1 \wedge B_2 \ldots \wedge B_m$, where $B_1, \ldots, B_m$ can be m concepts or states.

If there are c portions of objects in set T to meet either F or G, then

$$\text{Rule}: F(T) \Rightarrow G(T) \tag{IV.2.16}$$

with belief degree factor $0 \leq c \leq 1$ meeting object set T. For given object set T, the goal of data mining is to discover all rules to meet some constraints, where the constraints include syntactic constraints and support constraints.

Syntactic constraint is to indicate the constraint on predicates and methods such that only some predicates and methods appear in the rule. While the support constraint is to consider the number of objects in T to support the rule. The rule support is defined as the number of objects meeting the rule's premise and conclusion in object set T. In what follows, we united the main goals of data mining into a rule framework.

1. Classification—The object set consists of a labeled training set in which the label is assigned according to which class the corresponding tuple belongs to. The tuple's attributes represent the features of the object. The goal of classification is to find the rule describing the feature of each class in the training set, i.e., the conclusion form of all rules is $t.\text{label\_method} = k$, where

k fetches the value of the label. Note that there is a synaptic constraint whose label cannot appear in the premise of the rule.

2. Association—Here we have an object set consisting of each transaction event of customers. For example, according to each item, there is a two-value method. True/false represents whether the item appears in the transaction or not. The association rule is confined by either synaptic constraint or support constraint.

2.5.2 Problem Representation and Encoding

In order to express the rule extracted by neural networks explicitly, one should encode the input in advance. If the attributes of the input take discrete values, then assign some digits of binary code to the discrete value. If the attribute value is a continuous value, we discrete it at first into n subfields. Then the number of subfields is encoded. Finally, we can encode all attributes of inputs in this manner. There are two kinds of encoding methods. The first one is to encode each attribute completely, i.e., the enumerating method. The second method is the thermometer encoding method that can be described as follows: for an attribute whose value field $\in [0\ 10]$ with its subfield interval being 2, the value falling into $[0, 2]$ is encoded as 00001, while the value falling into $[2, 4]$ is encoded as 00010 and so on.

2.6 Convergent Network Method for Rule Extraction

2.6.1 First Convergence of the Main Network

2.6.1.1 Network Training

The training network (TrN for short) is a three-layer feedforward neural network whose first layer is an input layer with input vectors $\subset R^{D_{ITr}}$, second layer is a hidden layer with vector space $\subset R^{D_{HTr}}$, and last layer is an output layer belonging to $R^{D_{OTr}}$ space, where the subscripts *ITr*, *HTr*, and *OTr* indicate the number of neurons of the input, hidden, and output layers. Then, in the TrN, the input vector is $\mathbf{X}^{Tr} = (x_1^{Tr}, x_2^{Tr}, \ldots, x_{D_{ITr}}^{Tr})^T$, the output vector is $\mathbf{O}^{Tr} = (o_1^{Tr}, o_2^{Tr}, \ldots, o_{D_{OTr}}^{Tr})^T$, the weighting matrix from the input layer to the hidden layer is $\mathbf{W}^{Tr} = [w_{ij}^{Tr}]$, where w_{ij}^{Tr} represents the ith row and jth column entry of matrix $\mathbf{W}^{Tr}$, $i = 1, \ldots, D_{HTr}$, $j = 1, \ldots, D_{Itr}$, and the weighting matrix from the hidden layer to the output layer is $\mathbf{V}^{Tr} = [v_{ij}^{Tr}]$, where v_{ij}^{Tr} is the ith row and jth column entry of matrix $\mathbf{V}^{Tr}$, $i = 1, \ldots, D_{OTr}$, $j = 1, \ldots, D_{HT_r}$.

Thus, according to basic theory of neural networks, the output of hidden layer is

$$\mathbf{H}^{Tr} = \sigma(\mathbf{W}^{Tr}\mathbf{X}^{Tr}) \tag{IV.2.17}$$

The network output is

$$\mathbf{O}^{Tr} = \varphi(\mathbf{V}^{Tr}\mathbf{H}^{Tr}) = \varphi\lfloor\mathbf{V}^{Tr}\sigma(\mathbf{W}^{Tr}\mathbf{X}^{Tr})\rfloor \tag{IV.2.18}$$

According to practical requirements, the activation functions σ and φ of Equations IV.2.17 and IV.2.18 can be sigmoid functions, hard limiter functions, saturation linear functions, hyper tangent functions, or Gaussian functions.

The cost function for training is chosen as MSE, which can be expressed as

$$\min_{\mathbf{W},\mathbf{V}} E = (\mathbf{O}^{Tr} - \mathbf{T})^T(\mathbf{O}^{Tr} - \mathbf{T}) \tag{IV.2.19}$$

On the other hand, the training algorithm of the network can be chosen from the well-documented algorithms such as the BP algorithm and quasi-Newton algorithm as well as the author-proposed SLNN algorithm.[7,21]

2.6.1.2 Hidden Nodes Pruning Algorithm

When a connection weight from a hidden node to an output node is very small, it has a smaller contribution than that with a bigger connection weight. In order to raise the representative efficiency of the network, we enforce a penalty fluctuation factor ζ $(0 < \zeta < 1)$ to the training network again. Once a connection weight is smaller than a preset threshold, we can delete it from the network. Specifically, this hidden node pruning algorithm can be divided into the following two phases:

1. *Weight penalty phase*: When a new observation comes, if $\alpha_j < \alpha_{pu}$, then $\alpha_j = \zeta\alpha_j$ ($j = 1, \ldots, D_{HTr}$), where α_{pu} represents the threshold of penalty fluctuation, and α_j is a network connection weight, such as w_{ji} or v_{lj}.
2. *Weight pruning phase*: If $\alpha_j < \alpha_{pr}$, then $\alpha_j = 0$, ($j = 1, \ldots, D_{HTr}$), where α_{pr} is referred to as the pruning threshold of hidden nodes.

In addition, if all connection weights from the jth hidden node are zeros, then this hidden node can be deleted from the network and the number of hidden nodes decreases by one.

2.6.1.3 Convergent Network

The learned network (LN for short) is a network of trained TrN whose input vector space $\subset R^{D_{IL}}$, hidden vector space $\subset R^{D_{HL}}$, and output belong to $R^{D_{OL}}$ space. Then, for LN, we have

$$\text{Input vector: } \mathbf{X}^L = (x_1^L, x_2^L, \ldots, x_{D_{IL}}^L)^T \tag{IV.2.20}$$

$$\text{Output vector: } \mathbf{O}^L = (o_1^L, o_2^L, \ldots, o_{D_{OL}}^L)^T \tag{IV.2.21}$$

The connection weight matrix from the input layer to the hidden layer is

$$\mathbf{W}^L = [w_{ij}^L], \tag{IV.2.22}$$

where w_{ij}^L is the ith and jth column entry of matrix $\mathbf{W}^L$, $i = 1, \ldots, D_{HL}, j = 1, \ldots, D_{IL}$.
The connection weight matrix from the hidden layer to the output layer is

$$\mathbf{V}^L = [v_{ij}^L], \tag{IV.2.23}$$

where v_{ij}^L is the ith and jth column entry of matrix $\mathbf{V}^L$, $i = 1, \ldots, D_{OL}, j = 1, \ldots, D_{HL}$.

$$\text{Output vector of hidden layer: } \mathbf{H}^L = \sigma(\mathbf{W}^L \mathbf{X}^L) \tag{IV.2.24}$$

$$\text{Output vector of network: } \mathbf{O}^L = \varphi(\mathbf{V}^L \mathbf{H}^L) = \varphi[\mathbf{V}^L \sigma(\mathbf{W}^L X^L)] \tag{IV.2.25}$$

2.6.2 Network's Second Convergence

2.6.2.1 Network Mapping

Rule extraction starts after completing the network training. Based on LN, we establish a rule extraction network, i.e., REN for short. In REN, we can define

$$\text{Input vector: } \mathbf{X}^R = (x_1^R, x_2^R, \ldots, x_{D_{IR}}^R)^T \tag{IV.2.26}$$

$$\text{Output vector: } \mathbf{O}^R = (o_1^R, o_2^R, \ldots, o_{D_{OR}}^R)^T \tag{IV.2.27}$$

The connection weight matrix from the input layer to the hidden layer is

$$\mathbf{W}^R = [w_{ij}^R], \qquad (IV.2.28)$$

where w_{ij}^R represents the ith and jth entry of matrix $\mathbf{W}^R$, $i = 1, \ldots, D_{HR}$, $j = 1, \ldots, D_{IR}$. The connection weight matrix from the hidden layer to the output layer is

$$\mathbf{V}^R = [v_{ij}^R], \qquad (IV.2.29)$$

where v_{ij}^R represents the ith and jth entry of matrix $\mathbf{V}^R$, $i = 1, \ldots, D_{OR}$, $j = 1, \ldots, D_{HR}$.
The network's transfer property is

$$\mathbf{H}^R = \sigma(\mathbf{W}^R \mathbf{X}^R) \qquad (IV.2.30)$$

$$\mathbf{O}^R = \varphi(V^R \mathbf{H}^R) = \varphi[\mathbf{V}^R \sigma(\mathbf{W}^r \mathbf{X}^R)] \qquad (IV.2.31)$$

where $\mathbf{H}^R \in R^{D_{HR}}$.

Definition 6: The mapping relation between REN and LN is defined as $\Psi : LN \to REN$ i.e.,

$$\Psi : \begin{cases} \mathbf{X}^R = f(\mathbf{W}^L) \\ \mathbf{W}^R \leftarrow g(\mathbf{X}^L) \\ \mathbf{V}^R = \mathbf{V}^L \\ D_{HR} = D_{HL} \end{cases} \qquad (IV.2.32)$$

where function f is a multi-dimensional mapping, i.e., $f : R^{M \times N} \to R^{1 \times MN}$. Specifically, $\forall \mathbf{X} \in R^{M \times N}$, then $\exists \mathbf{Y} \in R^{1 \times MN}$, one has $\mathbf{Y} = f(\mathbf{X})$, where

$$\mathbf{Y} = \{y_k = f(x_{ij}) | y_k = x_{\lfloor \frac{k}{N} \rfloor k \ \lfloor \frac{k}{N} \rfloor N}\} \qquad (IV.2.33)$$

Definition 7: Assume $\mathbf{X} \in R^N$, and $\mathbf{I}$ is the unit matrix. Define a substitution multiplication of $\mathbf{X}$ and $\mathbf{I}$ as

$$\mathbf{X} \otimes \mathbf{I} = \begin{bmatrix} \mathbf{X} & 0 & 0 & 0 \\ 0 & \mathbf{X} & 0 & 0 \\ 0 & 0 & \ldots & 0 \\ 0 & 0 & 0 & \mathbf{X} \end{bmatrix} \qquad (IV.2.34)$$

The function g in Equation IV.2.32 is a mapping, i.e., $g: R^{D_{IL}} \to R^{D_{HL} \times (D_{IL} D_{HL})}$. According to the mapping of Equation IV.2.32, one can obtain

$$D_{HR} = D_{HL}$$
$$D_{OR} = D_{OL} \qquad (IV.2.35)$$

For $\forall \mathbf{X} \in R^{D_{IL}}$, then $\exists \mathbf{Y} \in R^{D_{HL} \times (D_{IL} D_{HL})}$, we have $\mathbf{Y} = f(\mathbf{X})$. In concise matrix notation, we have $\mathbf{Y} = \mathbf{X} \otimes \mathbf{I}$.

Assertion: The convergence solution of REN is corresponding to a valid solution of LN such that the extrema obtained by an optimization process gave out a specific rule of the system.

Because $\mathbf{O}^R = \varphi[\mathbf{V}^R \sigma(\mathbf{W}^r \mathbf{X}^R)]$, and according to the mapping relation in Equation IV.2.32, we obtain

$$
\begin{aligned}
\mathbf{O}^R &= \varphi[\mathbf{V}^L \sigma(g(\mathbf{X}^L) f(\mathbf{W}^L))] \\
&= \varphi[\mathbf{V}^L \sigma((\mathbf{X}^L \otimes \mathbf{I}) f(\mathbf{W}^L))]
\end{aligned}
\tag{IV.2.36}
$$

According to the monotonic property of the mapping f in Equation IV.2.32 and the definition of substitution multiplication $\otimes$ in Equation IV.2.34, we have the following solution relation for the two mapping-pair networks:

$$
\mathbf{O}^R \to \varphi[\mathbf{V}^L \sigma(\mathbf{W}^L \mathbf{X}^L)] = O^L
\tag{IV.2.37}
$$

2.6.2.2 Rule Extraction Algorithm

In summary, we can re-express the rule extraction algorithm of convergent networks as follows.

Step 1:   Randomly initialize the nonzero entries of matrix $\mathbf{W}^R$.

Step 2:   Build the same objective function and a proper error criterion of LN network. Search the optimal $\mathbf{W}^R$ by the LMS algorithm subject to confining the entry value of $\mathbf{W}^R$ as $w_{ij}^R \in [0, 1]$ and disadjusting $\mathbf{V}^R$ and $\mathbf{X}^R$.

Step 3:   When the training of REN ends, obtain one rule in hypothesis space.

Step 4:   If the goal is reached, stop the extraction of rules. Otherwise, go to Step 1.

2.6.2.3 Estimation of Minimal Rules

In order to obtain a stop criterion for the above rule production algorithm, we must estimate the minimal rules implied in a data set. Here we propose a method to determine the lower limit of rules based on the difference of belief degree (DBD)[7]. It is well known that the rules built on pattern discovery are an advanced description of patterns. As a result, the estimation of minimal patterns is just an estimation of minimal rules. The belief degree is an important index to describe the significance of patterns, but the difference of belief degree represents the unpredictability of patterns.[30] The computation of belief degree is by the Bayesian method, frequency method, Cyc method, and statistical method,[31] where the Bayesian estimation method can be used to compute the belief degree for any problem. Here, the belief degree of proposition α is defined as a prior probability $P(\alpha|\zeta)$ under prior event ζ. When new event E occurs, according to the Bayesian rule, belief degree $P(\alpha|E, \zeta)$ can be formulated as

$$
P(\alpha|E, \zeta) = \frac{P(E|\alpha, \zeta)P(\alpha|\zeta)}{P(E|\alpha, \zeta)P(\alpha|\zeta) + P(E|-\alpha, \zeta)P(-\alpha|\zeta)}
\tag{IV.2.38}
$$

The main difficulty in the above computation is that conditional probability $P(E|\alpha, \zeta)$ is hard to obtain in practice. Here, we make use of $P(\hat{\mathbf{x}}|\overline{\mathbf{x}}, \text{cov}(\overline{\mathbf{x}}))$ to estimate $P(E|\alpha, \zeta)$, where $\mathbf{x} = (x_1, \ldots, x_n)$ is the characteristic parameter set of data. For example, in experiment 2 of the next section, $\mathbf{x} = $ (attribute 1, $\ldots$, attribute 9). $\overline{\mathbf{x}} = (\overline{x}_1 \ldots, \overline{x}_n)$ is a mean estimation of $\mathbf{x}$ on the basis of event ζ, while $\hat{\mathbf{x}} = (\hat{x}_1, \ldots, \hat{x}_n)$ is an estimation of $\mathbf{x}$ on the basis of event E and $\text{cov}(\mathbf{x})$ is the covariance of $\mathbf{x}$.

In experiment 2 (including 9 attributes) below, 500 training samples are randomly generated. We sorted out 283 samples belonging to group A and drew two-dimensional sample scattering plots for every two attributes in Figure IV.2.8. It is intuitively seen from the age–salary data scattering plot that there are at least three patterns related to this attribute, but the data scattering plot of other attribute-pairs are the same as age–loan (as shown in Figure IV.2.9). It can be seen from Figure IV.2.9 that the relationship of this attribute-pair is a random relation.

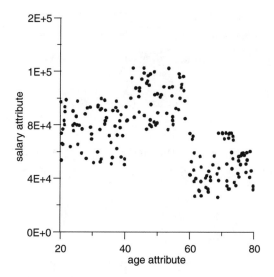

FIGURE IV.2.8   Scatter plot of attribute-pair age–salary.

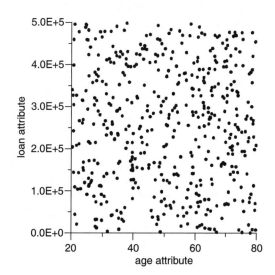

FIGURE IV.2.9   Scatter plot of attribute-pair age–loan.

For the above data, we computed the DBD of their adjacent items and plotted them in Figures IV.2.10 and IV.2.11. The difference in the two patterns' DBD is great. Therefore, it is very reasonable to estimate minimal patterns (rules) by using this kind of DBD index. When setting up an appropriate threshold, we can detect three patterns from Figure IV.2.10. In Figure IV.2.11, the detection system always demonstrates a big change in DBD for each input data, which is a typical behavior of random patterns.

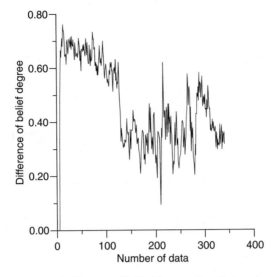

FIGURE IV.2.10   Difference of belief degree of attribute-pair age–salary.

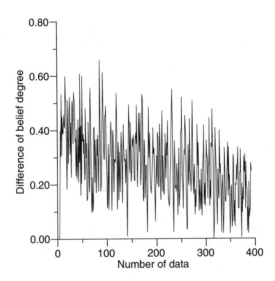

FIGURE IV.2.11   Difference of belief degree of attribute-pair age–loan.

2.7   Experiments and Analyses

2.7.1   Experiment 1: XOR Problem

In this experiment, TrN has 2 input nodes, 3 hidden nodes, and one output node. In order to make the network convergence fast, we employ the following cross-entropy function as our error function:

$$E(w, v) = - \sum_{i=1}^{S} \sum_{op=1}^{D_{OTr}} (t_{op}^i \log s_{op}^i + (1 - t_{op}^i) \log(1 - s_{op}^i)) \qquad \text{(IV.2.39)}$$

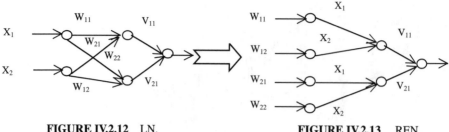

FIGURE IV.2.12   LN.   **FIGURE IV.2.13**   REN.

where S is the number of training samples. In the meantime, in order to make use of the pruning algorithm conveniently, we added a penalty function P into the error function in Equation IV.2.39.

$$P = \varepsilon_1 \left(\sum_{m=1}^{D_{HTr}} \sum_{l=11}^{D_{ITr}} \beta(w_l^m)^2 / 1 + \beta(w_l^m)^2 + \sum_{m=1}^{D_{HTr}} \sum_{p=1}^{D_{OTr}} \beta(V_p^m)^2 / 1 + \beta(V_p^m)^2 \right)$$

$$+ \varepsilon_2 \left(\sum_{m=1}^{D_{HTr}} \sum_{l=11}^{D_{ITr}} (w_l^m)^2 + \sum_{m=1}^{D_{HTr}} \sum_{p=1}^{D_{OTr}} (V_p^m)^2 \right) \qquad (IV.2.40)$$

where ε_1 and ε_2 are two positive constants to balance two terms and error E.

After the training process (final overall error $E + P = 0.0020$), we obtain LN with two hidden nodes whose network structure is shown in Figure IV.2.12. According to Equation IV.2.32, the REN obtained from the LN mapping is also shown in Figure IV.2.13. After that, we train the REN again with the update $\Delta W = -\alpha(\frac{\partial E}{\partial W} + \frac{\partial P}{\partial W})$, where W takes values from interval $[0, 1]$.

For a given target signal $O = 0$, when REN converges, the obtained weight matrices are $\begin{bmatrix} 1 & 1 & 0 & 0 \\ 0 & 0 & 1 & 1 \end{bmatrix}$ (overall error $E + P = 0.0032$) and $\begin{bmatrix} 0 & 0 & 0 & 0 \\ 0 & 0 & 0 & 0 \end{bmatrix}$ (overall error $E + P = 0.0027$).

Thus, rule 1 is $(x_1 = 1 \wedge x_2 = 1) \vee (x_1 = 0 \wedge x_2 = 0) \Rightarrow O = 0$.

For a given target signal $O = 1$, when REN converges, the obtained weight matrices are $\begin{bmatrix} 0.7176 & 0.0841 & 0 & 0 \\ 0 & 0 & 0.7176 & 0.0841 \end{bmatrix}$ (overall error $E + P = 2.7135e\text{-}004$) and $\begin{bmatrix} 0.0129 & 0.7791 & 0 & 0 \\ 0 & 0 & 0.0129 & 0.0779 \end{bmatrix}$ (overall error $E + P = 2.5929e\text{-}004$).

Thus, Rule 2 is $(x_1 = 1 \wedge x_2 = 0) \vee (x_1 = 0 \wedge x_2 = 1) \Rightarrow O = 1$.

It is obvious that Rules 1 and 2 are just the rules of XOR.

2.7.2 Experiment 2: The Problem of a Typical DM Function

Here we adopted the data mining function defined by Agrawal, Imielinski, Swami,[32] and cited and used many times by Lu, Setiono, and Liu,[17] which included 9 attributes, as shown in Table IV.2.4.

The function is defined as:

$$IF((age < 40) \wedge (50000 \leq salary \leq 100000)) \vee ((40 \leq age < 60)$$

$$\wedge(75000 \leq salary \leq 125000))$$

$$\vee((age \geq 60) \wedge (25000 \leq salary \leq 75000)),$$

THEN belong to group A. Otherwise belong to group B.

TABLE IV.2.4 Attributes and Value Ranges of Function 2 of Typical DM Functions

Attribute	Value range	Partition Levels of Value Range
E-level (education level)	[1 4] discrete value	4
Car	[1 20] discrete value	20
Zip code (zip code of the town)	[1 9] discrete value	9
Salary	[20,000 150,000] continuous	7
Commission	[10,000 75,000] continuous	7
Age	[20 80] continuous	7
H-year (age of the house)	[1 30] continuous	10
Loan (total amount of loan)	[1 500,000] continuous	10
H-value (value of house)	[0.5 k 10,000 1.5k 1,000,000] continuous (k = zipcode)	14

We randomly chose 500 samples to train TrN. In this experiment, the value of each attribute is encoded by using a thermometer-coding method. The recognition rate of the trained network is up to 96%. After network mapping and convergence, we can obtain the following rules:

Rule 1: $((20 \leq age < 30) \wedge (65000 \leq salary \leq 107000)) \Rightarrow$ Group A
Rule 2: $((50 \leq age < 60) \wedge (42000 \leq salary \leq 85000)) \Rightarrow$ Group A
Rule 3: $((70 \leq age < 80) \wedge (20000 \leq salary \leq 100000)) \Rightarrow$ Group A
Rule 4: $((80 < age) \wedge (20000 \leq salary \leq 65000)) \Rightarrow$ Group A
Rule 5: $((30 \leq age < 60) \wedge (65000 \leq salary \leq 107000)) \Rightarrow$ Group A
Rule 6: $((20 \leq age < 80) \wedge (65000 \leq salary \leq 107000)) \Rightarrow$ Group A
Default rule $\Rightarrow$ Group B

It can be seen from the extracted rules that rules 1–4 are just a subset of rules expressed by the original function except for minor boundary changes, which is probably interfered with by data noise and unrelated attributes. These rules are obtained just after 20 times of convergent computation. In our experiment, each network convergence produced a rule. Since different coding methods result in different sparse spaces of solution, some infeasible solutions need to be deleted sometimes. Rules 5 and 6 are obtained by averaging the salary of this age level. The error is generated by inaccurate partitions of attribute salary. According to the estimation of minimal rules, this experiment has at least three rules. This experimental result met this estimation and supported our estimation algorithm once again.

2.7.3 Experiment 3: Extraction of Soil Fertility-Ranking Evaluation Rules

This experiment is used to mine soil fertility-ranking evaluation rules from a soil database.

It was shown that the following 10 main factors (U) are the principal factors set to fix on soil fertility ranking:

$$U = (U_1, U_2, \ldots , U_{10})$$
$$= (\text{organism matter, all nitrogen, alkli resolving nitrogen, quick-result}$$
$$\text{phosphor, quick-result Kalium, soil natural production,}$$
$$\text{Ph value, soil CEC, weight, physical clay})$$

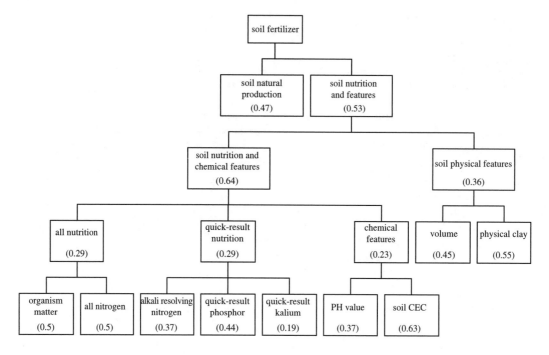

FIGURE IV.2.14   Experts' scoring weights for soil fertility ranking.

At the same time, we divided each fertility-ranking factor into four classes as the evaluation set (V) of fertility ranking:

$$V = (\; C_1 \quad C_2 \quad C_3 \quad C_4 \;)$$
$$= \text{(high, middle, low, extremely low)}$$

The hierarchical relationship of ten fertility-ranking factors and each factor's weight issued by domain experts are listed in Figure IV.2.14.

In a database, soil fertility ranking is divided into four levels, i.e., high, middle, low, and extremely low. The training set is formed by randomly choosing 500 tuples in a completely coding manner. The number of levels of $U_1, U_2, \ldots, U_6$ is four while $U_7, U_8, \ldots, U_{10}$ have 8 levels, respectively. The accuracy of the first-time convergence of the network is 92%. Through the second convergence of the mapping network, 36 rules are generated. Twenty of the 36 rules are listed in Table IV.2.5 for illustration.

2.8   Concluding Remarks

In this chapter, we discussed the neural network methods for chaotic pattern discovery and rule extraction in KDD, which is a very promising technique and has received extensive attention recently.

The chaotic pattern is a class of useful patterns existing in databases but was ignored and always treated as random fluctuations in the past, which is the embodiment of the natural complexity. In this chapter, we proposed a new chaotic pattern discovery net (CPDN) based on radial basis functional neural networks and applied it to discover and predict chaotic patterns. The CPDN can dynamically distribute the resource of the net according to the increase of input information and recycle the redundant resource efficiently. Therefore, it can not only determine whether there exists the pattern in a database, but it can also find and predict the chaotic pattern. For rule extraction from a database,

TABLE IV.2.5   Soil Fertility-Ranking Evaluation Rule by Two-Time Network Convergence (Partial)

Organic Matter	All Nitrogen	Alkali Dispelling Nitrogen	Quick Result Phosphor	Quick Result Kalium	Soil Natural Productivity	PH Value	Soil CEC	Weightiness	Physical Grume	Soil Fertility Grade
L	L	L				I	I	I		
I	I	L				H		I		
		L		H		H		L	H	
L	L		L			I	L	H	I	
			L			I		L	H	
I	I	I		L		L	L	I	L	L
I	I	I	I	L		I	H	I		L
L	L	L	H	I		I	H			L
				I				H	I	L
H	H	I	I		I	I	I	I	I	I
L	L	I	I	L	I	L	L	L	H	I
L	L	L	I	L	I	I	L	I	L	I
H	I	L	I		I	L	H	H	I	I
	I	H			H	L		I	H	I
L	H	L	I	I	H	I	H	H	H	H
L	I	H	H	L	H	H	I	L	H	H
I	I	H	I	I	H	H	I	L		H
H	I	H			H	H		I	L	H
L	I	H		I	H	H	I	I	L	H

Note: In the table, L means low, I means intermediate, H means high, and blank cells should be filled with extremely low. According to the evaluation and assessment of agricultural experts, 90% of the obtained rules are very efficient and valid.

we proposed a new rule extraction method based on two-times convergence of neural networks. This method can not only extract rules from a database efficiently, in particular for high-dimensional input space, but can also estimate the minimal rules by using the difference of belief degree. By experimenting on benchmark data and a real database, the validation of our proposed method was illustrated and justified. We believe that the neural network method of data mining will attract more attention in many communities and find increasingly practical applications in the near future.

References

1. Froyland, J., *Introduction to Chaos and Coherence*, Institute of Physics Publishing Ltd., London, 1992.
2. Klosgen, W. and Zytkow, J.M., Knowledge discovery in databases terminology, in *Advances in Knowledge Discovery and Data Mining*, Fayyad, U. et al., Eds., AAAI Press, Menlo Park, CA, 1996.
3. Casdagli, M., Nonlinear prediction of chaotic time series, *Physica D*, 35, 335, 1989.
4. Elder, J.F., IV and Pregibon, D., A statistical perspective on knowledge discovery in databases, in *Advances in Knowledge Discovery and Data Mining*, Fayyad, U. et al., Eds., AAAI Press, Menlo Park, CA, 1996.
5. Zytkow, J.M., Database exploration in search of regularities, *J. Intelligent Inf. Syst.*, 2, 39, 1993.
6. Zytkow, J.M., Creating a discoverer: autonomous knowledge seeking agent, in *Machine Discovery*, Zytkow, J.M., Ed., Kluwer, Dordrecht, 1997, 253.
7. Deng, C., *Data mining method of neural networks for knowledge discovery and its application*, Ph.D. dissertation, University of Science and Technology of China, Hefei, 1999.
8. Quinlan, J.R., *C4.5: Programs for Machine Learning*, Morgan Kaufmann, Palo Alto, CA, 1993.
9. Quinlan, J.R., Comparing connectionist and symbolic learning methods, in *Computational Learning Theory and Natural Learning Systems*, Hanson, S.J., Drastall, G.A., and Rivest, R.L., Eds., MIT Press, Cambridge, MA, 1994, 445.
10. Andrews, R., Diederich, J., and Tickle, A.B., Survey and critique of techniques for extracting rules from trained artificial neural networks, *Knowledge Based Syst.*, 8(6), 373, 1995.
11. Tickle, A.B. et al., The truth will come to light: directions and challenges in extracting the knowledge embedded within trained artificial neural networks, *IEEE Trans. Neural Networks*, 9(6), 1057, 1998.
12. Jagielska, I., Matthews, C., and Whitfort, T., An investigation into the application of neural networks, fuzzy logic, genetic algorithms, and rough sets to automated knowledge acquisition for classification problems, *Neurocomputing*, 24, 37, 1999.
13. Chen, M.S., Han, J., and Yu, P.S., Data mining: an overview from a database perspective, *IEEE Trans. Knowledge Data Eng.*, 8(5), 866, 1996.
14. Omlin, C.W. and Giles, C.L., Rule revision with recurrent neural networks, *IEEE Trans. Knowledge Data Eng.*, 8(1), 183, 1996.
15. Fu, L.M., A neural-network model for learning domain rules based on its activation function characteristics, *IEEE Trans. Neural Networks*, 9(5), 787, 1998.
16. Tsukimoto, H., Rule extraction from prediction models, in *Proceedings of the Third Pacific Asia Conference, PAKDD '99*, Beijing, China, LNAI 1574, *Methodologies for Knowledge Discovery and Data Mining*, Springer-Verlag, Heidelberg, 1999, 34.
17. Lu, H.J., Setiono, R., and Liu, H., Effective data mining using neural networks, *IEEE Trans. Knowledge Data Eng.*, 8(6), 957, 1996.
18. Platt, J., A resource-allocating network for function interpolation, *Neural Comput.*, 3, 213, 1991.
19. Kadirkamanathan, V., A function estimation approach to sequential learning with neural networks, *Neural Comput.*, 5, 954, 1993.

20. Lu, Y. et al., A sequential learning scheme for function approximation using minimal radial basis function neural networks, *Neural Comput.*, 9, 461, 1997.
21. Deng, C. et al., Sequential learning neural networks and its application in agriculture, in *Proceedings of the 1998 IEEE International Joint Conference on Neural Networks (IJCNN '98)*, Alaska, 1998, 2118.
22. Nielsen, R.H., Kolmogorov mapping neural network existence theorem, *Proc. 1987 Int. Conf. Neural Networks*, 3, 11, 1987.
23. Piggio, T. and Girosi, F., A theory of networks for approximation and learning, AI Memo. no. 1140, AI Lab, Massachusetts Institute of Technology, Cambridge, MA, 1989.
24. Yaratman, E.J., Keeler, J.D., and Kowalski, J.M., Layered neural networks with Gaussian hidden units as universal approximations, *Neural Comput.*, 2, 210, 1990.
25. Park, J. and Sandberg, I.W., Approximation and radial basis function networks, *Neural Comput.*, 5, 305, 1993.
26. Tan, Y., Gao X.Q., and He, Z.Y., Neural network design approach of cosine-modulated FIR filter bank and compactly supported wavelets with almost PR property, *Signal Process.*, 69(1), 22, 1999.
27. Tan, Y. and He, Z.Y., Arbitrary FIR filter synthesis with neural network, *Neural Process. Lett.*, 8(1), 9, 1998.
28. Bruce, A.W. et al., Cooperative-competitive genetic evolution of radial basis function centers and widths for time series prediction, *IEEE Trans. Neural Networks*, 7(4), 869, 1996.
29. Langari, R., Wang, L., and Yen, J., Radial basis function networks, regression weights, and the expectation-maximization algorithm, *IEEE Trans. Systems Man Cybernetics—Part A: Syst. Humans*, 27(5), 613, 1997.
30. Silberschata, A. and Tuzhilin, A., What makes patterns interesting in knowledge discovery system, *IEEE Trans. Knowledge Data Eng.*, 8(6), 970, 1996.
31. Kachigan, S.K., *Statistical Analysis*, Radius Press, 1986.
32. Agrawal, R., Imielinski, T., and Swami, A., Database mining: a performance perspective, *IEEE Trans. Knowledge Data Eng.*, 5(6), 914, 1993.

3

Topological Relations in Spatial Databases

Eliseo Clementini
University of L'Aquila

Abstract. Various models for the representation of topological relations have been developed. The aim of this chapter is to propose a comprehensive view of state-of-the-art models for topological relations in spatial databases. A prominent role is assumed by the CBM (calculus-based method), which offers a small set of topological relations with high expressiveness and suitability to be embedded as operators of a spatial query language. The CBM is applicable to all kinds of complex geometric features and constitutes the user-oriented level (top level) of a hierarchy of topological operators. At lower levels (with more geometric details), other models are proposed, which are also applicable to geometric features with a broad boundary (features that are able to model spatial uncertainty). Eventually, a classification of all topological invariants of a spatial relation between two features offers the tools to describe the finest topological details.

3.1 Introduction

Spatial database systems have constantly increased in popularity during the last decade. The major motivation pushing research in spatial database systems is the large number of application domains they involve.[1] An incomplete list of these domains includes: geographical information systems (GIS), scientific databases, pictorial databases, and CAD. Spatial data management imposes a number of new requirements on database systems when compared with traditional ones. Among the many requirements, the formalization of spatial relations among physical objects has a central role, since they occur in the majority of spatial queries. The following three examples are typical spatial queries borrowed from the geographical context:

- Retrieve all the states *bordering on* Italy.
- Retrieve the names of the lakes bigger than 40 square miles and located *northwest of* Venice.
- Retrieve the three-star hotels *close to* the railway station in Rome.

The usual way of classifying spatial relations is in terms of two basic categories, namely, topological and metric (orientation and distance):

- Topological relations (e.g., adjacent, overlap, . . .) describe whether two objects intersect or not and, in the former case, how they intersect.
- Orientation relations (e.g., north-of, south-of, . . .) describe where an object is located with respect to a reference.
- Distance relations (e.g., very close, close, . . .) describe how far an object is with respect to a reference.

Within these categories, topological relations have been studied in more depth.[2–19] Among the available models for classifying topological relations, in this chapter, we will review the evolution and the application of the so-called calculus-based method (CBM).[2,14,18] The CBM allows us to model the topological relation between two spatial objects represented as 2D geometric elements (namely, points, lines, areas, and collections of those) in terms of five relations (i.e., in, overlap, cross, touch, and disjoint) and three boundary operators. The CBM enjoys three formal properties:

- *Completeness*—The five relations make a full covering of all the possible topological situations.
- *Exclusiveness*—It cannot be the case that two different relations hold between the same two objects.
- *Expressiveness*—The CBM is better able to distinguish among finer topological configurations than an entire category of previous methods, all based on point-set topology, namely: the four-intersection model,[4] the dimension extended method,[2] and *the 9-intersection model*.[5] The expressiveness of the CBM was investigated in Clementini and Di Felice.[13]

The previous properties, combined with the small size of the set of relations and operators, make the CBM a natural candidate for being part of a spatial query language. It is worthwhile to note that existing spatial query languages[20–26] offer some topological operators (for example, on, adjacent, and within in the case of MAPQUERY[21]), however none of the above query languages discusses issues like the expressive power of the topological operators offered or their completeness with respect to some predefined formal criteria. Recently, the CBM has been adopted as the standard way of defining topological operators inside a spatial extension of SQL by the OpenGIS Consortium.[27]

Most of the literature on topological relations used to deal with simple areas and lines (simple areas are two-dimensional point sets homeomorphic to a disk, and simple lines are one-dimensional features embedded in the plane with only two endpoints). Unfortunately, the variety and complexity of spatial entities cannot be modeled with simple geometric features. In order to become operational within a real environment, spatial models should be general enough to include complex objects. Based on dimensionality, a *complex feature* is either a *complex area feature* (that is, an area made up of several components possibly containing holes), a *complex line feature* (that is, a line with separations, self-intersections, and an arbitrary number of endpoints), or a *complex point feature* (that is, a set of points thought of as a single entity).

Let us put various contributions in a historical perspective. Egenhofer et al.[8] extended the four-intersection model[4] to cover the case of topological relations between areas with holes. Worboys and Bofakos[19] proposed a tree-based model for the representation of areas with holes and islands nested to any finite level. Clementini et al.[9] proposed a hierarchical model, based on the CBM, able to treat composite areas in topological queries. In Clementini and Di Felice,[14] an extension of the CBM for the case of complex features was given, and it was formally proved that the topological relations are mutually exclusive and complete. The CBM also applies for modeling topological relations between 3D features (a 3D implementation of the CBM was described by van Oosterom et al.[28]). The spatial model adopted by the OpenGIS Consortium[27] is based on complex features (multipoints, multipolylines, and multipolygons).

Recently, another extension to spatial data models was considered with the goal of treating uncertainty of spatial data. In essence, a new geometric model is needed that overcomes the limits of the current models of spatial databases, which traditionally are a collection of features with a crisp boundary, which are a rough approximation of geographic reality. The proposal of many research papers is to introduce broad boundaries replacing crisp ones.[15,16,18,29−31] The advantage of this approach is that it can be implemented on existing database systems at reasonable cost: the new model can be seen as an extension of existing geometric models. More effort is needed to define new operators applying to the new category of features in order to use them with profit. A first step in this direction took place defining approximate topological relations.[16,18] Clementini and Di Felice[18] verified that the CBM can also be applied to features with a broad boundary by replacing the definition of set intersection between crisp areas with set intersection between areas with a broad boundary. The CBM relations applied to features with a broad boundary maintain the formal properties of being mutually exclusive and providing a complete coverage of all the realizable geometric configurations. The CBM relations can be seen as the upper level of a hierarchy of topological operators that allow users to query uncertain spatial data independently of the underlying geometric data model.

Topological relations that have been considered by the CBM, the 9-intersection, and all other models discussed so far are based on the analysis of few topological properties (also called *topological invariants*) of the two features involved in the scene: essentially, the void or non-void intersection between parts of the features (the so-called *content invariant*). The topological relation between two features can be refined by analyzing all topological invariants that characterize the relation. Topological properties do not change after topological transformations, which include very common GIS transformations such as rotation, translation, scaling, and rubber sheeting. If two different scenes have the same topological invariants, they are topologically equivalent. Formal criteria to establish topological equivalence are extremely important both for recognizing a particular configuration of features and for checking if a scene has been transformed consistently. We will propose a set of topological invariants that is necessary and sufficient to characterize topological equivalence, taken from Clementini and Di Felice.[17] The proposed set of invariants to fully describe a topological relation is relevant for the definition of a spatial query language that is able to consider different levels of granularity in topological queries. By considering all topological invariants, the query language would be able to operate at the most detailed level of granularity. Operators at higher levels can be defined starting from the most basic. This process also involves the meaning of natural language spatial expressions (e.g., the road crosses the park), which can be defined by aggregating several similar topological equivalence classes that have a common property.

This chapter is organized as follows: in Section 3.2, we define complex geometric features. Section 3.3 introduces the CBM. In Section 3.4, the 9-intersection model for topological relations between features with a broad boundary is described. Section 3.5 introduces a hierarchy of topological operators. In Section 3.6, the topological invariants are discussed. Section 3.7 gives short conclusions.

3.2 Complex Features

The aim of this section is to introduce basic concepts of *point-set topology* related to the definition of geometric two-dimensional features, which is the usual model for representing 2D projections of

3D scenes. The definition of geometric features is purely topological since we disregard all the shape and metric properties and concentrate on the study of topological relations.

A *topological space* is generally described as a set of arbitrary elements (points) in which a concept of continuity is specified. A mapping of a topological space X onto a subset of a topological space Y in which the neighborhood relations between mapped points are preserved in both spaces is called a *continuous mapping*. These mappings are also called *topological transformations* and include translation, rotation, and scaling. Topological relations are those remaining invariant under topological transformations. The study of topological relations between features also depends on the embedding space, which we assume to be $\mathbb{R}^2$.

First, we extend the definitions of area, line, and point given in Clementini, Di Felice, and van Oosterom[2] in order to take into account complex areas having both separations and holes, complex lines having separations, more than two endpoints, possibly self-intersections, and, also, sets of points as a single complex point feature. Complex features are far more common than simple ones in reality.

In the following, we introduce the notion of complex geometric features (complex area, complex line, and complex point, respectively); for each feature, the notions of boundary, interior, closure, and exterior are given. All the definitions in this section are based on the concepts of open sets, closed sets, and continuity of point-set topology.[32] Since all the point-sets we refer to in this chapter are subsets of $\mathbb{R}^2$, usually we do not repeat it.

3.2.1   Complex Area Features

Let A be a two-dimensional point-set:

- The interior of A, A°, is defined as the union of all open sets contained in A.
- The closure of A, $\overline{A}$, is defined as the intersection of all closed sets containing A. If A is a closed set, then $\overline{A} = A$.
- The boundary of a closed set A, ∂A, is defined as the intersection between its closure and its interior.
- The exterior of a closed set A, $\overline{A}$, is defined as $\overline{A} = (\mathbb{R}^2 - \overline{A})$.

Throughout this chapter, we assume that areas are regular closed sets.

Definition 1: A point-set A is called regular closed if and only if $A = \overline{A^\circ}$.

Regular closed point-sets are usually assumed in spatial data modelling[19] because they allow getting rid of strange cases such as those shown at the top level of Figure IV.3.1.

The notion of complex area relies on the topological notions of connectedness and component.[32] A separation of A is a pair of disjoint non-empty open sets A_1 and A_2 whose union is A. A is connected if there exists no separation of A, disconnected otherwise, and A_1 and A_2 are called the components of A. Separations and components refer to interiors of areas, while spatial features are represented as closed sets, in order to permit an area to also contain its boundary.

Definition 2: A simple area is a regular, closed (non-empty), two-dimensional point-set, A, with a connected interior and connected exterior.

Let us now proceed to extend the above definition by taking holes into account (Figure IV.3.2). The exterior of an area with holes may be separated. Separations of the exterior imply that there exists one outer exterior (unbounded set) and $n > 0$ inner exteriors (bounded sets). The outer exterior will be denoted by A_0^- and the inner exteriors by $A_1^- \ldots A_n^-$. Their union makes up the entire exterior, i.e.,

$$A^- = \bigcup_{i=0}^{n} A_i^-$$

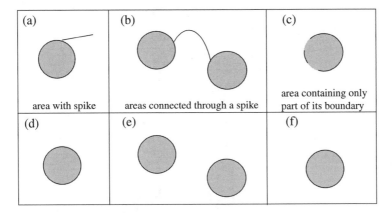

FIGURE IV.3.1 Examples of non-regular closed point-sets (cases (a), (b), and (c)) and their counterparts after "regularization" (cases (d), (e), and (f), respectively).

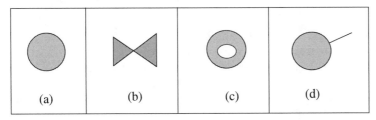

FIGURE IV.3.2 A valid (a) and three invalid ((b), (c) and (d)) simple areas.

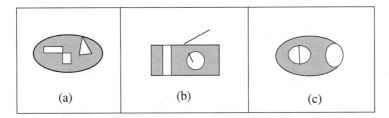

FIGURE IV.3.3 A valid (a) and two invalid ((b) and (c)) areas with holes.

Definition 3: An area with holes is a regular, closed (non-empty), two-dimensional point-set, A, with a connected interior such that the intersection of the closures of any two different exterior components is empty or equal to a finite set of points:

$$\forall i, j = 0 \ldots n, i \neq j : (\overline{A_i^-} \cap \overline{A_j^-} = \emptyset) \vee (\overline{A_i^-} \cap \overline{A_j^-} = \{p_1 \ldots p_k\})$$

According to Definition 3, case (a) in Figure IV.3.3 is allowed, while cases (b) and (c) are both not allowed since they do not correspond to regular closed point-sets.

If we relax the constraint that the interior of A, A°, is connected, we can define the components $A_1 \ldots A_n$ of A as the closure of the corresponding components of A°.

Definition 4: A complex area is a closed (non-empty) two-dimensional point-set A with components $A_1 \ldots A_n$, such that:

1. Each A_i is either a simple area or an area with holes.
2. $A_i^\circ \cap A_j^\circ = \emptyset \ \forall i \neq j$.
3. $(\partial A_i \cap \partial A_j = \emptyset) \vee (\partial A_i \cap \partial A_j = \{p_1 \ldots p_k\})$.

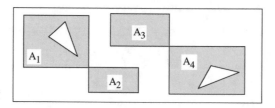

FIGURE IV.3.4 A complex area made up of four components.

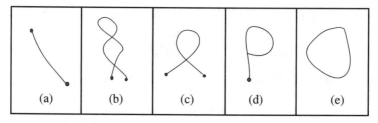

FIGURE IV.3.5 A simple line (a) and four lines with self-intersections (b–e).

Notice that a complex area is a regular closed point-set because, by definition, it is the union of a finite number of regular closed point-sets. Figure IV.3.4 shows an example of a complex area.

3.2.2 Complex Line Features

Definition 5: A simple line is a closed (non-empty) one-dimensional point-set, L, defined as the image of a continuous mapping $f : [0, 1] \rightarrow \mathbb{R}^2$, such that

$$f(t_i) \neq f(t_j), \; \forall t_i, t_j \in [0, 1], t_i \neq t_j.$$

The mappings of 0 and 1 through f are the two endpoints of L; these two points make up the boundary of a simple line in the plane; i.e., $\partial L = \{f(0), f(1)\}$.

By relaxing the constraint of no self-intersections in the interior, we get the notion of line with self-intersections. Figure IV.3.5 shows a simple line and four lines with self-intersections. The number of endpoints for a line with self-intersections can be either 2, 1, or 0. In detail, the number of endpoints of L is equal to:

- Two, if $f(0) \notin f((0, 1)) \wedge f(1) \notin f((0, 1))$; $\partial L = \{f(0), f(1)\}$
- One, if $f(0) \notin f((0, 1)) \wedge f(1) \notin f((0, 1))$; $\partial L = \{f(1)\}$ (vice versa if $f(0) \notin f((0, 1)) \wedge f(1) \in f((0, 1))$; $\partial L = \{f(0)\}$)
- Zero, if $f(0) = f(1)$; $(f(0) \in f((0, 1)) \wedge f(1) \in f((0, 1)))$; $\partial L = \emptyset$

Definition 6: Let $f_1, f_2, \ldots f_n$ be continuous mappings from the interval $[0, 1]$ to the plane. We call a complex line any closed (non-empty) one-dimensional point-set, L, defined as the union of the image of the functions $f_1, f_2, \ldots f_n$:

$$f_1([0, 1]) \cup f_2([0, 1]) \cup \cdots \cup f_n([0, 1])$$

Figure IV.3.6 shows a complex line made up of four components.

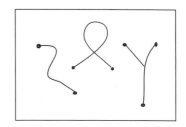

FIGURE IV.3.6 A complex line.

The boundary of L, i.e., the set of its endpoints, can be built subtracting from the set of endpoints of single f_i all the endpoints that connect with the lines, that is:

$$\begin{aligned}
\partial L = \ &\{f_1(0), f_1(1), f_2(0), f_2(1), \ldots, f_n(0), f_n(1)\} \\
&- \{f_i(0) | i \in 1 \ldots n, f_i(0) \in f_i([0, 1])\} \\
&\cup \{f_i(1) | i \in 1 \ldots n, f_i(1) \in f_i([0, 1])\} \\
&\cup \{f_i(0) | i, \exists j \in 1 \ldots n, i \neq j, f_i(0) \in f_i([0, 1])\} \\
&\cup (\{f_i(1) | i, \exists j \in 1 \ldots n, i \neq j, f_i(1) \in f_i([0, 1])\})
\end{aligned}$$

The closure of a line, $\overline{L}$, is the set of all points of L, endpoints included; therefore, $\overline{L} = L$. The interior of a line, L°, is the set difference between the closure of the line and its boundary: $L^\circ = \overline{L} - \partial L = L - \partial L$. The exterior of a line, L^-, is the set difference between the embedding space and the closure of the line: $L^- (\mathbb{R}^2 - \overline{L}) = (\mathbb{R}^2 - L)$.

3.2.3 Complex Point Features

Definition 7: A point is a (non-empty) zero-dimensional point-set, P, consisting of only one element.

As an obvious generalization of the previous concept, we introduce the notion of complex point features as follows.

Definition 8: A complex point is a (non-empty) zero-dimensional point-set, P, consisting of a finite number of distinct elements.

Throughout this chapter, we consider the boundary of a complex point feature P empty (i.e., $\partial P = \emptyset$); as a consequence, the interior of a complex point feature P is equal to the union of all the elements in P (i.e., $P^\circ = \bigcup_{i=1}^n P_i$).

3.2.4 Features with Broad Boundaries

Below we give the definition of features with broad boundaries, which differ from crisp ones with regard to the boundary definition. For an area with a broad boundary, we can define an inner boundary and an outer boundary, where the inner boundary is surrounded by the outer boundary. The closed annular area between the inner and outer boundary is the broad boundary of the original area.

Definition 9: An area with a broad boundary A is made up of two areas A_1 and A_2, with $A_1 \subseteq A_2$, where ∂A_1 is the inner boundary of A and ∂A_2 is the outer boundary of A.

Definition 10: The broad boundary ΔA of an area with a broad boundary A is the closed subset between the inner boundary and the outer boundary of A, i.e., $\Delta A = \overline{A_2 - A_1}$ or, equivalently, $\Delta A = A_2 - A_1^\circ$.

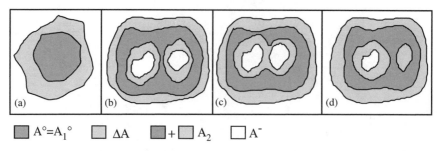

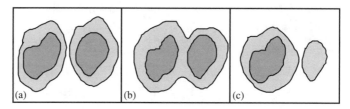

FIGURE IV.3.7 Areas with a broad boundary.

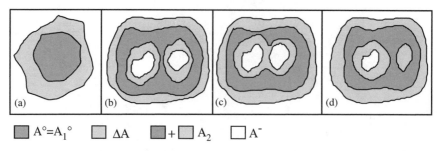

FIGURE IV.3.8 Composite areas with a broad boundary.

Definition 11: The interior, closure, and exterior of an area with a broad boundary A are defined as $A^\circ = A_2 - \Delta A, \overline{A} = A^\circ \cup \Delta A, A^- = \mathbb{R}^2 - \overline{A}$, respectively.

The interior and exterior of an area with a broad boundary are open sets, while the broad boundary is a closed set. Notice that both the exterior and broad boundary of an area A may have several components, because areas A_1 and A_2 may have holes. Figure IV.3.7. illustrates some cases that can arise: case (a) is an area without holes; case (b) is an area with two holes; case (c) is an area where A_1 has one hole and A_2 has two holes; case (d) is an area where A_1 has two holes and A_2 has one hole.

Now, we consider the most general case of composite areas with a broad boundary:

Definition 12: A composite area with a broad boundary A is made up of two composite areas A_1 and A_2, with $A_1 \subseteq A_2$, where ∂A_1 is the inner boundary of A and ∂A_2 is the outer boundary of A.

Definition 13: The broad boundary ΔA of a composite area with a broad boundary A is the closed subset between the inner boundary and the outer boundary of A, i.e., $\Delta A = A_2 - A_1$ or, equivalently, $\Delta A = A_2 - A_1^\circ$.

Definition 14: The interior, closure, and exterior of a composite area with a broad boundary A are defined as $A^\circ = A_2 - \Delta A, \overline{A} = A^\circ \cup \Delta A, A^- = \mathbb{R}^2 - \overline{A}$, respectively.

Figure IV.3.8 illustrates some configurations of composite areas: case (a) is an area with two components; case (b) is an area where A_1 has two components and A_2 has one component; case (c) is an area where A_1 has one component and A_2 has two components.

For lines embedded in $\mathbb{R}^2$, we can distinguish two kinds of broad boundaries, modeling either the position of the line or the position of its endpoints (Figure IV.3.9). The first kind of broad boundary is an area surrounding the whole line, while the second kind is made up of an area for each endpoint. We will refer to the first kind as a broad line and to the second one as a line with a broad boundary. A broad point is a special case of a broad line. The following definitions hold:

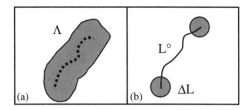

FIGURE IV.3.9 Broad lines (a) and lines with a broad boundary (b).

Definition 15: A broad line Λ is a simple area representing a family of positions that a simple line L_1 can assume under a continuous deformation. The interior of Λ is empty, while $\Delta\Lambda = \Lambda$.

Definition 16: A line with a broad boundary L is made up of a simple line L_1 and two simple areas A_1 and A_2, surrounding the two endpoints P_1 and P_2 of the line L_1, respectively.

Definition 17: The broad boundary ΔL of a line with a broad boundary L is the union of the two simple areas A_1 and A_2, that is, $\Delta L = A_1 \cup A_2$.

Definition 18: The interior, closure, and exterior of a line with a broad boundary L are defined as $L^\circ = L_1 - \Delta L, \overline{L} = L_1 \cup \Delta L$, and $L^- = \mathbb{R}^2 - \overline{L}$, respectively.

3.3 The Topological Relations of the CBM

In this section, we refer to generic complex features without the need to distinguish among areas, lines, and points. λ will be used to denote these features. Therefore, if λ is a point-set, then $\partial\lambda, \lambda^\circ, \overline{\lambda}, \lambda^-$, and dim (λ) will denote the boundary, the interior, the closure, the exterior, and the dimension of λ, respectively. The latter is a function, which returns the dimension of a point-set (0, 1, or 2) or nil $(-)$ for the empty set. In case the point-set consists of multiple parts, then the highest dimension is returned. The CBM, as it appeared in Clementini, Di Felice, and van Oosterom[2] for the case of simple features, was made up of five relations and three boundary operators. For the case of complex features, the CBM needs slight modifications to one relation (the cross) and to the boundary operators. For features with a broad boundary, as proven in Clementini and Di Felice,[18] the definitions are the same. In the following, we give all the definitions.

Definition 19: The touch relation (applies to area/area, line/line, line/area, point/area, and point/line groups of relations, but not to the point/point group):

$$\langle \lambda_1, \text{touch}, \lambda_2 \rangle \Leftrightarrow (\lambda_1^\circ \cap \lambda_2^\circ = \emptyset) \wedge (\lambda_1 \cap \lambda_2 \neq \emptyset).$$

Definition 20: The in relation (applies to every group):

$$\langle \lambda_1, \text{in}, \lambda_2 \rangle \Leftrightarrow (\lambda_1 \cap \lambda_2 = \lambda_1) \wedge (\lambda_1^\circ \cap \lambda_2^\circ \neq \emptyset).$$

Definition 21: The cross relation (applies to line/line, line/area, point/area, and point/line groups):

$$\langle \lambda_1, \text{cross}, \lambda_2 \rangle \Leftrightarrow (\dim(\lambda_1^\circ \cap \lambda_2^\circ) < \max(\dim(\lambda_1^\circ), \dim(\lambda_2^\circ)))$$
$$\wedge (\lambda_1 \cap \lambda_2 \neq \lambda_1) \wedge (\lambda_1 \cap \lambda_2 \neq \lambda_2).$$

Definition 22: The overlap relation (applies to area/area, line/line, and point/point groups):

$$\langle \lambda_1, overlap, \lambda_2 \rangle \Leftrightarrow ((\dim(\lambda_1^\circ) = \dim(\lambda_2^\circ) = (\dim(\lambda_1^\circ \cap \lambda_2^\circ))$$
$$\wedge (\lambda_1 \cap \lambda_2 \neq \lambda_1) \wedge (\lambda_1 \cap \lambda_2 \neq \lambda_2).$$

Definition 23: The disjoint relation (applies to every group):

$$\langle \lambda_1, disjoint, \lambda_2 \rangle \Leftrightarrow \lambda_1 \cap \lambda_2 = \emptyset.$$

Definition 24: The boundary operator b for a complex area A: the pair (A, b) returns the complex line ∂A, which is the union of several (either disjoint or intersecting in a finite number of points) circular lines (with no endpoints).

Definition 25: The boundary operators f, t for a complex line L: the pairs (L, f) and (L, t) return the two complex points $\{f_i(0), \forall i\} \cap \partial L$ and $\{f_i(1), \forall i\} \cap \partial L$.

3.4   The Topological Relations of the 9-Intersection

Binary topological relations between two features, A and B, in $\mathbb{R}^2$ can be classified according to the intersection of A's interior, boundary, and exterior with B's interior, boundary, and exterior. The nine intersections between the six feature parts describe a topological relation and can be concisely represented by the following 3×3 matrix M, called the *9-intersection*:[5]

$$M = \begin{pmatrix} A^\circ \cap B^\circ & A^\circ \cap \partial B & A^\circ \cap B^- \\ \partial A \cap B^\circ & \partial A \cap \partial B & A\partial \cap B^- \\ A^- \cap B^\circ & A^- \cap \partial B & A^- \cap B^- \end{pmatrix}.$$

By considering the values empty (0) and nonempty (1), we can distinguish between $2^9 = 512$ binary topological relations. For two simple areas with a one-dimensional boundary, only eight of them can be realized, and those are disjoint, meet, overlap, coveredBy, inside, covers, contains, equal. For two composite areas with a one-dimensional boundary, there are eight additional matrices that can be realized, totaling 16 relations.* Each set of relations provides complete coverage and is mutually exclusive.[4]

The 9-intersection model has been extended to simple areas with a broad boundary (non-composite and without holes).[15,16] The matrix M needs to be redefined as follows, having a broad boundary in place of a sharp boundary:

$$M = \begin{pmatrix} A^\circ \cap B^\circ & A^\circ \cap \Delta B & A^\circ \cap B^- \\ \Delta A \cap B^\circ & \Delta A \cap \Delta B & \Delta A \cap B^- \\ A^- \cap B^\circ & A^- \cap \Delta B & A^- \cap B^- \end{pmatrix}$$

Clementini and Di Felice,[15] showed that there are 44 realizable matrices for simple areas with a broad boundary.** In Figure IV.3.10, the 44 topological relations are illustrated. Notice that the numbering of relations is kept the same as in the mentioned paper[15] for compatibility.

---

*Currently, there are no names in the literature for these additional relations.

**The interested reader may refer to that paper[15] for understanding the geometric process that leads to exclude intersection matrices that do not correspond to a physical realization of a topological relation.

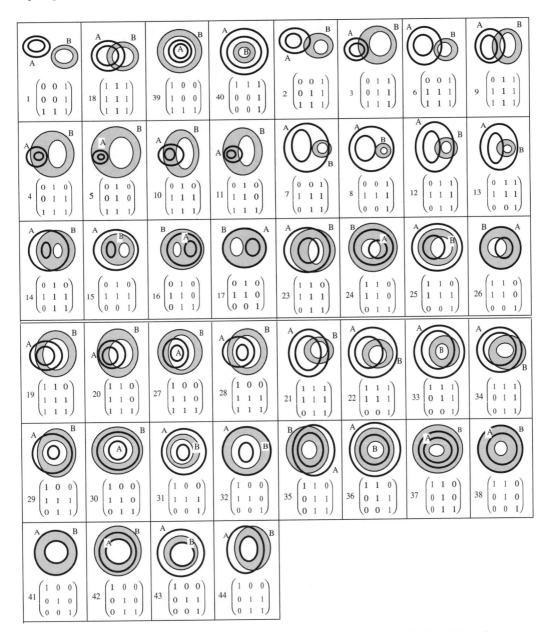

FIGURE IV.3.10 The 44 topological relations between simple regions with a broad boundary.

We now discuss the extension of the previous result to composite areas with a broad boundary. Clementini and Di Felice[15] discussed the geometric constraints that lead to 44 relations for simple areas with a broad boundary. Such constraints need to be replaced by a set of less restrictive constraints that allow for 56 possible relation matrices for composite areas with a broad boundary. In this analysis, areas with a broad boundary having an empty interior are excluded because they would generate matrices with an indeterminate A° or B°. Also, we assume that each component of an area has a non-empty interior.

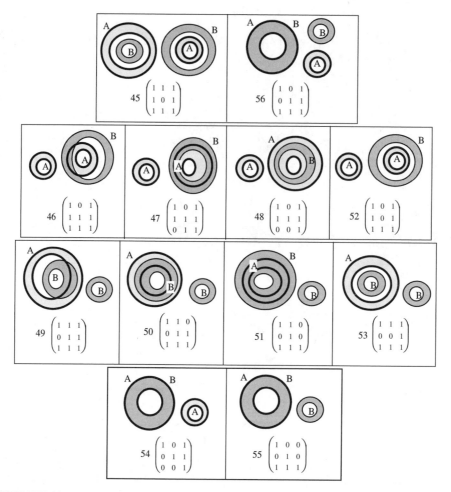

FIGURE IV.3.11   The 12 additional topological relations for composite regions with a broad boundary.

In detail, the following geometric constraints that must hold for simple areas with a broad boundary are no longer valid for composite areas (see Clementini and Di Felice[15]):

- If A's interior intersects with B's interior and exterior, then it must also intersect with B's boundary, and vice versa.
- If both boundaries intersect with the opposite interiors, then the boundaries must also intersect with each other.
- If A's broad boundary intersects with B's interior and exterior, then it must also intersect with B's broad boundary, and vice versa.

By removing the four constraints above, there are 12 additional relation matrices, besides the 44 already known, for composite areas with a broad boundary, whose geometric interpretations are given in Figure IV.3.11.

In Clementini and Di Felice,[16] spatial relations were organized in a graph having a node for each relation and an arc for each pair of matrices at a minimum topological distance. Such a distance is measured in terms of the number of different values in the corresponding matrices. This kind of graph has been called the *closest topological relation graph*.[33] Another way of interpreting the graph is to consider the arc between two relations as a smooth transition that can transform one relation to the other and vice versa. This kind of graph under the latter interpretation has been called the *conceptual neighborhood*[34] and is very useful for the analysis of deformations that can affect topological relations

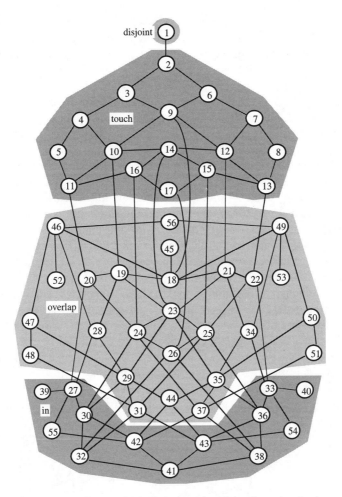

FIGURE IV.3.12   The topological relation graph and the clustering for the top level.

during motion or changes over time.* A sequence of elementary deformations corresponds to a path in the graph. The graph is extremely useful for spatial reasoning and for the process of grouping relations in clusters (as it will be clear in Section 3.5). Such clusters are meaningful to build more general operators, enabling users to communicate with the information system. The clustering process can be done in a complementary manner with a formal basis and with human subjects testing: it has been shown that both approaches reach a substantial agreement over the significant groupings.[35]

We have completed the graph for the new model for composite areas with a broad boundary: Figure IV.3.12 presents the new graph for the 56 relations, as well as a way of clustering them, which will be explained in Section 3.5.

3.5   Hierarchy of Topological Operators

In this section, we define a hierarchy of topological operators to be integrated in an SQL spatial extension such as that proposed by the OpenGIS Consortium.[27] The topological operators are hierarchically organized in three levels: the bottom level offers operators the ability to check for

---

*The same graph has also been called a *continuity network*.[3]

detailed topological relations between areas with a broad boundary using the model of Section 3.4. The intermediate and top level offer more abstract operators that allow users to query uncertain spatial data independently of the underlying geometric data model. The results of the application of an operator at one level can be refined by the corresponding operators at a lower level.

The OpenGIS model currently supports two kinds of topological operators: (1) a set of high level user-oriented operators (disjoint, touch, overlap, etc.) that have been extracted from the theoretical work of Clementini and Di Felice;[15] and (2) a low level geometric-oriented operator (relate) that is able to test for the specific topological relation holding among two features in terms of the corresponding 9-intersection matrix.

The model developed in Section 3.4 is able to extend the second kind of topological operators (the geometric-oriented ones) towards features with a broad boundary. Such a model has the advantage that users can test for a large number of spatial relations and fine tune the particular relation being tested. It has the disadvantage that it is a lower level building block and does not have a corresponding natural language equivalent. Users of spatial databases include, among others, SQL interactive users who may wish, for example, to select all the features "nearly overlapping" a query polygon. To address the needs of such users, a set of named spatial relation predicates for simple features has been recalled in Section 3.3, i.e., the calculus-based method (CBM).

The relations and the boundary operators of the CBM can be seen as the upper level of a hierarchy of topological operators having the relation matrices of the 9-intersection at the bottom level. The relations of the CBM are mutually exclusive and provide a complete coverage of all the realizable geometric configurations; as was proven by Clementini and Di Felice,[13] every topological relation expressed by a 9-intersection matrix can be translated into a boolean expression over the CBM relations and boundary operator.

We verified that the definitions of the CBM relations can be also applied to features with a broad boundary by replacing the intersection between features with a crisp boundary with the intersection between areas with a broad boundary. The set intersection between two features with a broad boundary is a new feature with a broad boundary $C = A \cap B$, which is defined analogously to ordinary point set intersection as follows:

$\cap$	B°	ΔB	B^-
A°	C°	ΔC	C^-
ΔA	ΔC	ΔC	C^-
$A-$	C^-	C^-	C^-

The CBM relations applied to features with a broad boundary maintain the formal properties of being mutually exclusive and providing a complete coverage of all the realizable geometric configurations. The proof of these properties would be straightforward, being similar to the work of Clementini and Di Felice,[13] and, therefore, it will not be repeated. The correspondence between the CBM relations and the 56 relations of the 9-intersection for composite areas with a broad boundary is shown in Figure IV.3.12: the CBM partitions the set of 56 relations into four clusters.

With only one exception (concerning the cluster induced by the relation disjoint), the number of geometric configurations falling inside a cluster is quite large. The CBM provides the user with a degree of resolution that is not able to capture the geometric details related to the presence of broad boundaries and separate components. It follows that an intermediate level of clustering of topological relations is needed to reach more detail than that offered by the CBM without going down to the bottom level of 56 relations. We give to each intermediate level cluster a name that corresponds to a topological operator.

Each of the previous CBM induced clusters can be subdivided into smaller ones following different criteria such as geometric patterns and topological distance. Specifically, we propose the splitting

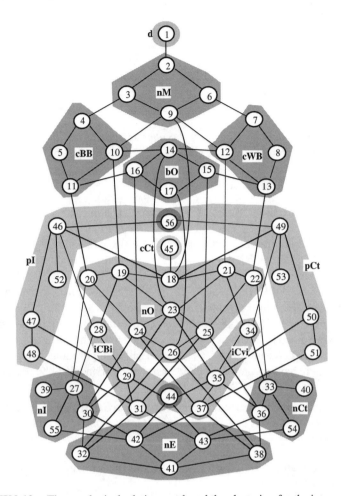

FIGURE IV.3.13   The topological relation graph and the clustering for the intermediate level.

of the touch cluster by identifying pattern matrices that characterize visually similar geometric configurations. The in cluster splits into three smaller clusters derived by aggregating the relations which are at distance 1 from specific relations that are valid also for simple areas with a crisp boundary. The overlap cluster is split into several smaller clusters following a combination of the criteria above. The intermediate level clusters are shown in Figure IV.3.13.

Below, each intermediate cluster is described by listing the numbers of relations (subsequently also called cases) they are composed of and the adopted criteria. The names given to the clusters are obtained by identifying a prototypical geometric configuration representing the cluster (see later in this section). However, we do not exclude that other possible names might be more descriptive for the operators in particular application contexts.

The intermediate level clusters coming from the touch relation are the following:

- *nearlyMeet* (**nM**) (cases 2, 3, 6, 9): $\begin{pmatrix} 0 & \delta & 1 \\ \delta & 1 & 1 \\ 1 & 1 & 1 \end{pmatrix}$

- *coveredByBoundary* (**cBB**) (cases 4, 5, 10, 11): $\begin{pmatrix} 0 & 1 & 0 \\ \delta & 1 & \delta \\ 1 & 1 & 1 \end{pmatrix}$

- *coversWithBoundary* (**cWB**) (cases 7, 8, 12, 13): $\begin{pmatrix} 0 & \delta & 1 \\ 1 & 1 & 1 \\ 0 & \delta & 1 \end{pmatrix}$

- *boundaryOverlap* (**bO**) (cases 14, 15, 16, 17): $\begin{pmatrix} 0 & 1 & 0 \\ 1 & 1 & \delta \\ 0 & \delta & 1 \end{pmatrix}$

The intermediate level clusters coming from the overlap relation are six, three of which concern composite areas; notice also that there are two cases (44 and 56) that, being symmetric, are shared by two different clusters:

- *nearlyOverlap* (**nO**) (cases 18–26): $\begin{pmatrix} 1 & 1 & \delta \\ 1 & 1 & \delta \\ \delta & \delta & 1 \end{pmatrix}$. Notice that this matrix gives rises to
  sixteen different configurations, seven of which are not realizable.
- *interiorCoveredByInterior* (**iCBi**) (cases 28, 29, 31, 44): case 29 plus other relations at distance 1 from it being part of the overlap cluster (except case 23)
- *interiorCoversInterior* (**iCvi**) (cases 34, 35, 37, 44): case 35 plus other relations at distance 1 from it being part of the overlap cluster (except case 23)
- *partlyInside* (**pI**) (cases 46, 47, 48, 52, 56): all configurations being part of the overlap cluster that need to have an area *A* with at least two components to be realized
- *partlyContains* (**pCt**) (cases 49, 50, 51, 53, 56): all configurations being part of the overlap cluster that need to have an area *B* with at least two components to be realized
- *crossContainment* (**cCt**) (case 45)

The in cluster splits into three intermediate level clusters derived by aggregating the relations which are at distance 1 from the relations 27 (coveredBy), 33 (covers), and 41 (equal) (valid also for simple areas with a crisp boundary), respectively:

- *nearlyInside* (**nI**) (cases 39, 27, 30, 55): case 27 plus other relations at distance 1 from it being part of the in cluster
- *nearlyContains* (**nCt**) (cases 40, 33, 36, 54): case 33 plus other relations at distance 1 from it being part of the in cluster
- *nearlyEqual* (**nE**) (cases 41, 42, 43, 32, 38): case 41 plus other relations at distance 1 from it

On average, the size of the clusters is 56 (relations)/14 (clusters) = 4. This size represents a reasonable trade-off from two opposite points of view: the need of enhancing the degree of granularity granted by the CBM and the need of keeping the number of operators low also at the intermediate level in order to encourage the user to use them. Table IV.3.1 collects information about the three-level hierarchy of topological operators.

The top level and the intermediate level can be used for queries that are independent of the geometric model used at the bottom level. While the top level is useful for a fast screening of the database, the intermediate level offers the possibility of refining the answer. The usability of the intermediate level operators is guaranteed by the fact that each corresponding cluster can be associated to a prototypical geometric configuration expressing, in a visual manner, the kind of results that the user can expect from that operator. In this way, queries on uncertain data are implemented as a small deviation from the reference prototype.

Prototypes for the intermediate level operators are shown in Figure IV.3.14. They are designed making use of the simplest and more intuitive shapes characterizing the clusters. Hence, simple crisp features are used for the operators disjoint, nearlyMeet, nearlyOverlap, and nearlyEqual. The prototypes for the operators coveredByBoundary, coversWithBoundary, boundaryOverlap, interiorCoveredByInterior, and interiorCoversInterior make explicit the broad boundary because the

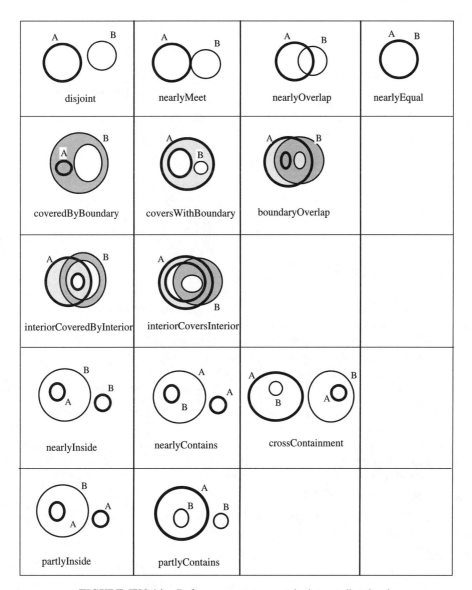

FIGURE IV.3.14 Reference prototypes at the intermediate level.

corresponding clusters refer to geometric configurations which do not exist in the case of features with crisp boundaries. Similarly, the prototypes for the operators nearlyInside, nearlyContains, cross-Containment, partlyInside, and partlyContains make explicit the existence of multiple components since the corresponding clusters include geometric configurations that are realizable for composite areas.

3.6 Topological Invariants

In the models for representing topological relations described so far, only a few general topological invariants are considered (mainly, the content invariant); such invariants provide a broad classification of topological relations. The introduction of other invariants allows finer topological distinctions.

TABLE IV.3.1   The Hierarchy of Topological Operators

Level					Total Number of Operators
Top level (CBM)	disjoint	touch	overlap	in (and reverse)	4
Intermediate level	disjoint	nearlyMeet coveredByBoundary coversWithBoundary boundaryOverlap	nearlyOverlap interiorCoveredByInterior interiorCoversInterior partlyInside partlyContains crossCointainment	nearlyInside nearlyContains nearlyEqual	14
Bottom level (9-Intersection)	1 relation matrix	16 relation matrices	26 relation matrices	13 relation matrices	56

3.6.1   Content Invariants

In the relation between two regular closed geometric elements, there are various point-sets to be considered, which can be empty or non-empty. This is generally called the *content invariant* and can be calculated for several sets: intersections, set differences, and symmetric differences.[6] The most convenient one seems to be the intersection, since it gives a comprehensive categorization of topological relations; further, it is extensively used in the literature.[4,5,36-38]

Content invariants are not independent from each other but can be hierarchically structured to form classes of topological relations. At the root of this hierarchy there is the following invariant:

Definition 26: Given two geometric features λ_1 and λ_2, the *feature intersection content invariant* is the intersection $\lambda_1 \cap \lambda_2$.

The feature intersection content invariant may assume the two values, empty ($\emptyset$) or nonempty ($\neg\emptyset$), and therefore leads to two classes of relations: disjoint and non-disjoint. The category of relations in which the features have some common parts can be further refined by considering intersections of boundary and interiors:

Definition 27: The *4-intersection content invariant*[4] is a 2×2 matrix of the intersections of the interiors and boundaries of the two features λ_1 and λ_2:

$$
M = \begin{pmatrix} \lambda_1^\circ \cap \lambda_2^\circ & \lambda_1^\circ \cap \partial\lambda_2 \\ \partial\lambda_1 \cap \lambda_2^\circ & \partial\lambda_1 \cap \partial\lambda_2 \end{pmatrix}
$$

Each intersection may be empty ($\emptyset$) or non-empty ($\neg\emptyset$), resulting in a total of $2^4 = 16$ combinations. Each case is represented by a matrix of values. It is possible to apply some simple geometric constraints to assess that not all combinations make sense for all areas, where the 4-intersection content invariant is able to recognize 8 different categories of relations. All 16 combinations are, instead, possible for lines.

Definition 28: The *9-intersection content invariant*[5] is a 3×3 matrix containing the empty/nonempty values for interior, boundary, and exterior intersections:

$$
\begin{pmatrix} \lambda_1^\circ \cap \lambda_2^\circ & \lambda_1^\circ \cap \partial\lambda_2 & \lambda_1^\circ \cap \lambda_2^- \\ \partial\lambda_1 \cap \lambda_2^\circ & \partial\lambda_1 \cap \partial\lambda_2 & \partial\lambda_1 \cap \lambda_2^- \\ \lambda_1^- \cap \lambda_2^\circ & \lambda_1^- \cap \partial\lambda_2 & \lambda_1^- \cap \lambda_2^- \end{pmatrix}
$$

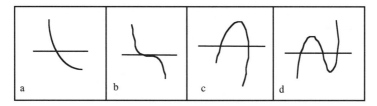

FIGURE IV.3.15 (a) Zero-dimensional intersection with one component, (b) one-dimensional intersection with one component, (c) zero-dimensional intersection with two components, and (d) zero-dimensional intersection with three components.

It is an extension of the 4-intersection that considers also the exterior of the features, besides interior and boundary. Excluding the impossible cases, we have 33 possible cases for line features. The 9-intersection allows for finer refinements of topological relations than the 4-intersection.

3.6.2 Dimension

A refinement of the content invariant is given by the dimension of each intersection set.

Definition 29: The *dimension invariant* is a function that returns the dimension of a given set S: $dim(S)$. If S has disconnected components, then the highest dimension is considered.

The dimension is useful for distinguishing whether the set intersection is 2, 1, or 0-dimensional. The dimension of the intersection cannot be higher than the lowest dimension of the two operands of the intersection. For line features, this means that the dimension of the intersection may be only 0 or 1 (Figure IV.3.15). The dimension is significant for the closures' intersection (or equivalently for the interiors' intersection), since the intersections involving boundaries can only be 0-dimensional.

The extension of the 4 and 9 intersections with the dimension has been also considered.[2,13] The topological configurations distinguishable with the 9-intersection extended with the dimension are 36.

3.6.3 Number of Intersections

Different configurations may have the same content and dimension of intersection sets, but each intersection set may be disconnected and made up of a different number of connected parts (Figure IV.3.15).

Definition 30: The *number* of connected components of a set A is a topological invariant and will be denoted as $\#(A)$.

$\#(A)$ may be any positive integer. We exclude cases of an infinite number of intersections.

Until now, we have discussed invariants that are meaningful for every geometric feature embedded in $\mathbb{R}^2$. To exploit all invariants for lines, we will concentrate on specific topological invariants characterizing the relation between pairs of simple lines. For the case of a non-void intersection between two simple lines, we are going to consider all connected intersection components. To find out all invariants, it is necessary to establish an order on the points belonging to the line, hence considering the line as an oriented feature.

3.6.4 Intersection Sequence

The intersection sequence describes the order in which the various components of the intersection between two lines L_1 and L_2 occur. Under a topological transformation, the intersection sequence must be preserved. Following the line L_1 from its first point and assigning numeric labels to the

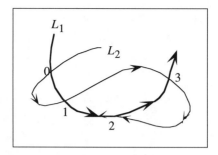

FIGURE IV.3.16   Intersection sequence.

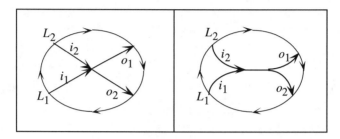

FIGURE IV.3.17   Incoming and outcoming arcs incident at an intersection component (for a zero-dimensional component (a), and a one-dimensional component (b)).

intersections until the last point is reached, the intersection sequence is a sequence of numbers established traversing the line L_2 and recording the labels that were previously assigned to L_1.

Definition 31: Let us consider two intersecting lines L_1 and L_2, images of the mappings f_1 and f_2, respectively. Their intersection has a number of connected components $m = \#(L_1 \cap L_2)$; let us label them with $0, 1, \ldots, m - 1$, following the order given by the line L_1 (from $f_1(0)$ to $f_1(1)$). The *intersection sequence* is an m-tuple $S(L_2) = \langle k_0, k_1, \ldots, k_{m-1} \rangle$, which is a permutation of the m-tuple $\langle 0, 1, \ldots, m - 1 \rangle$ and is obtained by traversing the line L_2 from $f_2(0)$ to $f_2(1)$ and recording the labels previously assigned when traversing the line L_1. The intersection sequence is a topological invariant.

For example, the sequence of intersection of the configuration in Figure IV.3.16 is $S(L_2) = \langle 0, 1, 3, 2 \rangle$. In general, given two lines with m intersections, all the possible different sequences are $m!$.

3.6.5   Intersection Type

Each intersection has "local" characteristics that make it different from the others. Let us consider a neighborhood of an intersection component big enough to properly contain the component itself (Figure IV.3.17). In general, each line with respect to such a neighborhood has an incoming arc i and an outcoming arc o, however, if the intersection involves some endpoint, some of those arcs may be empty.

Definition 32: For each intersection k, let us consider a disc $I(k)$ as a neighborhood. We define in this way at most four arcs, incident at the intersection, two incoming i_1, i_2, and two outcoming o_1, o_2, belonging to the lines L_1 and L_2, respectively. Establishing a clockwise orientation on the boundary of $I(k)$, we find a sequence of arcs $T(k) = (a_1, a_2, a_3, a_4)$ that we call the *intersection type*. The intersection type is a topological invariant.

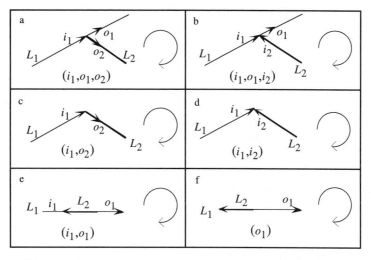

FIGURE IV.3.18 Intersection components with less than four arcs.

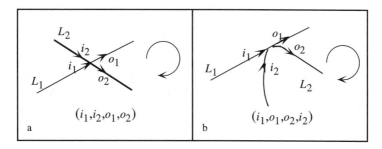

FIGURE IV.3.19 The crossing and touching intersection types.

As a convention, we start recording the arcs from i_1, but this is not restrictive since sequences of arcs are invariant to cyclic permutations. The arc a_i, $i = 1, \ldots 4$, may assume the values i_1, i_2, o_1, o_2. When some arcs are missing, the sequence can contain from zero to three values (Figure IV.3.18). If the sequence is empty, i.e., $T(k) = (\ )$, the lines are equal since the intersection has no incoming or outcoming arcs. If the intersection type contains only one arc, it means that a line is entirely contained in the other one with a common endpoint (Figure IV.3.18f). In the case of two arcs, each arc can belong to a different line (Figure IV.3.18c,d) or to the same line (Figure IV.3.18e). If a line is entirely contained in the other one and the intersection component is zero-dimensional, then the first line degenerates into a point.

Two important categories of intersections easy to differentiate are the *crossing* (Figure IV.3.19a), when the sequence of arcs of the two lines is alternated, and the *touching* (Figure IV.3.19b), when the two lines remain in the same part of the plane. The intersection type is able to distinguish the two categories.

The number of different intersection types in the case of four arcs can be evaluated by fixing the first arc in the sequence as i_1 and considering the permutations of the remaining three arcs; as a result we have $3! = 6$ possible values. In the case of three arcs, we have to consider the combinations of four values taken in groups of three; in this way, we obtain four groups of three arcs. For each group, which is invariant to cyclic permutations, we have to fix an arc and take into account the permutations of the remaining two. Therefore, the number of different intersection types is given by:

$$\binom{4}{3} \cdot 2! = 8.$$

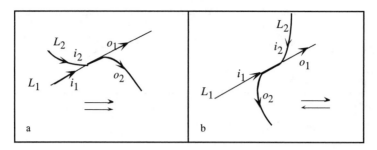

FIGURE IV.3.20   Collinearity sense.

In the case of two arcs, all possible combinations are:

$$\binom{4}{2} = 6.$$

For only one arc, the number of intersection types is 4.

3.6.6   Collinearity Sense

It is also necessary for one-dimensional intersections to distinguish whether the segments that make these components are traversed following the same orientation in the two lines (Figure IV.3.20a) or the reverse orientation (Figure IV.3.20b).

Definition 33: Given an intersection component k, with $dim(k) = 1$, between two lines L_1 and L_2, images of the mappings f_1 and f_2, respectively, there exists an interval $[t_1, t_2] \subseteq [0, 1]$ and an interval $[u_1, u_2] \subseteq [0, 1]$, such that $f_1([t_1, t_2]) = f_2([u_1, u_2])$. If $f_1(t_1) = f_2(u_1)$ and $f_1(t_2) = f_2(u_2)$, then the *collinearity sense* is positive ($CS(k) = +1$). If $f_1(t_1) = f_2(u_2)$ and $f_1(t_2) = f_2(u_1)$, then the collinearity sense is negative ($CS(k) = -1$). The collinearity sense is a topological invariant.

For convenience, we define $CS(k) = 0$ if $dim(k) = 0$. In this way, the CS invariant replaces the dimension.

3.6.7   Link Orientation

Given a sequence of intersections between two lines L_1 and L_2, for each pair of consecutive intersections $\langle h, k \rangle$, the part of line L_2 that is between the two intersections will be called a *link* (denoted with $L_2(h, k)$).

Definition 34: Given two lines L_1 and L_2 and a link L_2 (h, k) between two consecutive intersections $\langle h, k \rangle$, consider the cycle obtained traversing the link L_2 (h, k) and coming back to h traversing the line L_1. If such a cycle is counterclockwise, then the *link orientation* $LO_{L_2}(h, k)$ assumes the value l (left); if it is clockwise, $LO_{L_2}(h, k)$ assumes the value r (right). The link orientation is a topological invariant.

Notice that the orientation of the cycle is determined by L_2 (h, k) and disregards the orientation of the line L_1. The values l and r come from the following consideration: if the cycle is counterclockwise, L_2 (h, k) leaves the region bounded by the cycle on the left; if it is clockwise, L_2 (h, k) leaves the region on the right. Figure IV.3.21 shows a counterclockwise and a clockwise link orientation. Analogously, we could also consider the orientation of links belonging to the line L_1 with respect to L_2.

3.6.8   The Classifying Invariant for Simple Lines

To describe a scene involving two simple lines, we can use all the invariants discussed so far. As a matter of fact, not all invariants are necessary since some of them are implicit in others. The content invariants are able to express conditions on the location of the endpoints of the two lines: the same

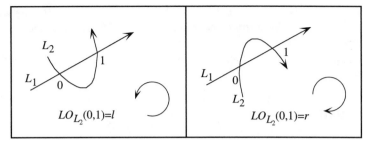

FIGURE IV.3.21 Link orientation.

information comes from the knowledge of the number of intersections and each intersection type. The number of intersections is implicitly given by the intersection sequence. The dimension of an intersection component falls in the definition of collinearity sense.

Given two simple lines L_1 and L_2, with m intersections, the *classifying invariant (CI)* for the topological relation between them is defined as a table $CI(L_1, L_2)$ made up of four columns and m entries. The four columns give the intersection sequence, the collinearity sense, the intersection type, and the link orientation, respectively. The generic entry contains the label of the intersection component k_i, the collinearity sense $CS(k_i)$, the type $T(k_i)$, and the link orientation $LO_{L_2}(k_i, k_{i+1})$. The link orientation is undefined in correspondence of the intersection k_{m-1}.

The general structure of $CI(L_1, L_2)$ is the following:

$S(L_2)$	CS	T	LO_{L_2}
k_0	$CS(k_0)$	$T(k_0)$	$LO_{L_2}(k_0, k_1)$
k_1	$CS(k_1)$	$T(k_1)$	$LO_{L_2}(k_1, k_2)$
$\vdots$	$\vdots$	$\vdots$	$\vdots$
$\vdots$	$\vdots$	$\vdots$	$LO_{L_2}(k_{m-2}, k_{m-1})$
k_{m-1}	$CS\ k_{m-1}$	$T(k_{m-1})$	—

If there are no intersections between the two lines ($m = 0$), then the classifying invariant is an empty table and will be denoted $CI(L_1, L_2) = \emptyset$.

For example, the scene in Figure IV.3.16 has the following CI:

$S(L_2)$	CS	T	LO_{L_2}
0	0	(i_1, i_2, o_1, o_2)	l
1	0	(i_1, o_2, o_1, i_2)	r
3	0	(i_1, i_2, o_1, o_2)	r
2	-1	(i_1, o_1, i_2)	—

Clementini and Di Felice[17] proved the necessity and the completeness of the classifying invariant. This result allows to conclude that if two scenes have the same classifying invariant, then they are topologically equivalent, while if two scenes have two distinct classifying invariants, they are topologically different. We refer to that paper also for a description of the classifying invariant for complex lines.

3.7 Conclusions

In this chapter, we have provided a global view of the models for representing topological relations in spatial databases. Such models have been developed during the last decade by the research community and, only in part, have been absorbed by database technology. The standard for spatial data recently established by the OpenGIS Consortium incorporates the CBM relations and the 9-intersection relations for complex features with crisp boundary. The models for topological relations between features with broad boundary and the topological invariants still have to be translated inside spatial database technology.

Acknowledgments

The work in this chapter was supported by M.U.R.S.T. under project *Rappresentazione ed elaborazione di dati spaziali nei Sistemi Informativi Territoriali.*

Portions of this chapter are reprinted from Clementini, E. and Di Felice, P., A model for representing topological relationships between complex geometric features in spatial databases, *Inf. Sci.,* 90(1–4), 121, 1996 and Clementini, E. and Di Felice, P., A spatial model for complex objects with a broad boundary supporting queries on uncertain data, *Data Knowledge Eng.,* 37(3), 285, 2001; with permission of Elsevier Science; and from Clementini, E. and Di Felice, P., Topological invariants for lines, *IEEE Trans. Knowledge Data Eng.,* 10(1), 38, 1998. © IEEE 1998; with permission.

References

1. Güting, R.H., An introduction to spatial database systems, *VLDB J.,* 3(4), 357, 1994.
2. Clementini, E., Di Felice, P., and van Oosterom, P., A small set of formal topological relationships suitable for end-user interaction, in *Advances in Spatial Databases—Third International Symposium, SSD '93*, Abel, D.J. and Ooi, B.C. Eds., Springer-Verlag, Berlin, 277, 1993.
3. Cui, Z., Cohn, A.G., and Randell, D.A., Qualitative and topological relationships in spatial databases, in *Advances in Spatial Databases—Third International Symposium, SSD '93*, Abel, D.J. and Ooi, B.C., Eds., Springer-Verlag, Singapore, 296, 1993.
4. Egenhofer, M.J. and Franzosa, R.D., Point-set topological spatial relations, *Int. J. Geographical Inf. Syst.,* 5(2), 161, 1991.
5. Egenhofer, M.J. and Herring, J.R., Categorizing Binary Topological Relationships Between Regions, Lines, and Points in Geographic Databases. Technical Report of the Department of Surveying Engineering, University of Maine, Orono, ME, 1991.
6. Herring, J.R., The mathematical modeling of spatial and non-spatial information in geographic information systems, in *Cognitive and Linguistic Aspects of Geographic Space*, Mark, D. and Frank, A., Eds., Kluwer Academic, Dordrecht, 313, 1991.
7. Kainz, W., Egenhofer, M.J., and Greasly, I., Modelling spatial relations and operations with partially ordered sets, *Int. J. Geographical Inf. Syst.,* 7(3), 214, 1993.
8. Egenhofer, M.J., Clementini, E., and Di Felice, P., Topological relations between regions with holes, *Int. J. Geographical Inf. Syst.,* 8(2), 129, 1994.
9. Clementini, E., Di Felice, P., and Califano, G., Composite regions in topological queries, *Inf. Syst.,* 20(7), 579, 1995.
10. Pullar, D.V. and Egenhofer, M.J., Toward the definition and use of topological relations among spatial objects, in *Third International Symposium on Spatial Data Handling*, Marble, D.F., Ed., International Geographical Union, Sydney, Australia, 225, 1988.
11. Egenhofer, M.J. and Sharma, J., Topological relations between regions in R^2 and Z^2, in *Advances in Spatial Databases—Third International Symposium, SSD '93*, Abel, D.J. and Ooi, B.C., Eds., Springer-Verlag, Singapore, 316, 1993.
12. Clementini, E., Sharma, J., and Egenhofer, M.J., Modelling topological spatial relations: strategies for query processing, *Comput. Graphics,* 18(6), 815, 1994.

13. Clementini, E. and Di Felice, P., A comparison of methods for representing topological relationships, *Inf. Sci.,* 3(3), 149, 1995.
14. Clementini, E. and Di Felice, P., A model for representing topological relationships between complex geometric features in spatial databases, *Inf. Sci.,* 90(1–4), 121, 1996.
15. Clementini, E. and Di Felice, P., An algebraic model for spatial objects with indeterminate boundaries, in *Geographic Objects with Indeterminate Boundaries*, Burrough, P.A. and Frank, A.U., Eds., Taylor & Francis, London, 155, 1996.
16. Clementini, E. and Di Felice, P., Approximate topological relations, *Int. J. Approximate Reasoning,* 16, 173, 1997.
17. Clementini, E. and Di Felice, P., Topological invariants for lines, *IEEE Trans. Knowledge Data Eng.,* 10(1), 38, 1998.
18. Clementini, E. and Di Felice, P., A spatial model for complex objects with a broad boundary supporting queries on uncertain data, *Data Knowledge Eng.,* 37(3), 285, 2001.
19. Worboys, M.F. and Bofakos, P., A canonical model for a class of areal spatial objects, in *Advances in Spatial Databases—Third International Symposium, SSD '93*, Abel, D.J. and Ooi, B.C., Eds., Springer-Verlag, Singapore, 36, 1993.
20. Tsurutani, T., Kasakawa, Y., and Naniwada, M., ATLAS: a geographic database system, *ACM Comput. Graphics,* 14(3), 71, 1980.
21. Frank, A.U., MAPQUERY: data base query language for retrieval of geometric data and their graphical representation, *ACM Comput. Graphics,* 16(3), 199, 1982.
22. Smith, T. et al., KBGIS-II: a knowledge-based geographical information system, *Int. J. Geographical Inf. Syst.,* 1(2), 149, 1987.
23. Roussopoulos, N., Faloutsos, C., and Sellis, T., An efficient pictorial database system for PSQL, *IEEE Trans. Software Eng.,* 14(5), 639, 1988.
24. Güting, R.H., Geo-relational algebra: a model and query language for geometric database systems, in *Advances in Database Technology—EDBT '88*, Schmidt, J.W., Ceri, S., and Missikoff, M., Eds., Springer-Verlag, Venice, Italy, 506, 1988.
25. Egenhofer, M.J., Spatial SQL: a query and presentation language, *IEEE Trans. Knowledge Data Eng.,* 6(1), 86, 1994.
26. Cardenas, A.F. et al., The knowledge-based object-oriented PICQUERY+ language, *IEEE Trans. Knowledge Data Eng.,* 5(4), 644, 1993.
27. OpenGIS Consortium, *OpenGIS Simple Features Specification for SQL*, 1998.
28. van Oosterom, P. et al., Integrated 3D modelling within a GIS, in *Advanced Geographic Data Modelling: Spatial Data Modelling and Query Languages for 2D and 3D Applications*, Molenaar, M. and Hoop, S.D., Eds., Netherlands Geodetic Commission, Delft, The Netherlands, 80, 1994.
29. Cohn, A.G. and Gotts, N.M., Representing spatial vagueness: a mereological approach, in *Principles of Knowledge Representation and Reasoning: Proceedings of the 5th International Conference (KR96)*, Morgan Kaufmann, San Francisco, CA, 230, 1996.
30. Erwig, M. and Schneider, M., Vague regions, in *Advances in Spatial Databases—Fifth International Symposium, SSD '97*, Scholl, M. and Voisard, A. Eds., Springer, Berlin, 298, 1997.
31. Schneider, M., Modelling spatial objects with undetermined boundaries using the Realm/ROSE approach, in *Geographic Objects with Indeterminate Boundaries*, Burrough, P. A. and Frank, A.U., Eds., Taylor & Francis, London, 141, 1996.
32. Munkres, J.R., *Topology: A First Course,* Prentice-Hall, Englewood Cliffs, NJ, 1975.
33. Egenhofer, M.J. and Al-Taha, K., Reasoning about gradual changes of topological relationships, in *Theories and Models of Spatio-Temporal Reasoning in Geographic Space*, Frank, A.U., Campari, I., and Formentini, U., Eds., Springer-Verlag, Berlin, 196, 1992.
34. Freksa, C., Temporal reasoning based on semi-intervals, *Artif. Intelligence,* 54, 199, 1992.

35. Mark, D.M. and Egenhofer, M.J., Modeling spatial relations between lines and regions: combining formal mathematical models and human subjects testing, *Cartography and GIS*, 21(3), 195, 1994.
36. Hernández, D., Maintaining qualitative spatial knowledge, in *Spatial Information Theory: A Theoretical Basis for GIS—European Conference, COSIT '93*, Frank, A.U. and Campari, I., Eds., Springer-Verlag, Berlin, 36, 1993.
37. Huang, Z. and Svensson, P., Neighborhood query and analysis with GeoSAL, a spatial database language, in *Advances in Spatial Databases—Third International Symposium, SSD '93*, Abel, D.J. and Ooi, B.C., Eds., Springer-Verlag, Singapore, 413, 1993.
38. Smith, T. and Park, K., Algebraic approach to spatial reasoning, *Int. J. Geographical Inf. Syst.*, 6(3), 177, 1992.

4

Intelligent Image Retrieval for Supporting Concept-Based Queries: A Knowledge-Based Approach

Jae Dong Yang
Chonbuk National University

4.1 Introduction

Currently, as a large volume of image databases has been emerging in a wide variety of Web application areas such as interior design, cyber museums, education on demand, and medical diagnosis, interests in retrieving relevant images from the databases are ever increasing. To get relevant images, users may issue Boolean queries against text documents with appropriate search terms, expecting the images are linked by hyperlinks. However, the users would soon become disappointed after finding that the result is far from their expectations.

Obviously, for the direct access of the required images without such hyperlinks, text retrieval functionality alone is no longer sufficient in information retrieval systems such as Lycos, Inktomi, and Google. The simplest solution is to use text annotations of images produced manually. However, since knowledge encoded even in one image is equivalent to thousands of words in general, it seems intolerable for such a volume of annotations to be entered manually with great tedium and prohibitive cost.[20] Furthermore, since images are decidedly multidisciplinary in nature, it is almost inevitable that the perspective of one indexer will differ from that of other indexers.[4]

To effectively attack this problem, image retrieval systems capable of indexing and searching for images based on the characteristics of their content are necessary. Photobook,[20] QBIC,[15] Virage,[3] and Visualseek,[21] as well as other developed methods[14,22] are attempts to index images based on color, texture, shape, or the combination of them. To be specific, Photobook provides a set of interactive tools that allows content-based search with appearance, 2D shape, and a combination of texture models such as reaction-diffusion, Markov random field, and a new wold model.

Aslandogan et al.,[2] Forsyth,[16] and Papadias and Sellis[19] proposed content-based image retrieval techniques exploiting spatial relationships to capture the semantics of images. The SCORE system[2] uses extended E-R diagrams to query image databases as well as index them. Since a user poses his/her query by an appropriate E-R diagram, images would be an answer of the query if their E-R diagrams match it to a certain degree. E-R diagram matching, however, is problematic since it incurs the NP complete problem. Symbolic arrays proposed by Papadias and Sellis[19] are used to capture the spatial structure of a spatial concept, where elements of the array denote the meaningful components of the concept. For example, a symbolic array for a spatial concept "continent" represents the spatial relationships between continents such as Africa and America, each of which has, in turn, its own symbolic array which specifies its spatial relationships between its countries. However, its purpose is mainly to represent the relative position of spatial concepts rather than capturing some semantics to which their aggregation leads.

Forsyth[16] proposed a new representation for peoples and animals, called a body plan. It is a sequence of grouping rules, constructed to mirror the layout of body segments in people and animals. For example, in the case of horses, it first collects body, neck, and leg segments. It then constructs pairs that could be views of a body–neck pair or a body–leg pair. From these pairs, it attempts to construct triples. This approach may be considered an attempt to semantically use the spatial relationships. However, it is basically a method adopted for a quite different purpose in that it tries to precisely recognize objects rather than capture concepts.

2D strings[1,7] and triple indexing,[9–12] which refines the 2D strings, are other approaches attempting to represent image structure with the spatial relationships. To be specific, Chang[9] indexes an image with triples specifying the relationships between its component objects. The inverted file to be used for triple matching is implemented with a well-designed hashing function. The triple structure is a novel index mechanism in that it enables us to simply describe the spatial structures of images, guaranteeing fast retrieval time. However, they do not support conceptual queries since they only exploit the relationships syntactically. In other words, they fail to notice that the relationships between objects in an image can derive the semantics of that image, if properly used. For example, a table could be labeled differently, depending on some objects near it. If it is near a sink, it would be a dining table, or a desk if it is near a bookshelf. Without handling the semantics, to extract images conceptually coinciding with users' intent may be a great burden. For example, suppose a user wants to retrieve some picture that includes a dining table, cupboard, and a dishwasher, which turns out to be "kitchen." To retrieve it, a user has to explicitly and thoroughly enumerate the objects in his/her query because the systems cannot recognize the concept "kitchen."

Concept-based image retrieval[24,26] is an attempt to solve the problem by making the conceptual queries possible. Yang and Yang[24] deal with concepts which are mainly detected by a predefined configuration of constituent objects. The configuration is formed according to objects' relative spatial positions. To detect the concepts, a knowledge base called a concept rule base is used in this approach. The concepts are basically composite objects formed from the aggregation of constituents—for example, the concept "kitchen" is a composite object. A detailed explanation of this approach will be given in Section 4.3.1.

Yang[26] refines the previous approach by considering another kind of concept which can be captured using generalization. For example, suppose that a user wants to retrieve a picture where a natural scene and a city appear simultaneously at its left- and right-hand side, respectively. While Yang and Yang's approach[24] may detect the concept "city," which is the aggregation of buildings, cars, people, etc., it does not capture the other concept "natural scene," which is the generalization of mountains, a lake, or a river. To capture the concept, another knowledge base called the concept thesaurus is used and it adopts a new data structure, a fuzzy term predicate for fuzzy matching between terms which are the specialization or generalization of each other. In addition to detecting concepts, this approach also extends the triple structure to accommodate inexact matching between spatial relationships. In Sections 4.3.2 and 4.3.4, it will be explained in greater detail.

From now on, we refer to objects in an image as concepts if they are used as the labels of semantics. The semantics are derived from the aggregation of more than one object with their spatial relationships or the generalization of an object. For example, a table and a bookshelf above the table can be reduced into (or labeled as) the concept "study table" together with their spatial configuration, and the concept can, in turn, be labeled as another generic concept, "furniture." Objects other than concepts are primitive objects. The primitive objects may be recognized by manual or automatic labeling procedures, which involve segmentation, physical image feature extraction, etc.

Note the difference between the concepts and semantics. Generally, semantics is the interpretation of image features such as color histogram, texture, and shapes as well as the spatial relationships. As mentioned, concepts are the labels of the semantics.

This chapter is concerned with an intelligent image retrieval model capable of capturing concepts buried in images by aggregation and generalization. As a uniform construct to retrieve images based on the two kinds of concepts, we introduce an extended triple structure which includes object identifiers (oids) and an angle value denoting the spatial relationship between them. The name, *XY* coordinates, color histogram, and other properties of an object appearing in an image may be obtained by referencing the corresponding oid. To successfully detect concepts by the aggregation or generalization of other concepts or primitive objects, we adopt two kinds of knowledge base: the concept rule base and the concept thesaurus. The former is used to detect the aggregation with rules defined by the logical connective of triples, whereas the latter is exploited to capture the generalization with is-a hierarchies.

This chapter proceeds as follows. Section 4.2 briefly reviews related image indexing techniques for intelligent image retrieval. Section 4.3 explains our techniques for concept-based image retrieval, which are uniformly developed on the extended triple indexing. The triple structure is extended to accommodate concepts which are captured by aggregation and generalization. In addition to that, a series of k-weight functions is used as membership functions to enumerate the compatibility between the eight directions and the corresponding angle degrees. In Section 4.4, we develop an image retrieval system which fully supports the proposed technique, and conclusions follow in Section 4.5.

4.2 Overview of Related Techniques toward Intelligent Image Retrieval

Classical image indexing techniques to exploit spatial relationships may be classified into iconic indexing based on 2D strings or triple indexing,[7,9–12] E-R diagrams,[2] and symbolic arrays.[19] They commonly represent the structure of an image in terms of the spatial relationships between its constituent objects. In this section, we explain the 2D strings and the triple indexing technique including the SMR (spatial match representation) scheme. These are selected as preliminaries to our approach, since icon indexing is closely related to it. Representing images in terms of icons, i.e., iconic images, is the basic principle of icon indexing, which makes an image a picture itself. It is generally conceived that using iconic images is well suited to hold visual information and to allow different levels of abstraction and management.[8]

4.2.1 2D String

Chang introduced a promising new way of representing image pictures based on orthogonal projections: the 2D string.[7] It represents the spatial relationships between object $o_i, i = 1, \ldots, n$ by $r \in \{=, <, >, :\}$. For example, the 2D string for the iconic image (or iconic picture) in Figure IV.4.1 is $(u, v) = (re = r < s = w, r = w < re = s)$.

The symbol $<$ denotes the left-right spatial relation in string u and the below-above spatial relation in string v. The symbol $=$ denotes "at the same spatial location as." $(u, v) = (o_1 r_1 o_2 r_2 \ldots r_{n-1} o_n, o_{p(1)} r_1 o_{p(2)} \ldots r_{n-1} o_{p(n)})$ is the general form of 2D strings where $P: \{1, 2, \ldots, n\} \rightarrow \{1, 2, \ldots, n\}$ is a permutation and the two strings u and v represent the spatial relationships of the X and Y projections, respectively, of the objects in the image. Therefore, the 2D string can be seen to be the symbolic projection of picture p along the x- and y-axes. The 2D string is not only a compact representation of the real image, but also an ideal representation suitable for formulating picture queries—these are also formulated with 2D strings, making the retrieval of pictorial information a problem of 2D subsequence matching.

As an extension of the 2D string to facilitate spatial reasoning in two-dimensional scenes, Chang[9] and Chang and Lee[10] made the 2D string structure more readable by transforming a 2D string into equivalent triples. The detailed explanation follows.

4.2.2 Triple Indexing

The triple-based indexing represents an iconic image as a set of ordered triples $\langle o_i, o_j, r_{ij} \rangle$. The triples are used to encode a pair-wise spatial relationship r_{ij} between objects o_i and o_j, where r_{ij} is one of eight directions, i.e., east, northeast, north, etc. Since $\langle o_i, o_j, r_{ij} \rangle$ implies $\langle o_j, o_i, r'_{ji} \rangle$ for r'_{ji}, which is the inverse of r_{ij}, only $n*(n-1)/2$ triples are enough to represent the iconic image having n constituent objects. For example, in Figure IV.4.1, let w, r, s, and re be working table, radio, speaker, and receiver, respectively. Then, the iconic image p_1 is indexed by $\{\langle w, re, \text{northwest}\rangle, \langle r, w, \text{east}\rangle, \langle r, re, \text{north}\rangle, \langle w, s, \text{north}\rangle, \langle r, s, \text{northeast}\rangle, \langle re, s, \text{east}\rangle\}$.

Note that o_i and o_j are not used as object identifiers in Chang[9] and Chang and Lee.[10] This means that the properties of the objects cannot be accessed from o_i and o_j; they are no more than icons.

The generated triples are then inserted into an inverted file for indexing p_1. For example, p_1 would be retrieved as an answer to the query "retrieve images containing $\langle r, s, \text{northeast}\rangle$" by inverted file lookup. The inverted file used by Chang[9] is given in Figure IV.4.2. Its structure was originally proposed by Cook and Oldehoeft.[13]

In the following subsection, we provide the SMR (spatial match representation) scheme as an extension of the triple structure. This scheme was proposed to remedy the drawback that image icons do not fully express the difference of objects in sizes and shapes, since the icons are approximately labeled by the center of gravity and the relative position of objects with respect to other ones.

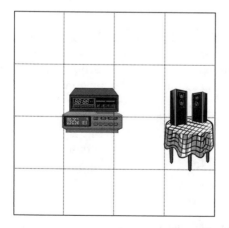 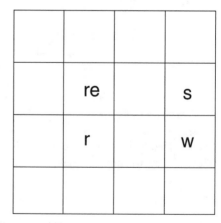

FIGURE IV.4.1 Iconic image p_1.

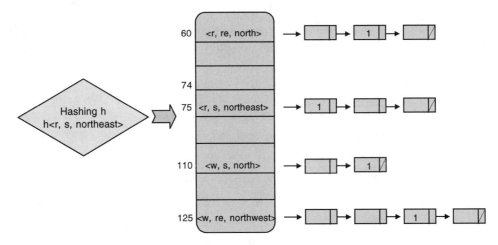

FIGURE IV.4.2   Inverted file constructed by a hashing function.

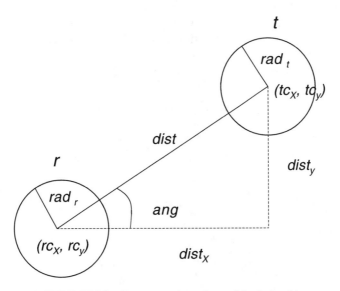

FIGURE IV.4.3   Representation of a spatial relationship.

4.2.3   SMR Scheme

In order to identify an object in an image, this scheme uses MBC (minimum bounding circle), which is a circle surrounding the object. In Figure IV.4.3, (rc_x, rc_y), (tc_x, tc_y) are the projected center points of the reference object r and the target object t on the X- and Y-axis, respectively. Rad_r and rad_t are their respective radius values. *Dist* is the distance between their center points defined by $dist = \sqrt{dist_x^2 + dist_y^2}$. Additionally, *ang* is an angle given as a degree, which is calculated counterclockwise.

Six positional operators are now defined such as FA (far away), DJ (disjoint), ME (meet), OL (overlap), IN (include), and SA (same).

With the operators, the spatial relation $r(r, t)$ is specified as given below, where *MF* is a multiflier for deciding FA and DJ. $r(r, t)$ is used in a spatial string $\langle r, t, r(r, t), ang \rangle$ with which image indexing is performed.

$r(r, t)$

$\quad$ = FA iff $dist >= (rad_r + rad_t) {}^* MF$

$\qquad$ DJ iff $(rad_r + rad_t) < dist <= (rad_r + rad_t) {}^* MF$

$\qquad$ ME iff $dist == (rad_r + rad_t)$

$\qquad$ OL iff $(dist < (rad_r + rad_t)) \wedge (((rad_r <= rad_t) \wedge (rad_r + dist > rad_t)$

$\qquad\quad \vee (rad_t > rad_r) \wedge (rad_t + dist > rad_r)))$

$\qquad$ IN iff $(rad_r > rad_t) \wedge (rad_t + dist <= rad_r)$

$\qquad$ SA iff $(rad_t == rad_r) \wedge (dist = 0)$

For obtaining results in the order of similarity to a user query, the scheme provides a weighting formula which ranks the results based on the angle measured between the two objects and their positional operator. Let s and s' be spatial string in an iconic image p and an iconic query image p' denoted by $s = (r, t, r(r, t), ang)$, $s' = (r', t', r'(r', t'), ang')$, respectively. Then, Equation IV.4.1 calculates the similarity between s and s' with $\alpha \in [0, 1]$, iff:

1. $r = r'$ and $t = t'$
2. Angle difference, i.e., $|ang - ang'|$ is less than a threshold $\theta > 0$.

$$\alpha = \left(1 - \frac{|ang - ang'|}{\theta}\right) * \delta \qquad\qquad (IV.4.1)$$

$\quad$ where

$$\delta = \frac{1}{1 + sim\_dist(r(r, t), r'(r, t))}.$$

$Sim\_dist(r(r, t), r'(r, t))$ means the similarity distance between $r(r, t)$ and $r'(r, t)$.

$\quad$ Since an iconic image has multiple objects in general, similarity between p and p' can be computed according to the following formula:

$$w_{pp'} = \frac{\sum_{i=1}^{n} \alpha_i}{n}$$

where n is the number of spatial strings in p' and α_i is the similarity between the ith spatial string of p and the corresponding string of p'.

4.3　Techniques for Concept-Based Image Retrieval

The techniques explained in this section focus on the detection of concepts by exploiting two knowledge bases: the concept rule base and the concept thesaurus. Section 4.3.1 treats composite concepts mainly captured by aggregation. The patterns of the composite concepts are defined by a series of concept rules in the concept rule base; the concept rules implement aggregation. The material in this section is partially taken from Yang and Yang.[24] Section 4.3.2 extends Section 4.3.1 to accommodate the other concepts detected by generalization. The generalization is defined by the concept thesaurus, which is formed by a collection of fuzzy membership functions. We also provide a query evaluation mechanism uniformly supporting the two concepts in user queries. This is described in Section 4.3.3. Section 4.3.4 shows a way of inexact matching between spatial relationships by introducing k-weight functions. The materials in Sections 4.3.2 and 4.3.4 are extended from Yang,[26] and Section 4.3.3 partially uses the query evaluation mechanism of Yang[25] in quite a different context—image databases disallowing indefinite labeling.

4.3.1 Detecting Concepts by Aggregation

We begin by defining our triple structure, which specifies spatial relationships between its two object identifiers in an image.

Definition 1: Let P be the whole set of images in an image retrieval system, O_p be the set of identifiers of all objects possibly occurring in an image $p \in P$ and, $D = \{\text{east, west, south, north, southeast, southwest, northeast, northwest}\}$ be the set of directions. Then, a triple t is defined as follows.

$$t = \langle o_i, o_j, r_{ij} \rangle \text{ where } o_i, o_j \in O_p \text{ and } r_{ij} \in D.$$

Additionally, a triple set used for indexing an image p is denoted by $T_p = \{t\}$.

Example 1: Figure IV.4.4 shows the image p_2 where the objects it contains are iconized into objects (object names): b, b' (chair), c (table), d (tableware), e (pineapple), f (banana), g (sink), and h (cupboard). Note that we assign more than one icon to c, considering the physical area spanned by it. Otherwise, it could incur the problem that their spatial relationships with respect to other objects depend on the location of their icons. For instance, if c is located on 31, the corresponding tuple would be $\langle c, d, \text{north} \rangle$, whereas it would be $\langle c, d, \text{northwest} \rangle$ if it moves to 32. Treating them as one conceptual object for representing the table allows us to fix its direction with respect to other objects, which naturally coincides with our intuition. The triple set T_{p2} is, hence, represented by

$$T_{p2} = \{\langle b, c, \text{east} \rangle, \langle c, d, \text{north} \rangle, \langle c, e, \text{north} \rangle, \langle b, e, \text{northeast} \rangle,$$
$$\langle c, e, \text{north} \rangle, \langle b', c, \text{north} \rangle, \langle b', d, \text{north} \rangle,$$
$$\langle b, b', \text{southeast} \rangle, \dots \}.$$

Definition 1 specifies triples in which object identifiers participate. However, since object equality is known to be extremely difficult to implement,[17] the definition may need a mapping from the object identifiers to conceptual linguistic terms (or simply terms). The following definition is provided to associate object identifiers with the terms.

Definition 2: A name function f_{name} is defined as follows.

$$f_{name} : O \rightarrow N, \text{ where } N \text{ is a set of terms.}$$

FIGURE IV.4.4 Iconic image p_2.

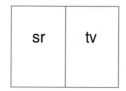

FIGURE IV.4.5   An iconic query.

In addition to such an association, the name function is crucial for making our triple structure more extensible—it allows the specification of two distinct objects that share the same name. For example, suppose that another radio r' appears in Figure IV.4.1, replacing working table w. Then r and r' need to be clearly distinguished. If we want to reduce r' and the speaker s into a composite concept, audio_set, together with their spatial relationship north; r should not be used instead of r' to capture the concept.

Our triple t may now be redefined with this name function.

Definition 3:   A triple t for an image p is defined as

$$t = \langle f_{name}(o_i), f_{name}(o_j), r_{ij} \rangle \text{ where } o_i, o_j \in O_p \text{ and } r_{ij} \in D.$$

For notational simplicity, $f_{name}(o), o \in O_p$ would be referred to as $o.name$ and the object identifier o as an object interchangeably.

To clarify the basic limitation of the extant triple indexing techniques, let us consider the query, "retrieve an image containing a study room (sr) to the left of TV (tv)," as shown in Figure IV.4.5. Apparently, it is beyond their retrieval power since the study room is not a primitive object to be recognized by an automatic labeling. To attack this limitation, we first introduce a concept rule base to capture concepts by aggregation and a mechanism of generating triples whenever the rule base detects them in an image.

To exploit concept rule base C_R in image indexing, two procedures are needed: rule specification and spatial relationship determination. In the former, rules are specified to define the intra-structure of a concept. This structure may serve as a pattern on which concepts are detected. The latter is needed to generate triples for concepts since a spatial relationship should be established between a newly recognized concept and other primitive objects in an image indexed by the triples. This will be explained later.

4.3.1.1   Concept Rule Specification

C_R consists of a set of concept rules defined by the logical connective of triples. The following is an example of C_R to capture concepts buried in p_2, where "," denotes AND and " | " denotes OR, respectively.

> kitchen $\leftarrow$ $\langle$kitchen-set, dining-room, below$\rangle$|$\langle$kitchen-set, dining-room, above$\rangle$;
>
> kitchen-set $\leftarrow$ $\langle$sink, cupboard, location$\rangle$;
>
> dining-room $\leftarrow$ $\langle$chairs, table, location$\rangle$|
>
> $\qquad$ $\langle$chairs, dining-room, location$\rangle$;
>
> study-room $\leftarrow$ $\langle$chairs, table, location$\rangle$, $\langle$chairs, bookshelf, location$\rangle$
>
> $\qquad$ |$\langle$chairs, table, location$\rangle$, $\langle$table, bookshelf, location$\rangle$;
>
> chairs $\leftarrow$ $\langle$chair, chairs, location$\rangle$;
>
> below $\leftarrow$ south | southwest | southeast;
>
> above $\leftarrow$ north | northwest | northeast;
>
> location $\leftarrow$ north | south | west |east | northeast | southeast| northwest | southwest;

Formally, it is defined as follows.

Definition 4: Let T_p be a set of triples for an image p. Then the conjunction of triples t_c is defined as follows.

$$t_c = t_1 \text{ AND } t_2 \text{ AND } \ldots \ t_s \text{ where } t_i \in T_p, i = 1, 2, \ldots, s.$$

In particular, when denoting that a triple t is involved in t_c, $t_c(t)$ will be used.

Definition 5: Let $n \in N$ be a concept name and T_C be a set of conjunction of triples. Then, a concept rule is defined as follows.

$$n \leftarrow t_{c1}|t_{c2}|\ldots|t_{cm} \text{ where } t_{ci} \in T_C, i = 1, \ldots, m.$$

Definition 6: Let O be a set of object identifiers (or simply objects) in an image database. Then, the interface of the concept rule base C_R is implemented by the following function.

$$C_R : T_C \rightarrow N, \ i.e., \ C_R(t_c) = n \in N \text{ for } t_c \in T_C \text{ where } N$$

$$\text{is a set of terms and } f_{name}^{-1}(n) = \{c\} \in 2^O.$$

For instance, a triple, ⟨sink, cupboard, north⟩, is interpreted as a concept name, kitchen-set, by the above rules. It is denoted by C_R (⟨sink, cupboard, north⟩) = kitchen-set.

Conversely, we may want to translate a concept c into an equivalent $t_c \in T_C$. $C_R^{-1}(c.name)$ could then be used for the purpose. It may also be used to check if c is a concept or a primitive object. c would be a primitive object if $C_R^{-1}(c.name) = \phi$ for $c \in O_p$.

If the rule base given in Definition 6 is not designed appropriately, procedures detecting concepts could be quite complicated. The reason is that to extract all concepts in an image containing n constituents, $n*2^n C_R$ calls are required. The rule base may be implemented by a composite event detection algorithm,[6] CFG (context free grammar), PDA (push down automata), as well as rules, which turn out to be the same mechanisms. To be specific, suppose C_R is constructed based on a composite hierarchy structure where primitive objects in terminal nodes are reduced into concepts (or a composite object) along with their spatial relationships, and the concepts are, in turn, reduced into higher level concepts. Then, concepts detected at previous calls may be directly exploited to capture higher level concepts at next calls. Figure IV.4.6 illustrates the nested semantics of the composite concept "kitchen."

In Figure IV.4.6, the concept "kitchen" may be viewed as the aggregation of "kitchen-set" and "dining-room," which may, in turn, be concepts—it may take two composite diagrams, comp_d_1 and comp_d_2, as its instances, each posing different spatial configurations which include two composite sub-diagrams, i.e., sub-patterns "kitchen-set" and "dining-room." In the sequel, the composite diagram comp_d_1 may, in turn, take d_1, d_2, and d_3 as its instances where possible configurations instantiate the two composite sub-diagrams of each. "Kitchen" may be directly used as a query if we translate each composite diagram into appropriate $t_c \in T_C$ by using C_R^{-1}(kitchen), if top down query evaluation is used. For example, the composite diagram d_2 may be translated into ⟨kitchen-set, dining-room, south⟩, and its two composite sub-diagrams are translated into ⟨sink, cupboard, north⟩ and ⟨table, chairs, south⟩, respectively. Conceptually, images containing objects which match these composite diagrams would be the answer of the query. However, since the top-down query evaluation incurs query evaluation delay due to query reformulation, here we adopt bottom-up query evaluation—every concept is detected and then the generated triples involving concepts are inserted into the inverted file prior to query evaluations. The triples in a user query may, therefore, match its direct counterpart in the inverted file. The detailed explanation will be given in Section 4.3.3.

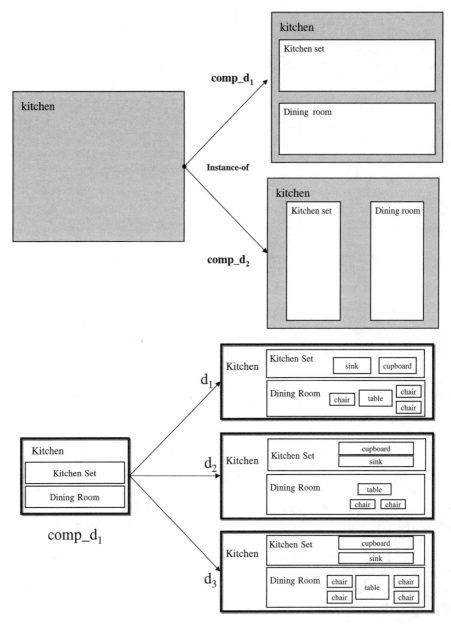

FIGURE IV.4.6   (a) Two possible instances of the concept "kitchen," and (b) the configuration of each instance of comp_d_1.

After they are identified, concepts are uniformly dealt with like primitive objects in an image. However, unlike primitive objects, once a concept is newly detected, the direction determination procedure should take place to generate the corresponding triples which specify spatial relationships between the concept and any other objects in the image. For determining the spatial relationships, we may need to obtain a set of objects participating in the concept. This set is indispensable to the procedure since the spatial relationships can be calculated from the directions of each constituent of the concept with respect to the other objects. For example, if we denote the constituent set of $c.name = $ kitchen by $c.comp$, then $c.comp = \{c_1, c_2 | c_1.name = $ kitchen-set, $c_2.name = $ dining-room$\}$. A more formal definition for $c.comp$ follows.

Definition 7: Let c be a concept. Then, a set of objects, $c.comp$, constituting c is defined by

$$c.comp = \{o_i, o_j | C_R(t_c) = c.name \text{ and } t_c(\langle o_i.name, o_j.name, r_{ij}\rangle) \text{ for } t_c \in T_C\}.$$

In addition to $c.comp$, the direction determination procedure may also need a set for recursively checking if a primitive object belongs to $c'.comp$ for each concept $c' \in c.comp$. The set is mainly used to prevent the procedure from considering an object directly and indirectly participating in c. The following is a definition to obtain the set.

Definition 8: Let $c.(recurse)comp \in 2^O$ be a set of all objects participating in a concept c. Then it is obtained as follows.

1. $c.comp \subseteq c.(recurse)comp$.
2. For $\forall o \in c.\ comp$,

$$o \in c.(recurse)\ comp \quad \text{if } C_R^{-1}(o.name) = \phi$$
$$o.(recurse)\ comp \subseteq c.(recurse)\ comp \text{ otherwise}$$

3. $c.(recurse)\ comp$ does not contain any other element.

Example 2: In Figure IV.4.4, consider $c.comp = \{c_1, c_2 | c_1.name = $ kitchen-set, $c_2.name = $ dining-room$\}$ for a concept c. First, $c.comp \subseteq c.(recurse)\ comp$. Second, since C_R^{-1} (kitchen-set) $\neq \phi$, $c_1.(recurse)comp \subseteq c.(recurse)\ comp$. Now that $C_R^{-1}(g) = \phi$, $C_R^{-1}(h) = \phi$ and g , h $\in c_1.comp$, $g, h \in c_1.(recurse)comp$. Similarly, we obtain $b, c \in c_2.(recurse)comp$ for $b, c \in c_2.comp$. Hence, $c.(recurse)comp = \{c_1, c_2, g, h, b, c\}$.

4.3.1.2 Spatial Relationship Determination for Concepts

We now define a subsumption relation between directions to determine spatial relationships between a newly identified concept and the other objects.

Definition 9: Let c be a concept, $o_i, o_i' \in c.comp$ and $\langle o_i, o_j, r \rangle, \langle o_i', o_j, r' \rangle \in T_p$ for all o_j satisfying $o_j \in O_p \wedge o_j \notin c.(recurse)comp$. Then for $r, r' \in D$ directly connected with each other, if r is not below r' in Figure IV.4.7, then we say r subsumes r'. It is denoted by $r \geq r'$.

$\geq$ is a partial order[5] explaining the implication of directions between the whole constituents of a concept and another object in an image. It is used when we need to get a representative direction which covers all the directions between them. For example, southeast subsumes south and east, since it may be regarded as south and east at the same time.

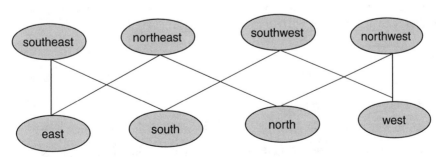

FIGURE IV.4.7 Ordering between directions.

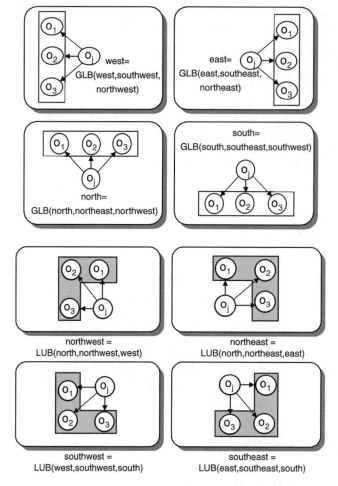

FIGURE IV.4.8 (a) Every GLB of concept-direction sets ; (b) every LUB of concept-direction sets.

Definition 10: Let c .comp $= \{o_1, o_2, \ldots, o_n\}$ for $c \in O_p$ and let R_c be a set of spatial relationships between $o_i \in c.comp, i = 1, \ldots, n$ and o_j such that $o_j \in O_p \wedge o_j \notin c.(recurse)comp$. Then R_c is given as follows.

$$R_c = \{r_{ij} | \langle o_i, o_j, r_{ij} \rangle \in T_p, i = 1, \ldots, n\}.$$

For referential simplicity, we will call R_c as a concept-direction set.

Definition 11: Let T_p^+ and T_p be sets of triples for $p \in P$ respectively. Then, T_p^+ is obtained by adding triples for a detected concept c to T_p. It is defined by

$$T_p^+ = T_p \cup \{\langle c, o, r \rangle | c, o \in O_p \wedge o \notin c.(recurse)comp \wedge r \in D\}$$

and

$r =$ GLB (R_c), when GLB (greatest lower bound) exists,

LUB (R_c), when GLB doesn't exist, but LUB (least upper bound) exists instead, $\emptyset$, else.

Figure IV.4.8 depicts GLBs and LUBs of every possible concept-direction set in an image. For example, given a concept-direction set $R_c = \{east, southeast, northeast\}$, GLB$(R_c) = east$

is a representative direction which is commonly subsumed by all the directions in R_c, whereas if R_c = {north, northwest, west}, $\text{LUB}(R_c)$ = northwest becomes a conclusive direction which subsumes all the directions in R_c.

We now provide a function, GenConceptTriple (T_p, c) for generating triples to be added into T_p^+. A spatial relationship needed in generating each triple is determined by $\text{GenDir}(R_c)$ and R_c with respect to an object $o \in O_p$ is obtained from GenConceptDirSet (T_p, o, c). We omit the specification of GenConceptDirSet() since it can be easily obtained by Definition 10.

Function GenConceptTriple (T_p, c)

        Input: T_p and a concept $c \in O_p$

        Output: TripleList $\subseteq T_p^+$

        TripleList = {};

Begin

        For each $o \in O_p \wedge o \notin c.(recurse)$ *comp*

        R_c = GenConceptDirSet (T_p, o, c);

        r = GenDir(R_c);

        TripleList = TripleList $\cup$ {$\langle c, o, r \rangle$};

End

        return (TripleList);

End

Function GenDir (R_c)

        Input : $R_c = \{r_1, \ldots, r_i, \ldots, r_m\}$

        Output: A direction in D

        // For notational convenience, we assume that each direction in R_c is internally converted into // a 4-bits array as shown in Figure IV.4.9. Result is then reconvertedinto the corresponding // direction r in D when returned—for example, 1010 is converted intonorthwest when // returned.

Begin

        $\text{GLB}(R_c)$ = AND$(r_1, \ldots, r_i, \ldots, r_m)$;

        If $\text{GLB}(R_c) \neq 0$ then return $\text{GLB}(R_c)$ else

                $\text{LUB}(R_c)$ = OR$(r_1, \ldots, r_i, \ldots, r_m)$;

        If $\text{LUB}(R_c) \neq 0$ then return $\text{LUB}(R_c)$;

        return (NULL)

End

Example 3: For the concept-direction set, R_c = {east, southeast, northeast} converted into R_c = {0100, 0101, 1100}, we get $\text{GLB}(R_c)$ = AND$(0100, 0101, 1100)$ = 0100 which denotes east. As another example, consider R_c = {north, northwest, west} internally represented by R_c = {1000, 1010, 0010}. Since $\text{GLB}(R_c)$ = AND$(1000, 1010, 0010)$ = 0000, $\text{LUB}(R_c)$ = OR$(1000, 1010, 0010)$ = 1010 becomes an alternative direction, i.e., northwest.

Example 4: In Figure IV.4.4, to determine the direction between *ks.name* = kitchen-set and *d.name* = tableware, let us obtain the concept-direction set R_{ks} = {south, southwest} between

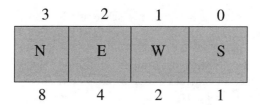

FIGURE IV.4.9   Internal 4-bits representation for spatial relationships.

$ks.comp = \{g, h\}$, and d by GenConceptDirSet(). Now, since GenDir() returns south, a triple list returned from GenConceptTriple() would include $\langle ks, d, south \rangle$, which is added into T_{p2}^+. It is quite natural that the location of the tableware (d) is viewed as the south side of kitchen-set (ks) once g and h are aggregated into one concept, i.e., ks.

In triples, since concepts as well as primitive objects are uniformly treated, we can employ the efficient inverted file structure sketched in Figure IV.4.2. The following is the definition for the inverted file.

Definition 12: An inverted file I_T is defined as a set of triples, each containing a triple t and the link for the corresponding set of images. It is given as follows.

$$I_T = \{\langle t, \{p\}\rangle | t \in T^+, p \in P\}$$

where P is a set of all images and $T^+ = \cup T_P^+$.

According to the definition, a tuple t_I in I_T is formed by attaching a triple t to a set of image oids $\{p\}$ indexed by it. A hash function would then assign an address to each tuple $t_I = \langle t, \{p\}\rangle$ for inserting it into I_T.

4.3.2 Detecting Concepts by Generalization

The technique introduced in Section 4.3.1 needs to be refined since it disallows concepts to be further generalized or specialized. To remedy the drawback, we may treat terms, $o_i.name$ and $o_j.name$ in Definition 3 as fuzzy sets characterized by their membership functions. Such terms are called fuzzy linguistic terms or, briefly, fuzzy terms. A formal definition follows.

Definition 13: A fuzzy term (or simply a term) F is a fuzzy set characterized by the membership function $\mu_F : N \to [0, 1]$, where N is a set of terms. $n \in N$ is a crisp term if there exists n' such that $\mu_n(n') = 1$, for $n' = n$, and $\mu_n(n') = 0$ otherwise. A term that is not crisp is referred to as a fuzzy term.

As an example, audio is a fuzzy term, since it can be interpreted as a fuzzy set, audio $= \{0.8/radio, 0.91/receiver, 0.84/speaker\}$. Radio, receiver, and speaker are crisp terms, provided that their membership functions are not defined. The concept thesaurus to be explained is a knowledge base defining the taxonomy of such terms.

4.3.2.1 Concept Thesaurus

A thesaurus is mainly used for replacing a term with its BTs (broader terms) or NTs (narrower terms), when the result of a query containing the term is unsatisfactory.[18] However, the thesaurus is adopted in our framework for quite a different purpose. Rather than suggesting such BT/NT terms, our thesaurus evaluates the conceptual distance between a term and its BT/NTs, interpreting it like a variable or a template. The evaluation is made by the integrated membership functions of fuzzy terms (or fuzzy sets) in the thesaurus, which has a hierarchical yet nested structure. Figure IV.4.10 shows the structure represented by a fuzzy graph specifying fuzzy membership values. Terms in leaf

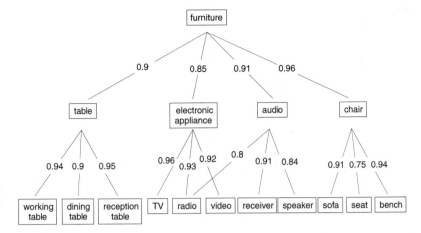

FIGURE IV.4.10   Membership hierarchy of fuzzy terms.

nodes denote crisp terms in N, and the other ones represent fuzzy terms, each taking lower level fuzzy terms as its members with degrees specified on the corresponding edges. Any degree between them is assumed 0 if unspecified.

For example, furniture is a fuzzy term taking table, electronic-appliance, audio, and chair as its members (or instances), and each of them is, in turn, a fuzzy term. So, furniture is a fuzzy set of second order. The instances of electronic-appliance, i.e., TV, radio, and video, are crisp. The corresponding membership functions are therefore given as furniture = {0.9/table, 0.85/electronic-appliance, 0.91/audio, 0.96/chair}, electronic-appliance = {0.96/TV, 0.93/radio, 0.92/video}, and so on. For using the thesaurus to match all the triples conceptually related with each other, we may need a composed membership function for obtaining membership values between two terms indirectly connected in the hierarchical thesaurus. For example, we may want to know the conceptual degree of closeness between furniture and radio. Definition 14 provides a function for computing such a degree.

Definition 14: Let F be a fuzzy term in N. Then,

$$\mu_F(a) = \max(\min(\mu_{ai}(a), \mu_F(a_i))), \forall a \in N$$

where N is a set of terms in the thesaurus and a_i is a fuzzy term such that $\mu_F(a_i) \neq 0$.

Since a_i in the above definition is a fuzzy term, this definition can also be applied to fuzzy terms which have more than two level distances. In other words, $\mu_{a_i}(a), \forall a \in N$ can be calculated by the recursive application of Definition 14 even if a is not directly connected with a_i.

Example 5: The degree of conceptual closeness between furniture and radio is calculated by

$$\mu_{furniture}(\text{radio})$$
$$= \max(\min(\mu_{electronic\text{-}appliance}(\text{radio}), \mu_{furniture}(\text{electronic-appliance}))),$$
$$\min(\mu_{audio}(\text{radio}), \mu_{furniture}(\text{audio})))$$
$$= \max(\min(0.93, 0.85), \min(0.8, 0.91))) = 0.85.$$

To conveniently deal with matching a term with other more general or specific terms, we now introduce the following data structure called the term predicate.

Definition 15: A fuzzy linguistic term predicate (or simply a term predicate) F corresponding to the term F in N is defined by

$$F : N \rightarrow [0, 1]$$

where

$$F(n) = 1 \ \text{if} \ n = F,$$
$$= \mu_F(n) \ \text{for each} \ n \neq F \ \text{and} \ n \in N.$$

For example, furniture(n) is a term predicate, whereas furniture is its corresponding term. Furniture(n) for some $n \in N$ is, therefore, interpreted as the degree of satisfaction of n to furniture.

The term predicates serve as templates which make it possible for terms to match each other based on generalization. The matching by the term predicates involves the concept hierarchy defining the generalization between terms.

We are now in a position to define our image retrieval system.

Definition 16: An intelligent image retrieval system I_RS is defined as the following 5-tuple.

$$I\_RS = \langle P, T, KB, Q, I_T \rangle$$

where

P is a set of all images.
T is a set of all triples.
KB is our knowledge base consisting of concept rule base and concept thesaurus.
Q is a set of queries, each formulated in terms of triples.
I_T is an inverted file.

4.3.3   Query Evaluation

Once concepts are detected by aggregation and the corresponding triples involving the concepts are generated, they are inserted into the inverted file prior to query evaluation. Hence, bottom-up query evaluation to be developed here may simply be a matching between the triples in a user query and its direct counterpart in the inverted file if only the aggregation is used. However, since the matching also considers term predicates, triples conceptually related by generalization would match each other with some degree instead of exact matching. For implementing $\wedge/\vee$ in a query Q, standard fuzzy min/max operator is used for conceptual simplicity. Now, the following definition is a first step toward such an evaluation.

Definition 17: Let each of $Qt_j, j = 1, \ldots, s$ be a triple used as a basic unit of a query. Then, the following disjunction of the s triples is called disjunctive query Q_D.

$$Q_D = \vee_{j=1}^{s} Qt_j.$$

Definition 18: Let $Q_{Di}, i = 1, \ldots, m$ be m disjunctive queries. Then, a conjunctive normal query Q (or simply query) is defined as follows.

$$Q = \wedge_{i=1}^{m} Q_D.$$

We now formalize a fuzzy match involving the term predicates.

Definition 19: Let d_T be a disjunction of term predicates. Then $|d\_T| = \{a \in N | d\_T(a) = 1\}$.

Definition 20: Let each of d_T_1 and d_T_2 be a disjunction of term predicates. Then, d_T_1 is more general with a degree $\alpha \in (0, 1)$ than d_T_2 iff min $(d\_T_1(a)$ for all $a \in |d\_T_2|) = \alpha$. It is denoted by $d\_T_2 \subseteq_\alpha d\_T_1$.

Example 6: Let $d\_T_1 =$ electronic-appliance and $d\_T_2 =$ radio $\vee$ video. Then, $d\_T_2 \subseteq_{0.92} d\_T_1$ since min$(d\_T_1(a)$ for all $a \in |d\_T_2| = \{$radio, video$\}) =$ min(electronic-appliance(radio), electronic-appliance(video)) $= 0.92 > 0$.

Definition 21: Let $Q = \vee_{j=1}^s \langle t_j^{(1)}, t_j^{(2)}, d \rangle$ be a query. Then, for $t = \langle t^{(1)}, t^{(2)}, d \rangle$ which has the same direction as that of Q,

$$t \subseteq_\alpha Q \text{ with } \alpha = \min(\alpha_1, \alpha_2) \in (0, 1]$$
$$\Leftrightarrow t^{(1)} \subseteq_{\alpha 1} \vee_{j=1}^s t_j^{(1)} \text{ and } t^{(2)} \subseteq_{\alpha 2} \vee_{j=1}^s t_j^{(2)}$$

Since Q is a disjunction of triples which share the common direction with $t \in T_p$, from now on we call it a cd-query for t. Additionally, if $t \subseteq_\alpha Q$ and $\alpha \geq \alpha'$ for all $t \subseteq_{\alpha'} Q, \alpha' > 0$, we denote it by $t \subseteq_{\alpha*} Q$.

The following proposition is provided for testing $t^{(1)} \subseteq_{\alpha 1} \vee_{j=1}^s t_j^{(1)}$ or $t^{(2)} \subseteq_{\alpha 2} \vee_{j=1}^s t_j^{(2)}$.

Proposition 1: Suppose $d\_T = \vee_{i=1}^m F_i$. Then, for a term predicate $F, F \subseteq_\alpha d\_T$ iff $\max(F_i(a/A), i = 1, 2, \ldots, m)$ for $a \in |F|) = \alpha > 0$ and $a \in N$.

Proof: This proposition can be directly proven by Definitions 19 and 20.

Example 7: Let $d\_T =$ electronic-appliance $\vee$ audio and $F =$ radio. Then, $F \subseteq_{0.93} d\_T$, since max (electronic-appliance (radio), audio(radio)) $=$ max$(0.93, 0.8) = 0.93 > 0$ by Proposition 1.

Consider next the condition that a triple $t \in T_p$ matches a query Q guaranteeing a predefined possibility. Detecting the condition may be crucial to obtain the following answer set of Q.

Definition 22: Suppose Q a cd-query for for $t \in T_p$. Then, for a threshold degree $\alpha \in [0, 1]$,

1. $I\_RS| -_\alpha Q(p)$ iff $t \subseteq_\alpha Q$,
2. $I\_RS| -_{\alpha*} Q(p)$ iff $t \subseteq_{\alpha*} Q$.

Definition 23: The answer set for a cd-query Q with a threshold degree $\alpha \in [0, 1]$ is defined as follows.

1. $\|Q\|_\alpha = \{p \in P | I\_RS| -_{\alpha'*} Q(p) \text{ and } I\_RS| -_\alpha Q(p) \text{ for } 0 < \alpha \leq \alpha' \leq 1\}$.
2. $\|Q\|_{\alpha*} = \{p \in P | I\_RS| -_{\alpha*} Q(p)\}$.

Number 1, above (Definition 23), is defined to compensate for the drawback of the answer set $\|Q\|_{\alpha*}$ that even $p \in P$ satisfying Q with a higher degree than α cannot be qualified for the set. On the contrary, $\|Q\|_\alpha$ may fail to obtain an answer set with the highest threshold degree α' avoiding $\|Q\|_{\alpha'} = \|Q\|_\alpha$ for $\alpha < \alpha' \leq 1$. Since $\|Q\|_\alpha$ can be easily obtained from $\|Q\|_{\alpha*}$ (but not vice versa), here we adopt number 2, above, as the definition of our answer set.

Example 8: Let the query Q_{D1} be "search for images where an electronic-appliance or audio is at the northeast of furniture." Then, $Q_{D1} = Q_{t1} \vee Q_{t2}$, where $Q_{t1} = \langle$electronic-appliance, furniture, east$\rangle$ and $Q_{t2} = \langle$audio, furniture, east$\rangle$. Now, for $t = \langle t^{(1)}, t^{(2)}, \text{east} \rangle \in T_{p1}$ where $t^{(1)} =$ radio, $t^{(2)} =$ working-table, $I\_RS| -_{0.9*} Q_{D1}(p_1)$ since $t^{(1)} \subseteq_{0.93*}$ electronic-appliance $\vee$ audio (see Example 7) and working-table $\subseteq_{0.9*}$ furniture. Hence, $p_1 \in \|Q_{D1}\|_{0.9*}$.

Note that $t \supseteq_\alpha Q$ for $t \in T_p$ is not a sufficient condition to be $I\_RS| -_\alpha Q(p), 0 < \alpha \leq 1$. For example, let the query Q be "search for images where a radio is at the east of table." If p' contains a speaker and it is indexed by $t = \langle \text{audio}, \text{table}, \text{east} \rangle \in T_{p'}$, then $t \supseteq_{0.7} Q$, which would lead to a wrong result, i.e., $I\_RS| -_{0.7} Q(p')$. Due to the functional limitation of extant image analyzers, it is not always possible for every primitive object to be clearly identified in image labeling. Related issues including indefinite labeling are found in the literature.[25]

Proposition 2: Let $Q_D = \vee_{j=1}^{s} Qt_j$ be a disjunctive query. Then, for all $p \in IDB$,

$$p \in \|Q_D\|_{\alpha^*}, \alpha = \max(\alpha_1, \alpha_2, \dots, \alpha_s) \in (0, 1]$$

iff there exists at least one cd-query Q_i satisfying $t_i \subseteq_{\alpha_{i*}} Q_i$, for at least one $t_i \in T_p, \alpha_i > 0$, $1 \leq i \leq s$.

Proof: ($\Leftarrow$) Without loss of generality, let us assume that such cd-queries are Q_1 and Q_2 for $t_1, t_2 \in T_p$ respectively. If $t_1 \subseteq_{\alpha_{1*}} Q_1$ or $t_2 \subseteq_{\alpha_{2*}} Q_2$, then $I\_RS| -_{\alpha_{1*}} Q_1(p)$ or $I\_RS| -_{\alpha_{2*}} Q_2(p)$, which implies $I\_RS| -_{\max(\alpha_1,\alpha_2)*} Q_1(p) \vee Q_2(p)$ for $\alpha_1, \alpha_2 > 0$ from Definition 22. Hence, according to Definition 23, $p \in \|Q_1 \vee Q_2\|_{\alpha^*}, \alpha = \max(\alpha_1, \alpha_2) > 0$. By the definition of cd-queries, we now conclude $p \in \|Q_D\|_{\alpha^*}, \alpha = \max(\alpha_1, \alpha_2) > 0$. The same is true even if $t_1 \subseteq_{\alpha_{1*}} Q_1$ and $t_2 \subseteq_{\alpha_{2*}} Q_2$. It is possible since p can be indexed simultaneously by t_1 and t_2, which are not necessarily the same with each other.

($\Rightarrow$) The proof of "if part" is similar to "only if part." It is omitted.

Theorem 1: Let $Q = \wedge_{i=1}^{n} Q_{D_i}$ be a query. Then for all $p \in IDB$,

$$p \in \|Q_{D_i}\|_{\alpha_i^*}, \alpha_i \in (0, 1], \quad i = 1, 2, \dots, n \Rightarrow p \in \|Q\|_{\alpha^*} \text{ with } \alpha = \min(\alpha_1, \alpha_2, \dots, \alpha_n).$$

Proof: If $p \in \|Q_{D_i}\|_{\alpha_i^*}$, then $I\_RS -_{\alpha_i^*} Q_{D_i}(p), \alpha_i > 0$ for all $i = 1, 2, \dots, n$. Hence, the following holds for $\alpha = \min(\alpha_1, \alpha_2, \dots, \alpha_n)$:

$$I\_RS| -_{\alpha^*} Q_{D_1}(p) \wedge Q_{D_2}(p) \wedge \cdots \wedge Q_{D_n}(p).$$

Accordingly, $p \in \|Q\|_{\alpha^*}$.

Theorem 2: Let $Q = \wedge_{i=1}^{n} Q_{D_i}$, and $Q_{D_i} = \vee_{j=1}^{Si} Qt_{ij}$ be a query. Then

$$p \in \|Q\|_{\alpha^*} \text{ with } \alpha = \min(\max(\alpha_{11}, \dots, \alpha_{1Si}), \dots, \max(\alpha_{i1}, \dots, \alpha_{is_i}), \dots, \max(\alpha_{n1}, \dots, \alpha_{ns_n}))$$

such that for $i, 1 \leq i \leq n, p \in \|Qt_{ij}\|_{\alpha_{ij}^*}, j = 1, \dots s_i$.

Proof: This theorem can be easily proven by Proposition 2 and Theorem 1.

Example 9: Let the query Q be $Q_{D1} \wedge Q_{D2}$ where $Q_{D2} = \langle \text{furniture}, \text{table}, \text{east} \rangle$. Then, since $p_1 \in \|Q_{D1}\|_{0.9^*}$ (see Example 8) and $p_1 \in \|Q_{D2}\|_{0.85^*}$ from min $(0.85, 0.94) = 0.85$, we get $p_1 \in \|Q\|_{0.85^*}$.

4.3.4 Fuzzification of Spatial Relationships for Inexact Matching

The set of eight directions D is not sufficient to precisely specify the spatial relationships between objects. To compensate for the drawback, we extend our triple structure by replacing the spatial relationship with an angle degree in [0, 359] or with one element of an ordered list of direction descriptors, $D = \{\text{east}, \text{northeast}, \text{north}, \text{northwest}, \text{west}, \text{southwest}, \text{south}, \text{southeast}\}$, in which the order of its elements is preserved. Additionally, when the order is important, the ith element of D is denoted by d_i. It is used when precisely specifying the corresponding membership function of each direction along with counterclockwise increasing angle.

Definition 24: Let P be the whole set of images in I_RS, O_p be the set of all objects possibly occurring in an image $p \in P$. Then, a set of triples for p, T_p is given by

$$T_p = \{t = \langle o_i.name \ldots \rangle \text{ for } o_i, o_j \in O_p | r_{ij} \in D \cup [0, 359] \text{ and}$$
$$r_{ij} \text{ is the relative direction of } o_j \text{ with respect to } o_i\}.$$

The ordered list of direction descriptors D is needed to define the corresponding k-weight function for each direction in D. Along with the order, the corresponding k-weight function interprets some angle into one of eight directions with some degree, when the angle is given as a spatial relationship. The following section is devoted to the development of this function.

4.3.4.1 K-Weight Functions

To support an inexact matching between directions, we introduce k-weight functions, which symmetric but not always convex. They are used as the membership functions of the fuzzy subsets defining the eight directions in terms of the numeric angle degree. A k-weight function[23] is defined as

$$f(x, k) = C*(a^2 - (x - b)^2)^k, \text{ for } |x - b| \leq a,$$

where k is positive and $C, a > 0$, and b are constants.

For each direction $d_i, i = 0, \ldots, 7$, the following membership function is provided, where east corresponds to $0°$ (i.e., $\mu_{east}(0) = 1$) and the angle increases counterclockwise—for example, northeast and north correspond to $45°$ and $90°$, respectively.

Definition 25: Let $b_o = 0$ or 360 and let $b_i = 45^*i, i = 1, \ldots, 7$. Then, membership functions μ_{d_i} for $d_i \in D, i = 0, \ldots, 7$ are defined as

$$\mu_{d_i}(x) = C^*(45^2 - (x - b_i)^2)^k, i = 1, \ldots, 7, \text{ if } 45^*(i - 1) \leq x \leq 45^*(i + 1)$$
$$= 0 \text{ otherwise}$$

$$\mu_{d_o}(x) = C^*(45^2 - (x - b_o)^2)^k, \text{ where } b_0 = 0 \text{ if } 0 \leq x \leq 45$$
$$b_o = 360 \text{ if } 45^*7 \leq x \leq 45^*8$$

$$= 0 \text{ otherwise}$$

For the normalization, C is set as $C = 1/2025^k$ for real number k. Since the shape of the functions, $\mu_{d_i}(x), i = 0, \ldots, 7$ is subject to the k value and the retrieval effectiveness of our framework may depend on the characteristic of the shapes, we need to analyze the versions of each of the functions, generated along with the k value. For example, consider the versions of $\mu_{northeast}(x)$ for $k = 1, 2, 2.5$, and 3 in Figure IV. 4.11.

At first, all of the three non-convex versions are obviously superior to the convex one in the precision of matching. In other words, they guarantee high precision by making themselves sharper, especially in their boundary angle areas. For example, at some point x near $45°$, each value of $\mu_{east}(x)$ and $\mu_{north}(x)$ is drastically lower than that of $\mu_{northeast}(x)$, while $\mu_{northeast}(x)$ is significantly lower than each of those near $0°$ and $90°$, respectively (see Figure IV.4.12). However, they may, in turn, compromise recall due to their relatively low membership values in comparison with the convex one. For example, given a non-convex version obtained by $k = 2$, an image indexed by $\langle$table, chair, $20°\rangle$ may not be retrieved as an answer of $\langle$table, chair, northeast$\rangle$, since $\mu_{northeast}(20°) = 0.48 < \alpha$ for the threshold value $\alpha = 0.5$. The threshold value is the degree of relevance of results to be accepted as answers.

On the other hand, the version of $k = 2.5$ has an interesting property that the sum of $\mu_{northeast}(x)$ and either of its neighbor functions, i.e., $\mu_{east}(x)$ and $\mu_{north}(x)$ at the corresponding angle range is approximately 1. For example, when $\mu_{northeast}(20°) = 0.39$, $\mu_{east}(20°) = 0.61$. It may lead to highest precision since it exactly coincides with our intuition. However, since its recall is even worse than that of the version of $k = 2$, the latter version is our choice as a trade-off. Nevertheless, it is not

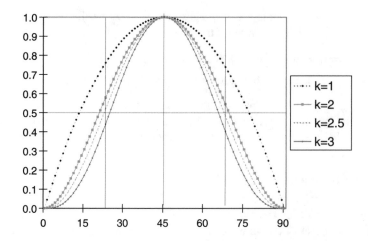

FIGURE IV.4.11   The four versions of $\mu_{northeast}(x)$ for $k = 1, 2, 2.5$, and 3. (From Yang, J.D., An image retrieval model based on fuzzy triples, *Fuzzy Sets Syst.*, 121(3), 89, 2001. With permission from Elsevier Science.)

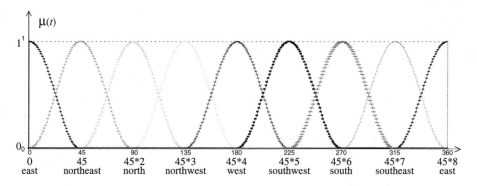

FIGURE IV.4.12   Membership functions for 8 directions. (From Yang, J.D., An image retrieval model based on fuzzy triples, *Fuzzy Sets Syst.*, 121(3), 89, 2001. With permission from Elsevier Science.)

clear which one is the best choice, since which one counts between precision and recall may largely depend on the characteristics of application domains.

We now get the following eight direction functions drawn by the version of $k = 2$.

Though the functions leave room for improvement, we avoid further detailed discussion here, since developing functions to maximize retrieval effectivenenss is far from our concern. They are introduced as an example of designing a membership function to enumerate compatibility between a direction and the corresponding angle degrees.

We now provide an example to incorporate the inexact matching into our query evaluation.

Example 10: Let a query be given as "search for images where an audio locates between the east and northeast side of furniture with an angle of $20°$." Then it is converted into $Qt_1 = \langle$ furniture, audio, $20° \rangle$. Now, for the stored triple, $t' = \langle f_{NAME}(r) = radio, f_{NAME}(s) = $ speaker, northeast$\rangle$ in Figure IV.4.1, since

$$\mu_{furniture} \text{ (radio)} = 0.85 \text{ (see Example 5),}$$

$$\mu_{audio} \text{ (speaker)} = 0.84 \text{ (see Figure IV.4.10) and}$$

$$\mu_{northeast}(20°) = \tfrac{1}{2025^2} * \{45^2 - (20° - 45)^2\}^2 = 0.48,$$

$$p_1 \in \| Qt_1 \|_{0.48*} \text{ with } \alpha = \min (0.85, 0.84, 0.48).$$

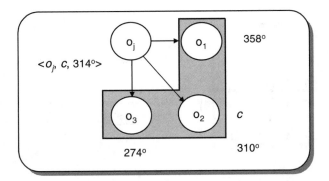

FIGURE IV.4.13 Averaging all the directions in R_c.

Besides the flexibility in matching between spatial relationships, inexact matching has another advantage—the matching can replace LUB/GLB with the average of the concept-direction set of a concept in determining a direction of the concept with respect to another object. For example, in Figure IV.4.8b, suppose the concept-direction set R_c is given as $R_c = \{358°, 310°, 274°\}$, which is approximately equivalent to {east, southeast, south}. Simply averaging every degree in R_c, it replaces the function GenDir(R_c); $\mu_{southeast}(314°) = 0.99$ implies that the average is interpreted into southeast with the degree 0.99. This is shown in Figure IV.4.13.

4.4   Implementation

In this section, we provide a prototypal system to demonstrate the feasibility of the proposed technique. It was implemented on top of Windows 2000 by using Visual C++ and Java. It consists of 5 modules: visual image indexer, concept rule base, concept thesaurus, inverted file, and query processor. The visual image indexer facilitates object labeling and the specification of relative position of objects. Concept rule base and concept thesaurus capture the concepts by aggregation and generalization respectively. The query processor enables users to formulate queries in terms of triples or object icons. A query is evaluated by matching the triples of the query with the inverted file. Shown in Figure IV.4.14 is the overall architecture of the system incorporating the 5 modules.

There are two approaches in implementing the query processor: top-down evaluation and bottom-up evaluation. In top-down evaluation, references to the concept rule base are confined to concepts appearing in the triples of user queries. Any other references for trying to detect concepts in images is not made. The inverted file, therefore, does not include any triple containing concepts. When evaluating a query, every triple containing concepts is translated into more than one concept-free triple which are semantically equivalent to it. The concept rule base is used in this translation. Target images may be retrieved by searching the inverted file with the concept-free triples. On the contrary, in bottom-up evaluation, every concept is detected and then the generated triples involving concepts are inserted into the inverted file prior to query evaluations. The triples in a user query may match their direct counterparts in the inverted file. Currently, our query processor adopts the bottom-up evaluation for not compromising the user response time, avoiding query evaluation delay due to query reformulation. However, the bottom-up evalution also has the drawback that concept detection is time consuming when images contain too many objects. Judgement on which one is better may depend on the characteristic of application domains.

The concept rule base and the concept thesaurus may need their dedicated editors for facilitating rule base or thesaurus construction. For example, as shown in Figure IV.4.15, the thesaurus editor can directly construct fuzzy term hierarchies in the concept thesaurus.

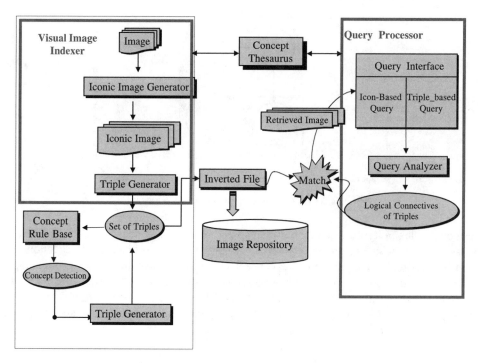

FIGURE IV.4.14   Overall system architecture.

With this concept thesaurus, the fuzzy term matcher performs a concept-based match between fuzzy terms in a query and the counterparts in extended triples of target images. It also enumerates the conceptual distance between them according to Definition 14 and the associated k-weight functions in Definition 25. Recall that the queries are assumed to use fuzzy terms in the thesaurus. The same is true in the triples, which are generated from an image indexer.

Figure IV.4.16 shows the image indexer for manually labeling objects in an image. The reason we adopt manual labeling is to avoid compromising the recall and precision of our method due to the poor object recognition ratio of the state-of-the-art image processing technologies. Unfortunately, it is widely conceived that object labeling without human intervention would not be made in the near future.

In Figure IV.4.16, a dialog box is displayed, waiting for a domain knowledge engineer to enter the name of a pointed object in an image whenever he or she draws an MBR (minimum bounding rectangle) surrounding the object. Once the needed names are entered, for example, television and sofa, the corresponding triples are automatically generated. The triple viewer window shows them, where either an angle degree or one of eight directions specifies a spatial relationship between a pair of objects. The relationship is calculated by drawing lines between the centers of their MBRs.

In labeling objects, a term mismatch problem may be addressed due to the lack of the other relationships except BT/NT in our thesaurus. For example, what if a user tries to retrieve images containing TV instead of the term television? It may be left as further research to cover the relationships such as "synonym of" and "part of" without compromising the simplicity of our framework.

Finally, Figure IV.4.17 is the interface of the query processor showing the answer images, when the evaluated query is ⟨electronic-appliance, furniture, east⟩ ∨ ⟨audio, furniture, 82°⟩. You can see images where TV is captured as an electronic-appliance or home appliance, speaker as audio,

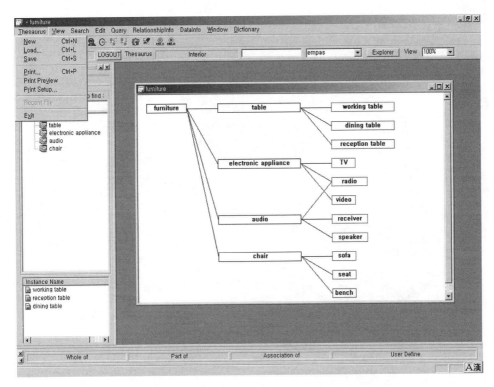

FIGURE IV.4.15   Concept thesaurus editor.

sofa as furniture, etc. As mentioned earlier, the fuzzy term matcher is repetitively called by this query processor whenever any fuzzy term matching is needed in the query triples—for example, when electronic-appliance needs to match any other term, say, TV, with the evaluation of their conceptual closeness. Practically, such term matching may imply the reformulation of queries. For each reformulated query, the query processor would search for the inverted files to get answers, assigning appropriate ranking according to Theorem 2.

4.5   Conclusions

In this chapter, we introduced an extended version of triples as a framework capable of matching images based on concepts as well as indexing them. The concepts are the aggregation and generalization of primitive objects in an image or other concepts. For the flexible matching between spatial relationships, the k-weight function was adopted for the precise specification of spatial relationships with angle degrees. We also showed a way of query evaluation supporting such a concept-based matching. To demonstrate the feasibility of our model, a prototype system was developed.

The contribution of this chapter is twofold: to show a systematic way of incorporating a concept-based match facility into the conventional triple framework, making its spatial relationship flexible, to provide a formal specification of a query evaluation capable of greatly enhancing recall in comparison with other ones based on the triple framework.

Since the extended triple is a powerful yet conceptually simple framework in its matching capability, we expect that it would be easily applied as well to the other content-based image retrieval systems which do not use the triples. For example, it may serve to enhance the functionality of the systems such as QBIC or Photobook, if it is used to filter images conceptually not relevant to user queries, before retrieving images based on colors, shapes, and texture or vice versa.

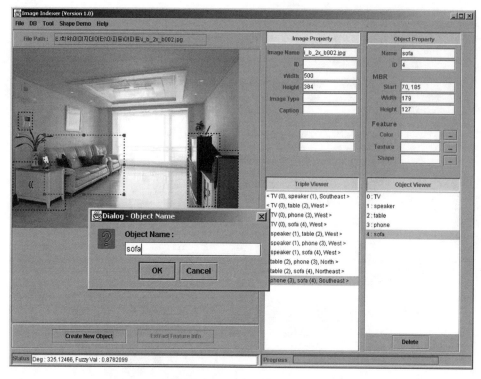

FIGURE IV.4.16   Image indexer.

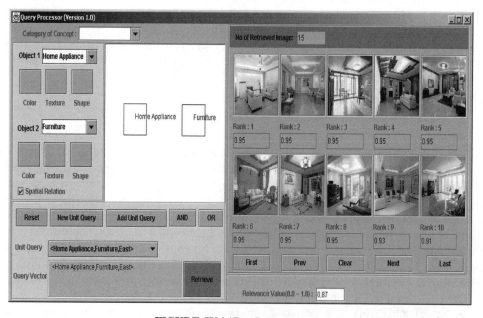

FIGURE IV.4.17   Query processor.

References

1. Arndt, T. and Kuppanna, R., A prototype multimedia database system incorporating iconic indexing, in *Intelligent Image Database System, Series on Software Engineering and Knowledge Engineering*, vol. 5, World Scientific Publishing Co., Singapore, 1996.
2. Aslandogan, Y.A. et al., Design, implementation and evaluation of SCORE (a system for content based retrieval of pictures), Proceedings of the 11th International Conference on Data Engineering, 280, 1995.
3. Bach, J.R. et al., The Virage image search engine: an open framework for image management, *Proc. Storage Retrieval Still Image Video Databases, SPIE*, 2670, 76, 1996.
4. Baxter, G. and Anderson, D., Image indexing and retrieval: some problems and proposed solutions, *New Lib. World*, 96(1123), 4, 1995.
5. Busby, R.C. and Kolman, B., *Discrete Mathematical Structures for Computer Science*, Prentice Hall, Upper Saddle River, NJ, 1997.
6. Chakravarthy, S. et al., Composite events for active databases: semantics, context and detection, Proceedings of the 20th International Conference on VLDB, 606, 1994.
7. Chang, S.K., Shi, Q.Y., and Yan, C.W., Iconic indexing by 2D string, *IEEE Trans. Pattern Anal. Mach. Intelligence*, Pami-9, 413, 1987.
8. Chang, S.K. et al., An intelligent image database system, *IEEE Trans. Software Eng.*, 14(5), 681, 1988.
9. Chang, C.C., Spatial match retrieval of symbolic pictures, *J. Inf. Sci. Eng.*, 7, 405, 1991.
10. Chang, C.C. and Lee, S.Y., Retrieval of similar pictures on pictorial databases, *Pattern Recognition*, 24, 675, 1991.
11. Chang, J.W., Kim, Y.J., and Chang, K.J., A spatial match representation scheme for indexing and querying in iconic image databases, Proceedings of ACM CIKM, 169, 1997.
12. Chang, J.W., Kim, Y.J., and Jin, K.S., Spatial match representation and retrieval for supporting ranking in iconic image databases, *IEEE Int. Conf. Multimedia Expo*, 1, 315, 2000.
13. Cook, C.R., and Oldehoeft, R., A letter-oriented minimal perfect hashing function, *ACM SIGplan Notices*, 17, 18, 1982.
14. Corridoni, J.M., Bimbo, A.D., and Magistris, S.D., Querying and retrieving pictorial data using semantics induced by colour quality and arrangement, Proceedings of the International Conference on Multimedia Computing and Systems, Hiroshima, Japan, June, 219, 1996.
15. Flickner, J. et al., Query by image and video content: the QBIC system, *Computer*, 28(9), 24, 1995.
16. Forsyth, D.A., *Finding People and Animals by Guided Assembly*, ICIP97, IEEE, New York, 1997.
17. Kim, W., *Introduction to Object Oriented Databases*, MIT Press, Cambridge, MA, 1990.
18. Larsen, H.L. and Yager, R.R., The use of fuzzy relational thesauri for classificatory problem solving in information retrieval and expert systems, *IEEE Trans. Syst. Man Cybernetics*, 23(1), 31, 1993.
19. Papadias, D. and Sellis, T., Spatial reasoning using symbolic arrays, Proceedings of the International Conference on GIS—From Space to Territory Theories and Method of Spatial-Temporal Reasoning in Geographic Space, Pisa, Italy, 1992.
20. Pentland, A., Picard, R.W., and Sclaroff, S., Photobook: tools for content-based manipulation of image databases, *Int. J. Comput. Vision*, 1996.
21. Smith, J.R. and Chang, S.F., Visualseek: a fully automated content-based image query system, Proceedings of the International Conference on Image Processing, Lausanne, Switzerland, 1996.
22. Vinod, V.V., Murase, H., and Hashizume, C., Focused color intersection with efficient searching for object detection and image retrieval, Proceedings of Multimedia '96, 229, 1996.

23. Wand, M.P. and Jones, M.C., *Kernel Smoothing*, Chapman & Hall, New York, 1995.
24. Yang, J.D. and Yang, H.J., A formal framework for image indexing with triples: toward a concept-based image retrieval, *Int. J. Intelligent Syst.*, 14(6), 1999.
25. Yang, J.D., A concept-based query evaluation with indefinite fuzzy triples, *Inf. Process. Lett.*, 74, 2000.
26. Yang, J.D., An image retrieval model based on fuzzy triples, *Fuzzy Sets Syst.*, 121(3), 89, 2001.

5

The Techniques and Experiences of Integrating Deductive Database Management Systems and Intelligent Agents for Electronic Commerce Applications*

Chih-Hung Wu
Shu-Te University

Shing-Hwang Doong
Shu-Te University

Chih-Chin Lai
Shu-Te University

Abstract. Exact order analysis and merchandising are two main issues in electronic commerce applications. In this chapter, we present the techniques and experiences of integrating deductive database management systems (DDBMS) and intelligent agents in electronic commerce applications. DDBMS is employed for analyzing online orders and assisting marketing managers in extracting applicable historical plans. Intelligent agents are designed for autonomously and periodically collecting orders and for deduction

*Partially supported by the National Science Council, Taiwan, R.O.C., under grant NSC89-2218-E-366-002.

of rules. An experimental Web portal is built for online B2B transactions. The issues on design and implementation are discussed.

Key words: Electronic commerce, deduction rules, deductive databases, intelligent agents, artificial intelligence

5.1   Introduction

For the past few decades, information technologies have successfully improved the efficiency and effectiveness of business transactions. Traditional commerce concerns the establishment of business transactions where participants—customers, suppliers, intermediaries, brokers, etc.—meet at a certain place and a certain time. Due to the innovations on the Internet and World Wide Web (WWW), interactions among different participants in a traditional business transaction have been changed. The Web-based technologies have broken the restrictions on time and space. Web-based transactions have become a new competitive business tool. Electronic commerce (EC),[12,19,23,37] which refers to a modern business methodology that addresses the exchange of information, products, services, and payments in a virtual environment, is emerging as one of the most important activities on the Internet.

Different types of EC transactions such as business-to-customers (B2C) and business-to-business (B2B) have arisen and received a lot of attention. The following issues are benefits that EC brings to business organizations.

1. Cost reduction: EC provides online mechanisms to search, compare, and order products or services for customers that can reduce costs such as money, time, and effort expended in gathering information about product price, quality, and specifications. On the other hand, suppliers can reach and serve more customers at lower costs.[8]
2. Global convergence: The open nature of the Internet, on which all kinds of computers and communication media are connected, technically enables worldwide customers and suppliers to conduct commercial transactions anytime and anywhere in the world.
3. Competition increased: EC can help business organizations cut costs, interact with customers, and provide timeless and worldwide communication. Additionally, it can even help an organization outperform its competitors.

5.1.1   Requirements of EC Applications

The supporting information technologies for a successful EC application include the following:

1. Web server: With the powerful abilities of communication and integration of different types of contents and systems, the Web is an ideal platform for developing EC applications.
2. Database: A database is a vehicle for organizing, storing, and accessing large amounts of information on the Web (e.g., transactional records, product catalogs, etc.) and provides transactions, concurrency control, distribution, and crash recovery for business data.
3. Application program: Application programs implement business logic, which usually bridge between the Web server and the database. They are designed for interacting with the users and the backend information systems.
4. Security: Cryptographs are used to ensure the safety of communication, providing a secure transactional environment.[3,36]

The Web can help a business collect customers' personal preferences as well as purchase behaviors and history. Consequently, business corporations have to build up large network-based databases so that their products and services can be reached by the users and the behaviors of their customers can be analyzed. This information provides the foundation for a business to offer customers individually tailored products or services. The Web also improves competition of business by reducing the operational cost. EC applications are data-driven, i.e., all business operations and interactions with customers are triggered by data. Clearly, among the above supporting technologies, the database is a critical one. In such a rapidly changing and highly competitive environment, the following requirements for implementing EC applications must be considered:

1. Flexibility: EC transactions are customer-oriented in the sense that business rules should be adjustable for different customer segmentations.
2. Time-to-market: To outperform the competitors, EC applications have to quickly respond to customers' needs. Additionally, they have to promptly respond to the change of market by efficiently differentiating products or business and marketing rules.
3. Easy to change: The life cycles of new products are short in EC. Applications of EC should be able to change the operational parameters such as prices, supplying relationships, marketing strategies, and business logic in a fast and easy manner.

5.1.2   Restrictions of Current Implementation Methods

Most software development projects adopt the spiral model for fast prototyping. The gap between a prototype and the final system converges if the specifications are precisely given. Developing EC applications in traditional methods encounters some difficulties.

First, in most EC applications, the database is built before application programs are developed. As we have mentioned previously, the requirements and specifications of EC applications may change frequently. Constant modification to both application programs and the database is inevitable, yet time-consuming and costly. For example, to implement a new pricing rule may require the creation or modification of the schemas of the database and the associated access programs and then a thorough test of their correctness.

Second, most EC applications employ the relational database management system (RDBMS), which is a mature technique for managing static data with well-established relationships. Because of the changing characteristics of EC applications, determining all relationships among products, components, suppliers, and manufacturers and creating all corresponding schemas prior to the development of the application programs are almost impossible.

Moreover, most application programs utilize the standard structured query language (SQL) for access to the RDBMS. For sophisticated EC applications, recursion and inheritance are commonly employed to describe the relationships among data. Unfortunately, constructing access programs using SQL commands for such relationships is very complicated and expensive.

Finally, data in an RDBMS is closed, i.e., queries to an RDBMS succeed only if the data is already stored in the database. Access to data of recursive or inheritant relationships requires flattening all relationships and storing all related data in database. For such a large amount of data and relationships, to implement EC applications in an RDBMS is difficult. Even when all data can be flattened and stored in an RDBMS, the inefficiency of data processing and a huge amount of storage space are two tough questions.

Traditional approaches have received a lot of success, but, unfortunately, they are inflexible and inefficient in implementing such an online system in a competitive and changing market. Obviously, a DBMS providing flexible, customized, and quick response for EC applications is highly required.

5.1.3   The Proposed Method

Logic deduction is a unique ability of human intelligence and is one of the most important research topics in artificial intelligence.[38] Rule-based deduction that derives new conclusions with the condition-action model is the simplest yet most effective inference method. Forward-chaining reasoning (deriving new conclusions from conditions) and backward-chaining reasoning (confirming conclusions by verifying conditions) are two common inference mechanisms. Rule-based deduction is widely adopted in developing intelligent systems for its ease of use and its ability to capture various inference relationships, e.g., recursion, inheritance, and cross-reference. To cope with the disadvantages of RDBMS, the combination of rule-based deduction and DBMS, referred to as the deductive database management system (DDBMS), provides a flexible way for managing complex data relationships and eliminating the waste of storage space.

An agent is a piece of software that autonomously performs specific missions assigned by the users. Mobility, autonomy, and intelligence are the main features of agents. Systems built with agents benefit from completing their missions autonomously with high mobility, transparency, and expansibility, providing users with an easy method to develop and maintain application programs in a distributed environment.

We integrate the DDBMS with intelligent agents for EC applications. The restrictions of RDBMS in implementing EC applications can be eliminated by flexible, intuitive, and easy-to-use deduction rules. The complexity and inconvenience caused by the frequent modification of application programs are reduced by the use of intelligent agents. Deduction rules are embedded in the brain of agents for completing jobs. For most EC applications, these seem to be essential, especially when personalization and customization are required.

5.1.4 Organization

The rest of this chapter is organized as follows. We give brief introductions to DDBMS and intelligent agents in Section 5.2. A graphic description scheme is defined in Section 5.3 for representing the relationships among data and deduction rules. In Sections 5.4 and 5.5, we describe how the proposed techniques cope with two problems encountered in EC applications, namely, decomposition of online orders and deployment of new marketing strategies. The system architecture we designed and implemented is presented in Section 5.6. Section 5.7 discusses our experiences and conclusions.

5.2 Background

5.2.1 Deductive Database Management Systems

Even though RDBMSs have been successfully applied in many business operations, an RDBMS still has the following disadvantages: unable to represent complex data type, redundancy in data entities, less intuitive semantic meaning in SQL expression, and lack of recursive query power in SQL. Logic programming (LP) with rule-based deduction provides a more intuitive semantic meaning than SQL, and it also offers recursive query power effectively. The marriage of LP and RDBMS has created a powerful database management system—the deductive database management system (DDBMS).[15–17,22,26,29] While LP is used to write declarative and powerful queries, RDBMS is used to handle a larger volume of data than a single LP system such as Prolog can handle alone. A deductive database (DDB) can thus be divided into two parts: intensional database (IDB) and extensional database (EDB). EDB contains the facts stored in an RDBMS, while IDB contains the deduction rules used in the inference process.[9,11] In this way, we can minimize the facts to be stored in EDB and apply the rules in IDB to deduce the other facts. Therefore, a DDB can usually save more storage space than a regular relational database.

There are a couple of DDBMSs that are commonly in use nowadays, such as LDL,[7,40] CORAL,[30,32] NAIL!,[24,25] and XSB.[31,33] In this project, we use the XSB system because of its academic background. The XSB system has the following special features not commonly found in LP:

- Evaluation according to the well-founded semantics through full SLG resolution[6]
- A compiled Hilog[5,34] implementation
- Interfaces to C, Java, Oracle, and ODBC systems

XSB can be seen as an extension of the Prolog system. Through ODBC, we use the built-in predicates in XSB to interface an external RDBMS for extensional data. Intensional data such as deduction rules are created using Prolog-like syntax in XSB. For simplicity, a rule is represented in the form of $p\text{:-}\ q_1, q_2, \ldots, q_3$, where the terms $p, q_1, q_2, \ldots, q_3$ in the rule are represented in first-order logic without functional terms. Variables are represented as strings beginning with a capital letter; constants are strings beginning with a lower-case letter. The term p is called the conclusion or the head of the rule, while $q_1, q_2, \ldots, q_3$ are the conditions of the rule body. All conditions are connected by logic AND. A single term t can be viewed as a rule with logic TRUE as the only condition, i.e., $t\text{:-}TRUE$. For example, $ancestor(X,Y)\text{:-}father(X,Z),ancestor(Z,Y)$ and $a(p,q)\text{:-}b(p,X),c(X,q)$ are legal rules.

5.2.2   Intelligent Agents

An agent is a software piece that automates some operations for user. It can be a dumb agent for providing simple choices or a smart agent empowering all sorts of artificial intelligence techniques to help users do things faster and more accurately. Perception and action are the two steps that an intelligent agent performs to automate operations for the users.[4] The agent gathers information through its sensors and perceives the environment around it by a simple predefined taxonomy or deduction rules and facts. Then, it takes action through effectors, which are like human muscles.

Multiple agents can cooperate or compete in the same environment. For example, we may divide a complex job into simpler sub-jobs for cooperating agents to reduce the programming efforts and increase the functional stability. On the other hand, competing agents normally exist in an EC environment for buyers and sellers—the buyer's agent wants to get the best deal, while the seller's agent wants to get the best profit. Communication becomes an important issue in a multiagent system. To eliminate the ambiguity in interpreting terms, a common vocabulary called ontology is used in the communication. The DARPA-developed knowledge query and manipulation language (KQML)[39] is commonly used for communication. Programs and agents can use KQML to exchange information and knowledge. KQML messages are called performatives, each of which is used to perform a specified action. A performative encodes information at the content, message, and communication levels. For example, in the following KQML message,

```
(ask-ibm
        :sender bill
        :content (real price = ibm.getprice())
        :receiver nyse-server
        :reply-with ibm-price
        :language java
        :ontology NYSE-QUOTE)
```

the content parameter is the content level, the reply-with, sender, and receiver parameters define the communication level, and the performative name ask-ibm, language, and ontology parameters are in the message level.

5.2.3   An Experimental Web Portal for Business-to-Business Transactions

Throughout this paper, we demonstrate the effectiveness of deductive databases and intelligent agents using a simplified online Web portal for B2B transactions. The whole project is to provide a flexible and convenient environment for business users to purchase large volumes of products or components of consumer electronics. The portal connects transactional information for retailers, wholesalers, and upstream suppliers and manufacturers. Among many of the requirements of building this portal, the ones related to this research include the following:

- A mechanism periodically collecting valid orders and calculating the volume of products to be shipped to the customers. The insufficient volume of products has to be ordered from and supplied by the upstream suppliers or manufacturers.
- An assisting mechanism for the marketing manager who frequently needs to react to the change of market condition by deploying, in a very short time, new strategies for promoting new products.
- A flexible tool for easily reconfiguring the relationships among products, components, and the associated suppliers or manufacturers.

To achieve the best cost-effectiveness, two issues are considered: (1) reducing the operational cost, and (2) increasing the sales force. Batch processing for shipping orders, negotiating for purchasing a larger volume of product components with lower prices, eliminating personnel expenses by automation, etc., are approaches to reduce the operational cost. Exact and efficient analysis of orders is a key point for accomplishing the above goals. Additionally, merchandising and deploying effective promotion strategies can increase the sales force. Below we describe how the techniques of DDBMS and intelligent agents assist the managers of the B2B portal in decomposing the online orders and deploying new marketing strategies.

5.3 Data Semantics Network

Conceptually, a database can be viewed as a network connecting various data entities via user-defined relationships. To describe the relationships among data for deduction in a concise manner, we introduce the data semantics network (DSN). A DSN is a directed graph in which a circular node represents an entity involved in the underlying problem, and a labeled, directed link between two nodes represents a specific relationship between two entities. DSN represents the schema of a database as well as a structural concept given by users. For example, suppose we have the following part–subpart relationships.

- Product a is assembled by products d and e.
- Product b is assembled by products e and f.
- Product e is assembled by products k and l.

A DBMS can record the above relationships easily with the schema shown in Figure IV.5.1a. The DSN representation scheme for the example is given in Figure IV.5.1b.

DSN can also describe the structure of a concept. Suppose a marketing manager successfully deployed a cross-sale strategy as follows:

> *cross-sale-discount m-375-a: customers who purchase the mobile phone (nj8765k) together with an additional cell battery pack (the bb1226x series) and the headset (ac0217) can get a 15% discount.*

The above concept can be represented by the DSN in Figure IV.5.2.

Note that the entity-relationship diagram (ERD) used in developing databases also describes the relationships among entities. However, all relationships in ERD are eventually recorded in database tables, and the ones in DSN may be given by the users and excluded from the database. Conceptually, an ERD of a database is a DSN, but not vice versa. DSN is an auxiliary tool for recognizing and defining relationships among data entities.

PID	Component ID	STW	...
a	d	TPE	
a	e	KHH	
b	e	CYI	
b	f	TXG	
e	k		
e	l		
...	...		

Component ID	Stock	...
d	2	
e	1	
f	2	
k	1	
l	3	
...	...	

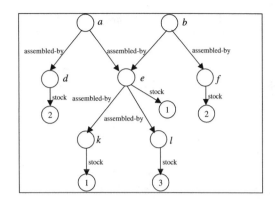

(a) Schema for the part—subpart relationships

(b) A DSN for the part—subpart relationships

FIGURE IV.5.1 The part–subpart relationship in the form of database schema and DSN.

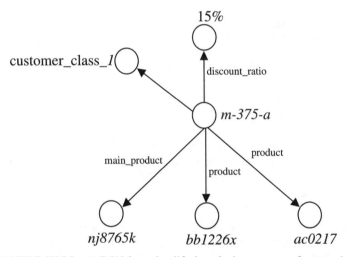

FIGURE IV.5.2 A DSN for a simplified marketing strategy of cross-sale.

5.4 Order Decomposition and Analysis

The items listed in an order usually include the identification number, volume, price, and total amount of purchased products. Since the portal provides consumer electronics of different scales, some products are assembled by independently functional components. To satisfy an order, we have to place orders for the functional components associated with the purchased products to the upstream suppliers or manufacturers before delivery. The administrator has to negotiate and contract with the suppliers and configure the most efficient way for quick response to the customers. He or she also needs to know the following information:

- The volumes of all products and associated components purchased by customers
- The volume of stocks in the warehouses for assembling the products
- The volume of stocks can be provided by the suppliers and manufacturers

TABLE IV.5.1   Table for the Supplying and In-Stock Relationships of Products

	Supplying Relationships						Volumes of Products in Stock				
PID	SID	PRIT	MRG	DSC%	DYTIME (hr.)	...	PID	STW	VOL	DST	RSPTIME (hr.)	...
i	s1	5000	1000	10	72		i	TPE	3000	350	48	
j	s2	22,000	500	15	48		j	TNN	700	75	36	
j	s3	20,500	700	12	48		j	KHH	500	50	48	
k	s2	300	6000	5	24		k	CYI	500	150	72	
l	s3	17,000	800	20	96		l	KHH	1280	50	48	
m	s1	8500	2000	10	120		m	PIF	650	100	60	
n	s1	950	15,000	15	184		n	TNN	2400	75	96	
n	s2	1150	9500	10	120		n	TPE	1000	350	80	
p	s3	6300	4000	25	168		p	TPE	100	350	96	
q	s1	4850	900	20	72		q	TXG	975	200	72	
q	s3	5500	1300	15	96		q	CYI	1200	150	72	
...	...	...	...	...	...	...	...	...	...	...	...	...

Note: PID is the product number; SID is the supplier; PRIT is the price per item; MRG is the margin of large volume; DSC is the discount for large volume; DYTIME is the response time to delivery; STW is the warehouse location; VOL is the volume of products in stock; DST is the distance to STW; RSPTIME is the response time to delivery.

Unfortunately, the market of consumer electronics is highly volatile and competitive. A product may have various or frequently changing ways to be assembled by different components. Moreover, suppliers and manufacturers may intentionally provide similar products or components with different but highly volatile prices. They may adjust the supplying items for the best selling items. A product or component provided by a supplier and manufacturer may become out-of-date and suspended from the supply chain. Clearly, the administrator of the portal needs to efficiently and effectively capture the information from the customers and suppliers for dynamically reconfiguring the status of products and components.

Intuitively, the relationships among products, components, suppliers, and manufacturers can be viewed as deduction rules. Gathering information from a database by these relationships can be viewed as an inference process of deduction. Compared with traditional access programs, deduction rules can be changed more easily. Suppose that the assembly relationships of products and components are as follows:

- Product a is assembled by products d and e.
- Product b is assembled by products e and f.
- Product c is assembled by products f and g and h.
- Product d is assembled by products i and j.
- Product e is assembled by products k and l.
- Product f is assembled by products m and n.
- Product h is assembled by products p and q.

The status of supplying these products and components is recorded in the database of Table IV.5.1.

Now we have five orders, each of which contains different volumes and types of products as follows.

- Order # 561226: product $a \times 50$, product $c \times 100$
- Order # 571029: product $a \times 150$, product $b \times 50$, product $c \times 30$
- Order # 850217: product $b \times 50$, product $c \times 120$

- Order # 870312: product $a \times 50$, product $b \times 150$, product $c \times 200$
- Order # 900829: product $a \times 100$, product $c \times 50$

A key task of the administrator is to analyze the orders and determine what, when, and for how many products or components the firm should place orders to the upstream suppliers or manufacturers. Cost-effectiveness and quick-response are two main considerations. In making such decisions, the administrator can construct declarative inference rules by referring to the corresponding DSN. For example, the calculation of the volume of products to be ordered from the upstream suppliers can be achieved by constructing the deduction rules to recursively query into the database. Figure IV.5.3a presents the corresponding DSN and inference rules.

With the recursive DSN set up as in Figure IV.5.3a and the order example in the last section, we implement the query operation in XSB with an ODBC link to an external database containing the facts of the DSN and the order example. These facts are depicted in Figure IV.5.4 as relational tables in the external database.

A sample LP code implementing the calculating rules is listed in Figure IV.5.5. By running the calculating rules, we can obtain that the qualities of the components i, g, j, k, l, m, n, p, and q are 350, 500, 350, 600, 600, 750, 750, 500, and 500, respectively.

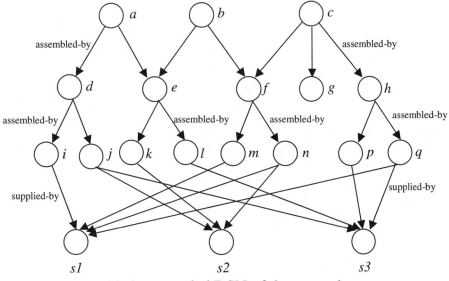

(a) An extended DSN of the example

```
/* Recursive def of part and subpart */
subpart(X,Y,C) :- parts(X,Y,C), \+ (Y == '999').
/* atom has '999' as subpart */
subpart(X,Y,C) :- parts(X,Z,C1), subpart(Z,Y,C2),
\+ (Y == '999'), C is C1*C2.
```

(b) Inference rules in the example

FIGURE IV.5.3 The calculation of the volume of products.

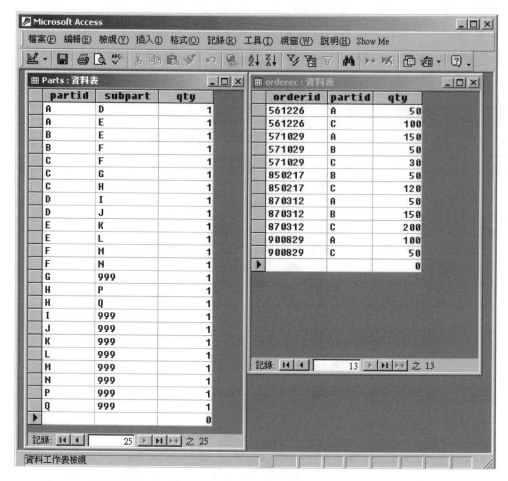

FIGURE IV.5.4   Sample facts recorded in the external database.

We can see that the recursive relation between part and subpart can be easily solved in logic programming by this simple definition in the XSB source code in Figure IV.5.3b. Once the total subpart numbers have been obtained in the XSB program, it becomes an optimization problem to compare the parts request with the available warehouse stock and make possible the placement of orders to the upstream suppliers. The Web portal can optimize the request with respect to the quickest response time or the cheapest order fee with another intelligent agent. Since the optimization problem is beyond the scope of this project, we leave out the details of this step.

5.5   Deployment of New Marketing Strategies

Another important task of EC is merchandising. Various methods of deploying marketing strategies have been proposed. Here we only discuss cross-sale. Cross-sale is one of the most effective and widely used strategies, which promotes products of high value, time-sensitivity, and high-homogeneity by combining as a purchasing package with other products of closely related characteristics. Setting up a marketing strategy requires analyzing the characteristics of the products, the purchasing behavior of customers, the strategies of competitors, and so on to predict its effectiveness. Clearly, this is a time-consuming and costly task. However, when promoting a new

```
/* Main utility */

sumpart(X) :- catpart(X,S), compa(X,S).

sumorder :- \+ clear_all_f, \+ setorders, printorder.

/* Recursive def of part and subpart */

subpart(X,Y,C) :- parts(X,Y,C), \+ (Y == '999').  /* atom has '999' as subpart */

subpart(X,Y,C) :- parts(X,Z,C1), subpart(Z,Y,C2), \+ (Y == '999'), C is C1*C2.

/* Supporting utility */

clear_all_f :- retract(f(_,_)), fail.                /* remove all facts concerning qty of atom */

setorders :- orderec(X,C), setorder(X,C), fail. /* for each order, decompose part into atoms, and insert facts with
qty */

setorder(X,C) :- catpart(X,S), compb(X,S,C).    /* case of a composite part */

setorder(X,C) :- parts(X,'999',1), assert(f(X,C)).  /* case of an atom part */

printorder:- setof(X, Y^f(X,Y), S), torder(S).

torder([X|Tail]) :- findall(C, f(X,C), S), list_sum(S, K), write('Total qty for subpart '),
        write(X), write(' is '), write(K), nl, torder(Tail).

catpart(X,S) :- setof(Y, C^bsubpart(X,Y,C), S).

bsubpart(X,Y,C) :- subpart(X,Y,C), parts(Y,'999',_). /* Y must be an atom */

compa(X,[Y|Tail]) :- comps(X,Y,K), write('Qty for subpart '), write(Y), write(' is '), write(K), nl, compa(X,Tail).

compb(X,[Y|Tail],C) :- comps(X,Y,K), J is K*C, assert(f(Y,J)), compb(X,Tail,C).

                        /* insert qty of atom into facts */

comps(X,Y,K) :- findall(C, subpart(X,Y,C), S), list_sum(S, K).

list_sum([],0).           /* sum the elements in a list */
list_sum([X|Tail],K) :- list_sum(Tail,J), K is X+J.
```

FIGURE IV.5.5   The code in implementing order-decomposition.

product, senior marketing managers can quickly figure out a prototypical strategy according to their experiences and historical data. For an inexperienced marketing staff, to mature a complete plan is difficult. He or she needs to thoroughly search from the archives for a plan that once successfully promoted similar products. In both cases, an historical plan retrieved from the archives partially confirms the correctness of the new marketing strategy. Such information can be used by human decision-makers for constructing a complete plan. The "what-if" analysis considers the results caused by assuming a set of conditions, e.g., "what will we gain from a strategy similar to an historical and successful one"? Historical data are usually retained and distributed in a data warehouse; therefore, the retrieval of historical plans needs to invoke a series of database queries. However, many uncertain or hypothetical factors exist in a prototypical strategy; composing sophisticated, yet complicated, SQL commands is unavoidable. Composing sophisticated SQL commands requires experienced IT professionals. Unfortunately, most marketing experts are not IT professionals and are usually frustrated by direct access to databases using plain SQL commands. The need exists for a flexible and easy-to-use mechanism for extracting similar marketing plans.

   The main characteristics and relationships of the promoted products, such as the specifications and combinations of the products, manufacturing cost, and targeting customer segmentation, can be described in the DSN. Refer to the marketing strategy *m-375-a* in Figure IV.5.2 again. The main

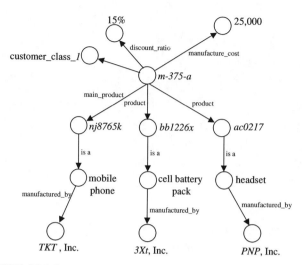

FIGURE IV.5.6   An extended DSN of the marketing strategy *m-375-a*.

features of the products *nj8765k, bb1226x*, and *ac0217* as well as their connecting relationships are described in Figure IV.5.6. Conceptually, finding applicable historical plans is analogous to identifying structurally similar DSNs. Using deduction rules, the two activities of marketing can be done easily.

5.5.1   Verification of New Marketing Plans

Generally, in designing a new strategy, marketing managers scratch a new plan and then verify its effectiveness by market survey. Questionnaire and comparison with historical plans are two common approaches for this purpose. A cross-sale strategy that successfully improves the sales force also successfully associates the relationships among the products or services. Referring to the structure of DSN in Figure IV.5.6, a successful marketing strategy can be viewed as a network connecting closely related products. According to marketing experience, a new strategy can succeed if it matches the key features of an historical strategy. In DSN, such key features reflect the structure of relationships. If a marketing manager conceives a new strategy, he or she can verify it by examining if there exist any structurally similar DSNs in the historical database. He or she can try some different combinations of products and evaluate the possibility of their success. By analogy, the process is to construct a set of deduction rules for the given DSN and to verify if similar rules can be confirmed. For example, a new cross-sale plan for products *ax*, *by*, and *cz* may describe the following relationships.

- *ax* is of feature p_1, *by* is of feature p_2, and *cz* is of feature p_3.
- The relationship between *ax* and *by* is r_1 and the relationship between *ax* and *cz* is r_2.

The manager verifies the plan by checking if the terms p_1, p_2, p_3, r_1, and r_2 ever appear together in the database with a similar DSN structure. Suppose an historical marketing plan *mj-99-ch-a* ever successfully promoted products *pm*, *qn*, and *rk*, where

- *pm* is of feature p_1, *qn* is of feature p_2, and *rk* is of feature p_3.
- The relationship between *pm* and *qn* is r_1 and the relationship between *pm* and *rk* is r_2.

In DSNs, the above two plans are not identical but are structurally similar, i.e., the terms p_1, p_2, p_3, r_1, and r_2 appear together at the same locations of the DSN structure. With this confirmation, the

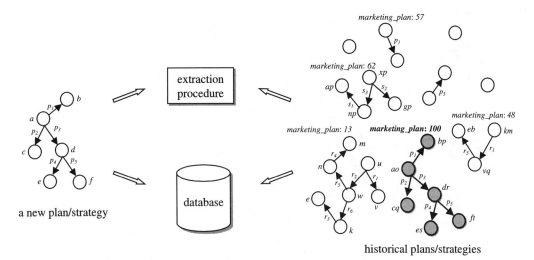

FIGURE IV.5.7   Confirming a new cross-sale strategy by DSNs.

marketing manager surely feels more confident with the success of the new plan. Figure IV.5.7 illustrates the concept of this process.

5.5.2   Extracting Applicable Marketing Plans

For marketing managers with less experience, to mature a plan is not easy. However, according to specific features of the new product, they can extract from the marketing archives a plan that promoted products with similar features and then modify the plan accordingly. In DSNs, this process is analogous to the extraction of a complete DSN containing a specific structure. The partial DSN represents the features of the new product given by the users. We can construct a set of deduction rules that represent specific nodes and links of DSN and query the database to investigate if confirming data exist. For example, suppose that we only know some features p_1, p_2, and p_3 of a new product az and intend to extract applicable plans. We investigate the database to see if any marketing plan contains any items having the features of az. Suppose an historical marketing plan cn-00-en-jp ever successfully promoted products mx and kz, where

- mx is of features p_1, p_2, and p_3 and kz is of feature p_4.
- The relationship between mx and kz is r.

Since the nodes and links representing the features p_1, p_2, and p_3 of az are included in the plan cn-00-en-jp, we could possibly promote az by cross-selling it with a product which is of feature p_4 and has a relationship r with az. Figure IV.5.8 depicts the concept of the above.

5.5.3   The Extraction Procedure

To extract a specific structure from a set of DSNs, we implement an extraction procedure and integrate it with DDBMS. In this procedure, the structure of the given DSN is represented as deduction rules; nodes and links of DSNs are stored in an external database. The extraction procedure consists of four main steps: rule construction, data generalization, recursive confirmation with backtracking, and statistics reporting. For the length of this chapter, we only give conceptual description of the procedure as follows.

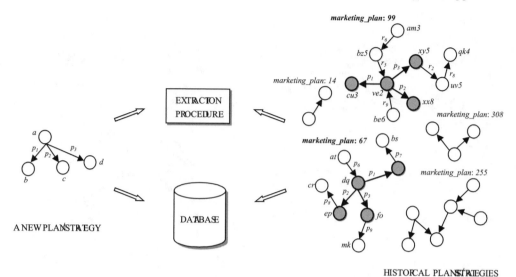

FIGURE IV.5.8   Finding applicable strategies with DSNs.

5.5.3.1 Rule Construction

Rule-based formalism is one of the effective schemes of knowledge representation. Users can use rules to record historical plans or to describe the features of new products. For example, the plan *mj-99-ch-a* mentioned previously can be described as follows:

$succ\_cross\_sale\_plan(mj\text{-}99\text{-}ch\text{-}a, pm, cross\_sale\_1, 15\%, home\_electronics, cus\_class\_1):\text{-}$
$p_1(pm), p_2(qn), p_3(rk), r1(pm, qn), r2(pm, rk).$

The rule head describes the record number, the strategy type, discount ratio, type of the promoting product, and the targeted customer segmentation; the rule body describes the main features of the associated products. The users can utilize rules for describing new products. For example, a new product *az* holding some features p_1, p_2, and p_3 can be described as below.

$promoting\_product(az, new, unclassified):\text{-} p_1(az), p_2(az), p_3(az), \ldots.$

Additionally, rules can describe general characteristics or taxonomy of products. Below are some examples:

- $communication\_device(X):\text{-} mobile\_phone(X)$
- $portability(high,X):\text{-} light\_weight(X), handheld(X), cordless(X)$
- $accessory\_of(X,Y):\text{-} battery\_pack(X,Y)$

The users can refer to the corresponding DSNs for constructing inference rules.

5.5.3.2 Data Generalization

The above rules describe the specific characteristics of products or historical plans. Unfortunately, exactly matching DSNs of applicable or reusable plans according to such characteristics is almost impossible. However, structural similarity is the most important so that we can focus on the relationships among products, not their specific characteristics. We consider two common generalization procedures in the following.

- Attribute generalization—generalizing specific attributes into variables. For example, the terms *strategy(cross_sale,nz1235)*, *mobility(nz1235,high)*, stating how the high-mobility product *nz1235* was promoted by cross-sale, can be generalized into *strategy(cross_sale,X)*, *mobility(X,high)*. Then, we can search for historical strategies that already promoted any high-mobility product *X*.

- Concept generalization—generalizing specific terms into a general concept. For example, a new strategy *customers who purchase the packet PC (lk86335m) together with an additional 32MB flash memory card (fmc32rc) and a wireless LAN module (w8325ls) can get 10% discount* is represented as

$$discount\_main(10\%, str\_341b, lk86335m):-$$
$$mobility(lk86335m, high), storage(lk86335m, fmc32rc),$$
$$interface\_card(lk86335m,w8325ls).$$

It is probably impossible to find an historical plan totally matching the new strategy. However, according to marketing experience, a promoting concept *customers would like to purchase consumer electronics of high-mobility together with their accessories* would be applicable. Then, we can use the following rule for investigating if the strategy can be applied to *lk86335m*.

$$discount\_main(S,ID,P):- mobility(P,high), accessory(P,Q),accessory(P,R).$$

According to the product hierarchy, the terms *mobility(lk86335m,high), storage(lk86335m,fmc32rc),* and *interface_card(lk86335m,w8325ls)* are generalized into *mobility(P,high), accessory(P,Q),* and *accessory(P,R)* with consistent variable substitution. Then, a much greater possibility exists of finding matching historical cases for the new product *lk86335m*.

The generalization procedure relies heavily on the taxonomy of products that is provided by the domain experts, e.g., the product designers or managers. With the structure of a given DSN, the generalization procedure searches applicable data and, if that search fails, generalizes the terms causing the failure according to the taxonomy and tries again. This procedure continues until all applicable records are found or no more data are to be searched. In order not to exhaustively search the database, the level of generalization is determined by the users.

5.5.3.3 Recursive Confirmation with Backtracking

All terms of deduction rules are queried to the database for finding satisfied data. Once the given rules fail to match data, the confirmation process continues with a backtracking mechanism for thoroughly finding all answers. Note that this sub-process is embedded in the generalization procedure, i.e., recursive confirmation with backtracking is performed with respect to the newly instantiated variables or generalized concepts.

5.5.3.4 Similarity Statistics Reporting

Finally, the extraction results are reported to the users. We rank and list the data that cause matching failures for giving users with a concept of percentage of the usability of the rules. A threshold is set for filtering out the results with lower similarity of DSN structure. The extraction procedure is illustrated in Figure IV.5.9.

5.5.4 Two Examples

5.5.4.1 Verification of New Marketing Plans

With the assumption in Section 5.5.1, we can list the rule for the marketing plan as follows:

$$marketing\_rule(ax, by, cz): - p_1(ax, v_1), p_2(by, v_2), r_1(ax, by), p_3(cz, v_3), r_2(ax, cz).$$

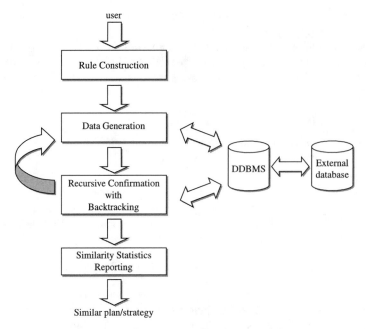

FIGURE IV.5.9   The extraction procedure.

Table: p_1

PID	VAL	NUMCITY	...
rk	v_1	NCR	...
pm	v_1	KHH	
mw	v_3	LGY	...
...	...	...	...

Table: r_1

PID-main	PID-acc	PROPERTY	...
pm	qn	A type_u	...
pm	yj	A type_j	
mw	io	R_fuu_k	...
...	...	...	...

Table: p_2

PID	VAL	DSC	...
qn	u_2	150	...
qn	v_2	553	
qj	u_2	369	...
...	...	...	...

Table: *cat*

PID_up	PID_down	Spec	...
pg	t_9	KHH	...
pm	t_1	JPQ	
t_1	t_2	CHX	
t_3	rk	TPE	
qj	io	UJA	
t_2	t_3	QXY	...
...	...	...	...

Table: p_3

PID	VAL	SIM	...
rk	v_3	12dae856a	...
an	u_3	458q9559b	
dr	i_{34}	58972685p	...
...	...	...	...

FIGURE IV.5.10   Database tables and records of the example.

The parameters v_1, v_2, and v_3 in p_1, p_2, and p_3 indicate the associated values of ax, by, and cz. First, this rule can be generalized into

$$marketing\_rule(X, Y, Z):- p_1(X, v_1), p_2(Y, v_2), r_1(X, Y), p_3(Z, v_3), r_3(X, Z)$$

since we like to verify if the relationships p_1, p_2, p_3, r_1, and r_2 hold on a set of consistent data. Suppose that the marketing manager employs r_2 to describe a recursive relationship between

$p1(rk,v1)$	1. $p1(rk,v1)$, $p2(qn,u2)$ ñ *failure with backtracking*
$p1(pm,v1)$	2. $p1(rk,v1)$,$p2(qn,v2)$, $r1(rk,qn)$ ñ *failure with backtracking*
$p1(mw,v3)$	3. $p1(rk,v1)$,$p2(qj,u2)$ñ *failure with backtracking*
$p2(qn,u2)$	4. $p1(pm,v1)$,$p2(qn,u2)$ ñ *failure with backtracking*
$p2(qn,v2)$	5. $p1(pm,v1)$,$p2(qn,v2)$,$r1(pm,qn)$,$p3(rk,v3)$,$r2(pm,rk)$
$p2(qj,u2)$	6. $p1(pm,v1)$,$p2(qn,v2)$,$r1(pm,qn)$,$p3(rk,v3)$,$cat(pm,rk)$ ñ *failure with*
$p3(rk,v3)$	*backtracking*
$p3(an,u3)$	7. $p1(pm,v1)$,$p2(qn,v2)$,$r1(pm,qn)$,$p3(rk,v3)$,$cat(pm,t1)$,$r2(t1,t2)$
$p3(dr,i34)$	8. $p1(pm,v1)$,$p2(qn,v2)$,$r1(pm,qn)$,$p3(rk,v3)$,$cat(pm,t1)$,$cat(t1,t2)$,
$r1(pm,qn)$	$r2(t2,rk)$ ñ *failure with backtracking*
$r1(pm,yj)$	9. $p1(pm,v1)$,$p2(qn,v2)$,$r1(pm,qn)$,$p3(rk,v3)$,$cat(pm,t1)$,$cat(t1,t2)$,
$r1(mw,io)$	$cat(t2,t3)$,$r2(t3,rk)$
$cat(pm,t1)$	10. $p1(pm,v1)$,$p2(qn,v2)$,$r1(pm,qn)$,$p3(rk,v3)$,$cat(pm,t1)$,$cat(t1,t2)$,
$cat(pg,t9)$	$cat(t2,t3)$,$cat(t3,rk)$
$cat(t1,t2)$	11. $\cdots$
$cat(t3,rk)$	12. $\cdots$
$cat(qj,io)$	13. $p1(pm,v1)$,$p2(qn,v2)$,$r1(pm,qn)$,$p3(rk,v3)$,$r2(pm,rk)$
$cat(t2,t3)$	

(a) The representation of data   (b) Running scenario

FIGURE IV.5.11 Scenario of running the program.

products, e.g., the position of products in catalog classification. The relationship $r2$ can be described as the following rules:

- $r2(X,Y):-\ cat(X,Y)$
- $r2(X,Y):-\ cat(X,Z),\ r2(Z,Y)$

We assume that the historical data regarding products, properties, and relationships are retained in an external RDBMS. For simplicity, we only list part of the associated tables and records in Figure IV.5.10. These data are loaded by XSB and represented as the logic predicates in Figure IV.5.11a.

Now, we consult the system by the query *marketing_rule(X,Y,Z)*. The scenario of running the program is listed in Figure IV.5.11b. Running the program for querying possible products X, Y, Z in the rule of *marketing_rule*, we confirm that the plan existed before in the historical data. Clearly, we can obtain $X = pm$, $Y = qn$, and $Z = rk$. The new marketing rule is supported by an historical plan which ever promoted pm together with qn and rk.

There are two points worth mentioning in this implementation.

1. When translated into an SQL statement, the body in the rule of *marketing_rule(X,Y,Z)* can become very complicated with several SQL join statements. From this viewpoint, we can see that LP provides a cleaner declarative method for stating our goals.
2. The relationship $r2$ can be implicitly and recursively defined by other relationships (*cat* in the example) in the original RDBMS tables. Current SQL 92 standards cannot handle this situation easily, but LP can describe and solve the problem efficiently.

5.5.4.2 Extracting Applicable Marketing Plans

Another example is for the case in Section 5.5.2. Again, we assume that the product az holds three unique features $p1$, $p2$, and $p3$, and we would like to know if any historical plan includes a product with these features. The associated rule can be briefly represented as follows:

$$known\_feature(az) :\text{-} \ p1(az), p2(az), p3(az).$$

The extraction of a plan is implemented in two steps. First, we search the database for the existence of products with terms $p1$, $p2$, and $p3$. Second, from the search results in the first step, we examine if an historical plan containing the found product with proper relations ever existed in the database. The historical data in the code are also loaded from an external database and represented as logic predicates in the following:

- $p1(mx)$
- $p2(mx)$
- $p3(mx)$
- $p4(kz)$
- $p5(lz)$
- $r(mx,kz)$
- $s(mx,lz)$

The rules implementing the above procedure are listed below.

- $findit(X,R,Y,P) :\text{-} known\_feature(X), R(X,Y), P(Y)$
- $known\_feature(X) :\text{-} p1(X), p2(X), p3(X)$

In the code, we are using the special feature of XSB regarding Hilog implementation to extract the functors. In the search for suitable solutions for the query *?- findit(X,R,Y,P)*, XSB first resolves $known\_feature(X)$ for $X = mx$. It then solves $R(mx, Y)$ for the functor R and variable Y and in terms of the functor P.

Here is the output from the program. Two solutions are obtained, i.e., $X = mx$, $R = r$, $Y = kz$, $P = p4$ and $X = mx$, $R = s$, $Y = lz$, $P = p5$. We can see that there are two historical marketing plans that match the known properties of the new product az. Therefore, we may promote az with something that has feature $p4$ and is related to az by r, or some other product with property $p5$ and related to az by s.

5.6 System Architecture

The two tasks discussed above, order-decomposition (Section 5.4) and marketing strategies deployment (Section 5.5), are designed and implemented using the techniques of DDBMS and Java-based agents. We integrate these techniques into the B2B portal and try to achieve two objectives: (1) intelligently and automatically collecting and analyzing customers' online orders; and (2) partially but flexibly assisting marketing managers in making decisions for deploying new marketing strategies. The system consists of the following four main functional blocks: external database, rule base and inference engine, intelligent agents, and user interface.

5.6.1 External Database

The external database is a relational database for recording plain data such as daily transactions prices, and specifications of products. The nodes and links of DSNs are also retained in the external database serving as a source of data for deriving conclusions from inference rules. During the process of deduction, intermediary data derived from inference rules are temporarily cached in the database. Derived data corresponding to the specified goals are recorded in the database by new record tables.

5.6.2   Rule Base and Inference Engine

Inference rules given by experienced managers or obtained from the results of data mining are retained in a rule base (namely, the intensional database, IDB). During the process of deduction, the inference engine resolves the condition elements of the inference rules and searches for satisfied facts or decomposes them into several SQL commands for enquiring the external database. Queried data are used to instantiate the other conditions iteratively, until all conclusions are obtained. As mentioned previously, inference of rules is goal-driven with backtracking.

Rules are partitioned into several groups, each of which serves a different purpose. Rules are also wrapped by agents. The rule base and the inference engine are implemented in XSB, with some Java-based components for connecting to the external database through ODBC and JDBC.

5.6.3   Agents

Since the portal is Web-based, operates 24 hours a day, 7 days a week, and orders may come from worldwide customers, automation is essential. Several issues need to be considered.

- **Collecting and Analyzing Orders.** As mentioned previously, collecting and analyzing orders are routines in running the portal. Customers come randomly and may have a variety of needs, such as the types, volumes, and delivery urgency of purchasing products. Orders are usually collected periodically. However, determining the period for collecting orders is a dilemma. Collecting and analyzing orders in a shorter period can respond quickly to the customers, but the firm may profit less due to the smaller total volume gathered in such a short period. A smaller volume means that we have to pay higher cost in frequent processing of shipping and logistics. Also, the orders of smaller volume to the upstream suppliers may get a less preferable discount ratio.
- **Changing the Supplying Relationships.** Once the relationships among products, components, suppliers, and manufacturers change, many relationships associated with the data entities may have to be changed accordingly. Some database tables as well as the related access programs have to be created or modified. Reconfiguration of supplying status is a costly task, and its complexity increases with varieties of the associated data entities.
- **Making Decisions in Deploying New Marketing Strategies.** For making an effective decision, successful and efficient access to the related decision models plays an important role. As we have presented previously, the extraction procedure can extract related DSNs using inference rules. However, the task of determining the levels of data generalization may be difficult for non-IT professional marketing managers. If the level is deep, the process may spend much more time extracting satisfied results in more iterations. On the contrary, if the level is shallow, the process iterates for fewer rounds to obtain less satisfied results.

Here we employ Java-based technologies to build agents serving these tasks. The roles and objectives of these autonomous agents are as follows.

- **OrderCollector**—designed for collecting orders. The knowledge about collecting orders is acquired from domain experts and considers some critical factors related to the products (e.g., types, market-values, logistic complexity, etc.), the customers (e.g., historical transaction records, financial status, purchasing potential, etc.), and the status of the working system (e.g., CPU loads, communication traffics, the number of online customers, etc.). The OrderCollector agent monitors the status of orders placed by online customers and determines the period of collecting and decomposing orders.

- **OrderAnalyzer**—designed for analyzing orders. The OrderAnalyzer agent analyzes the volumes, response time, delivery cost, etc., of ordered products, and determines if an order can be satisfied by the stock in a local warehouse or if we have to place orders to the upstream suppliers. The criteria and decision knowledge are deduction rules given by the administrator of the portal.
- **DBInterActor**—designed for interacting with the users and the external database. The DBInterActor agent takes care of prompting the users, gathering the results, creating new tables, and storing data back to the external database. Since there may be more than one external database, a collection of communication schemes are embedded in the DBInterActor for interacting with different types of databases.
- **MarketMaker**—designed for assisting the marketing managers for making decisions of cross-sale strategies. This agent receives from human managers the rules for extracting historical marketing plans and performs rule deduction. It communicates with the Generalizer and StatisticReporter agents for generalizing data and reporting the statistics of structural similarity.
- **Generalizer**—designed for data generalization. The Generalizer agent works with MarketMaker and provides suggestions to users about the level of data generation. It considers the amount of data entities in the external database and predicts the complexity of generalizing data.
- **StatisticReporter**—designed for showing the inference results to the users.

Knowledge in all agents is represented in rule-based formats. Agents are implemented in Java with the wrapping technologies for executing deduction rules in the brains. KQML is used as the mechanism for inter-agent communication. Figure IV.5.12 presents the collaboration of these agents in this system.

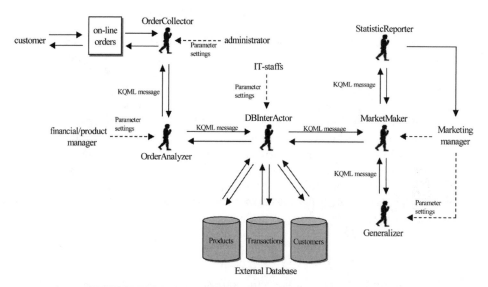

FIGURE IV.5.12   The collaboration of agents in this system.

5.6.4   User Interface

GUI is provided for the users to create and maintain the rule bases, set up the evaluation criteria, and manage the external databases. Once all operational factors and rules are properly set up, the system runs automatically.

5.6.5   Integration

The portal receives about 1000–1500 orders a day, and each order contains about 10–25 different items. Moreover, 35% of the relationships among products, elements, suppliers, and manufacturers change each week. The marketing manager deploys or modifies the cross-sale strategies within a period of about 5–9 days. In such a changing and competitive market, quick response to customers' orders is an essential key to success. Traditional implementation of such a system suffers from the considerably long time required for prototyping and modifying the information systems and related databases iteratively. A highly flexible information system supporting such a scenario is much needed. Figure IV.5.13 illustrates the main functionalities and the user interactions of the portal.

5.7   Discussion and Conclusions

We have presented how to integrate the techniques of DDBMS and intelligent agents into B2B electronic commerce applications. DDBMS is employed for analyzing online orders and assisting marketing managers in extracting applicable historical plans. Intelligent agents are designed for autonomously and periodically collecting orders and wrapping rule deduction. The lessons we learned in this project are discussed as follows:

1. Many deductive systems consider non-monotonicity,[10] i.e., knowledge revision on modification of data. Since data involved in this project are used for making decisions for completing transactions or merchandizing of the present time, revision on derived conclusions according to a modification of historical data is not needed. For example, an allied supplier that changes its shipping policy can affect our profits from now on, but not the ones in the past. Also, the change of the specifications of a product in an historical marketing plan does not need to revise a new plan. Non-monotonicity is not included in this project.
2. Our method can be viewed as a deductive case-based reasoning system.[1,21] We directly facilitate the external database for retaining cases and the inference rules for retrieving cases. However, we integrate different functionalities in DDBMS besides the retrieving of historical cases.
3. In this project, we only consider cross-sale as the marketing strategy. Our method is flexible and can support other marketing strategies by modifying the inference rules.
4. In retrieving historical marketing plans, the derived conclusions are for references. The techniques presented in this project are not intended to totally replace human decision-making. We provide decision-makers a flexible, easy-to-use, and automated tool for collecting and evaluating valuable references.
5. The efficiency of our system relies heavily on the structure of the external database. An ill-structured relational database contains many redundant and fragmented data pieces. To retrieve a complicated concept in an ill-structured relational database may require accessing the external database a large number of times. An improvement can be achieved by reorganizing the database using classification methods such as ID3 and C4.5.[27,28]
6. The content of the portal is highly changeable. Autonomous agents successfully reduce the cost and complexity of modifying and maintaining the content. Currently, only six types of agents are designed. We are designing other types of agents and integrating them in the portal.

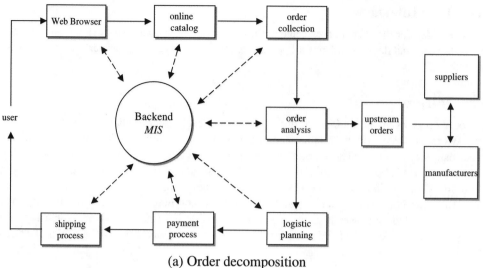

(a) Order decomposition

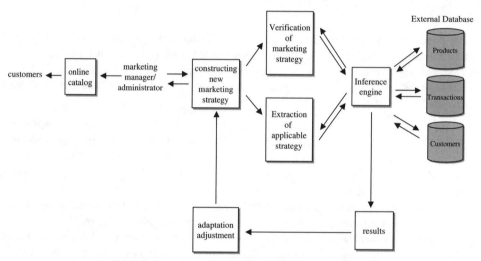

(b) Deployment of new marketing strategies

FIGURE IV.5.13   The functional diagram of the system.

References

1. Aamodt, A. and Plaza, E., Case-based reasoning: foundational issues, methodological variations, and system approaches, *AI Commun.*, 7, 59, 1994.
2. Andreoli, J., Pacull, F., and Pareschi R., XPECT: a framework for electronic commerce, *IEEE Internet Comput.*, 1, 40, 1997.
3. Bhimani, A., Securing the commercial Internet, *Commun. ACM*, 39, 29, 1996.
4. Bigus, J. and Bigus, J., *Constructing Intelligent Agents with Java*, John Wiley & Sons, New York, 1998.
5. Chen, W., Kifer, M., and Warren, D., HiLog: A first order semantics for higher-order logic programming constructs, *Proc. N. Am. Logic Programming Conf.*, 1989.

6. Chen, W., Swift, T., and Warren, D.S., Efficient top-down computation of queries under the well-founded semantics, *J. Logic Programming*, 24, 161, 1995.
7. Chimenti, D. et al., The LDL system prototype, *IEEE Trans. Knowledge Data Eng.*, 2, 76, 1990.
8. Conway, D.G. and Koehler, G.J., Interface agents: caveat merchator in electronic commerce, *Decision Support Syst.*, 27, 355, 2000.
9. Dahr, M., *Deductive Databases: Theory and Applications*, International Thomson Computer Press, 1997.
10. de Kleer, J., An assumption-based TMS, *Artif. Intelligence*, 28, 127, 1986.
11. Elmasri, B. and Navathe, S., *Fundamentals of database systems*, 3rd ed., Addison-Wesley, Reading, MA, 2000.
12. Fuller, F., *Getting Started with Electronic Commerce*, Harcourt College Publishers, Orlando, FL, 2000.
13. Gannoun, L. et al., Domain name exchange: a mobile-agent based shared registry system, *IEEE Internet Comput.*, 4, 59, 2000.
14. Genesereth, M. and Ketchpel, S., Software agents, *Commun. ACM*, 37, 48, 1994.
15. Grant, J. and Minker, J., The impact of logic programming on databases, *Commun. ACM*, 35, 66, 1992.
16. Han, J., Cai, Y., and Cercone, N., Data-driven discovery of quantitative rules in relational databases, *IEEE Trans. Knowledge Data Eng.*, 5, 29, 1993.
17. Heydecker, B.G., Small, C., and Poulovassilis, A., Deductive databases for transport engineering, *Transp. Res.* C, 3, 277, 1995.
18. Jain, A.K., Aparicio, M., IV, and Singh, M.P., Agents for process coherence in virtual enterprises, *Commun. ACM*, 42, 62, 1999.
19. Kalakota, R. and Whinston, A.B., *Frontiers of Electronic Commerce*, Addison-Wesley, Reading, MA, 1996.
20. Kosoresow, A.P. and Kaiser, G.E., Using agents to enable collaborative work, *IEEE Internet Comput.*, 85, 1998.
21. Leake, D.B., *Case-Based Reasoning: Experiences, Lessons, and Future Directions*, AIII Press, Menlo Park, CA, 1996.
22. Liu, M., Deductive database languages: problems and solutions, *ACM Comput. Surv.*, 31, 27, 1999.
23. Lucas, A., What in the world is electronic commerce?, *Sales Mark. Manage.*, 148(6), 24, 1996.
24. Morris, K. et al., YAWN!—Yet another window on NAIL!, *Database Eng.*, 1987.
25. Morris, K., Ullman, J.D., and Gelder, A.V., Design overview of the Nail! system, in *Proceedings of the International Conference on Logic Programming*, MIT Press, London 554, 1986.
26. Nussbaum, M., *Building a Deductive Database*, Ablex Publishing Corporation, 1992.
27. Quinlan, J.R., *C4.5 Programs for Machine Learning*, Morgan Kaufmann, Palo Alto, CA, 1993.
28. Quinlan, J.R., Induction of decision trees, *Machine Learning*, 1, 81, 1986.
29. Ramakrishnan, R. and Ullman, J.D., A survey of deductive database systems, *J. Logic Programming*, 23, 125, 1995.
30. Ramakrishnan, R. et al., The coral deductive system, *VLDB J.*, 3, 161, 1994.
31. Ramakrishnan, R. et al., XSB as an effective deductive database engine, Proceedings of the ACM SIGMOD International Conference on Management of Data, Minneapolis, MN, 442, 1994.
32. Ramakrishnan, R., Srivastava, D., and Sudarshan, S., CORAL: control, relations and logic, in *Proceedings of the International Conference on Very Large Databases*, 1992.
33. Rao, P. et al., XSB: a system for efficiently computing, in *Proceedings of the International Conference on Logic Programming and Non-Monotonic Reasoning*, Springer-Verlag, New York, 431, 1997.
34. Ross, K., Relations with relation names as arguments: algebra and calculus, *Proc. PODS*, 1992.
35. Sagonas, K., *The XSB System Version 2.3*, vol. 1, programmer's manual, 2001.
36. Schneier, B., *Applied Cryptography*, 2nd ed., Wiley, New York, 1996.

37. Shaw, M.J., Gardner, D.M., and Thomas, H., Research opportunities in electronic commerce, *Decision Support Syst.*, 21, 149, 1997.
38. Stuart, J.R. and Norvig, P., *Artificial Intelligence: A Modern Approach*, Prentice Hall, Upper Saddle River, NJ, 1995.
39. The KQML specifications, 1993. http://java.stanford.edu/KQMLSyntaxBNF.html.
40. Tsur, S. and Zaniolo, C., LDL: a logic-based data language, in *Proceedings of the International Conference on Very Large Databases*, Morgan Kaufmann Publishers, Palo Alto, CA, 33, 1986.
41. Turban, E. et al., *Electronic Commerce: A Managerial Perspective*, Prentice-Hall, Upper Saddle River, NJ, 2000.
42. Wooldridge, M. and Jennings, N.R., Agent theories, architectures, and languages: a survey, in *Intelligent Agents*, Wooldridge, M.J. and Jennings, N.R., Eds., Springer-Verlag, Berlin, 1, 1995.

6

A Method for Fuzzy Query Processing in Relational Database Systems

Shyi-Ming Chen
*National Taiwan University of Science
and Technology*

Yun-Shyang Lin
*National Taiwan University of Science
and Technology*

Abstract.   In recent years, many researchers focused on the research topic of fuzzy query processing in database systems. In this chapter, a method for fuzzy query processing in relational database systems is presented based on automatic clustering techniques and weighting concepts. It allows the query conditions and the weights of query items of users' fuzzy SQL queries to be described by linguistic terms represented by fuzzy numbers. Because the users can construct their fuzzy queries more conveniently, the existing relational database systems will be more intelligent and more flexible to the users.

6.1 Introduction

Zadeh[30] proposed the theory of fuzzy sets. The fuzzy set theory has been widely used to deal with vague and imprecise data. In traditional database systems, it is required that users' queries be specified precisely. Therefore, the users must fully know the database system structure. However, in many cases, the users may not have a precise view regarding the information stored in the database. Thus, the existing database systems lack friendliness and flexibility to the users. Fuzzy query languages based on fuzzy set theory provide a useful way to retrieve data from database systems,[3–12,17–20,26,28–29] where the users can express their queries in a form very similar to natural language to retrieve data from database systems. Bosc et al.[3] presented a relational database language for fuzzy querying. Bosc and Lietard[4] presented a method for database flexible querying. Bosc and Pivert[5] presented some approaches for relational database flexible querying. They also presented a relational database language for fuzzy querying.[6] Bosc et al.[7] presented a method for integrating fuzzy queries into an existing database management system. Bosc and Prade[8] presented an introduction to the fuzzy set and possibility theory-based treatment of flexible queries and uncertain or imprecise databases. Chang and Ke[9] presented the concept of a database skeleton and its application to fuzzy query translation. Chang and Ke[10] also presented a method for fuzzy query translation for relational database systems. Chen, Ke, and Chang presented a method for fuzzy query translation in database systems.[11] Chen and Jong[12] presented fuzzy query translation techniques for relational database systems. Kacprzyk[17] presented techniques for queries in DBMS based on fuzzy logic. Kacprzyk, Zadrozny, and Ziolkowski[18] presented a "human-consistent" database querying system, FQUERY III+, based on fuzzy logic with linguistic quantifiers. Kacprzyk and Ziolkowski[19] presented a method for database queries with fuzzy linguistic quantifiers. Kamel, Hadfield, and Ismail[20] presented a fuzzy query processing method using clustering techniques. Tahani[26] presented a conceptual framework for fuzzy query processing. Wong and Leung[28] presented a fuzzy database-query language. Yeh and Chen[29] presented a method for fuzzy query processing using automatic clustering techniques.

In this article, we present a method for fuzzy query processing in relational database systems by using automatic clustering techniques and weighting concepts. It allows the retrieval conditions and the weights of query items in the users' fuzzy SQL queries to be described by fuzzy terms represented by fuzzy numbers. Because the users can construct their query more conveniently, the existing relational database systems will be more intelligent and more flexible to the users.

The rest of this article is organized as follows. In Section 6.2, we briefly review the basic concepts of fuzzy sets and fuzzy number arithmetic operations.[13,21–23,30] In Section 6.3, we briefly review the automatic clustering algorithm.[29] In Section 6.4, we present a method for fuzzy query processing in relational database systems. The conclusions are discussed in Section 6.5.

6.2   Fuzzy Set Theory

In this session, we briefly review the theory of fuzzy sets and fuzzy number arithmetic operations found in the literature.[13,21–23,30]

6.2.1   Basic Concepts of Fuzzy Sets

Roughly speaking, a fuzzy set is a set with fuzzy boundaries. A fuzzy set A of the universe of discourse U, $U = \{u_1, u_2, \ldots, u_n\}$, can be characterized by a membership function μ_A, $\mu_A : U \rightarrow [0, 1]$, represented as follows:

$$A = \mu_A(u_1)/u_1 + \mu_A(u_2)/u_2 + \cdots + \mu_A(u_n)/u_n \qquad (IV.6.1)$$

where $\mu_A(u_i)$ indicates the grade of membership of u_i in the fuzzy set A, $\mu_A(u_i) \in [0, 1]$, and $1 \leq i \leq n$.

Definition 1:   Let A be a fuzzy set of the universe of discourse U, where $U = \{u_1, u_2, \ldots, u_n\}$ and $A = \sum_{i=1}^{n} \mu_A(u_i)/u_i$. Furthermore, let α be a threshold value, where $\alpha \in [0, 1]$. The α-level-cut $A_{\alpha\text{-level}}$ of the fuzzy set A is defined as follows:

$$A_{\alpha\text{-level}} = \sum_{i=1}^{n} \mu_{A\alpha\text{-level}}(u_i)/u_i \qquad (IV.6.2)$$

where if $\mu_A(u_i) \geq \alpha$, then let $\mu_{A_{\alpha\text{-level}}}(u_i) = \mu_A(u_i)$; if $\mu_A(u_i) < \alpha$, then let $\mu_{A_{\alpha\text{-level}}}(u_i) = 0$.

Definition 2: A fuzzy set A of the universe of discourse U is convex if and only if for all u_1, u_2 in U,

$$\mu_A(\lambda u_1 + (1 - \lambda)u_2) \geq \text{Min}(\mu_A(u_1), \mu_A(u_2)) \qquad \text{(IV.6.3)}$$

where $\lambda \in [0, 1]$. A fuzzy set of the universe of discourse U is called a normal fuzzy set if $\exists u_i \in U, \mu_A(u_i) = 1$. A fuzzy number is a fuzzy set which is both convex and normal.

The membership function of the fuzzy proposition "close to γ" can be expressed as follows:

$$\mu_{\text{close to } \gamma}(u) = \frac{1}{1 + (\frac{u-\gamma}{\beta})^2} \qquad \text{(IV.6.4)}$$

where the parameter β is the half-width of the curve at the crossover point and the crossover points are at $u = \gamma \pm \beta$. The membership function curve of the fuzzy number "close to γ" is shown in Figure IV.6.1.

Figure IV.6.2 shows a trapezoidal fuzzy number A of the universe of discourse U, where the trapezoidal fuzzy number A can be represented by $A = (a, b, c, d)$.

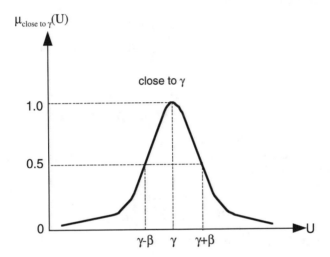

FIGURE IV.6.1   Membership function of the fuzzy number "close to γ."

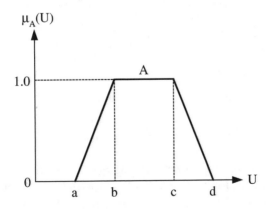

FIGURE IV.6.2   A trapezoidal fuzzy number.

6.2.2 Fuzzy Number Arithmetic Operations

In the following, we briefly review the simplified arithmetic operations of trapezoidal fuzzy numbers.

Definition 3: Let A and B be two trapezoidal fuzzy numbers, where

$$A = (a_1, b_1, c_1, d_1),$$
$$B = (a_2, b_2, c_2, d_2).$$

Then,

1. Fuzzy numbers addition $\oplus$:

$$A \oplus B = (a_1, b_1, c_1, d_1) \oplus (a_2, b_2, c_2, d_2)$$
$$= (a_1 + a_2, b_1 + b_2, c_1 + c_2, d_1 + d_2) \tag{IV.6.5}$$

2. Fuzzy numbers subtraction $\ominus$:

$$A \ominus B = (a_1, b_1, c_1, d_1) \ominus (a_2, b_2, c_2, d_2)$$
$$= (a_1 - d_2, b_1 - c_2, c_1 - b_2, d_1 - a_2) \tag{IV.6.6}$$

3. Fuzzy numbers multiplication $\otimes$:

$$A \otimes B = (a_1, b_1, c_1, d_1) \otimes (a_2, b_2, c_2, d_2)$$
$$= (a_1 \times a_2, b_1 \times b_2, c_1 \times c_2, d_1 \times d_2) \tag{IV.6.7}$$

4. Fuzzy numbers division $\oslash$:

$$A \oslash B = (a_1, b_1, c_1, d_1) \oslash (a_2, b_2, c_2, d_2)$$
$$= (a_1/d_2, b_1/c_2, c_1/b_2, d_1/a_2) \tag{IV.6.8}$$

Definition 4: Let the trapezoidal fuzzy numbers W_1 and W_2 be the weights of the attributes A_1 and A_2, respectively, where

$$W_1 = (a_1, b_1, c_1, d_1),$$
$$W_2 = (a_2, b_2, c_2, d_2).$$

Furthermore, let $\overline{W}_1$ and $\overline{W}_2$ be the normalized weights of W_1 and W_2, respectively. Then,

$$\overline{W}_1 = W_1 \oslash (W_1 \oplus W_2)$$
$$= (a_1, b_1, c_1, d_1) \oslash (a_1 + a_2, b_1 + b_2, c_1 + c_2, d_1 + d_2)$$
$$= \left(\frac{a_1}{d_1 + d_2}, \frac{b_1}{c_1 + c_2}, \frac{c_1}{b_1 + b_2}, \frac{d_1}{a_1 + a_2} \right), \tag{IV.6.9}$$

$$\overline{W}_2 = W_2 \oslash (W_1 \oplus W_2)$$
$$= (a_2, b_2, c_2, d_2) \oslash (a_1 + a_2, b_1 + b_2, c_1 + c_2, d_1 + d_2)$$
$$= \left(\frac{a_2}{d_1 + d_2}, \frac{b_2}{c_1 + c_2}, \frac{c_2}{b_1 + b_2}, \frac{d_2}{a_1 + a_2} \right). \tag{IV.6.10}$$

Definition 5: Let A be a trapezoidal fuzzy number, where $A = (a, b, c, d)$ and $0 \leq a \leq b \leq c \leq d \leq 1$, then the defuzzified value $DEF(A)$ of the trapezoidal fuzzy number A is calculated as follows:[13,23]

$$DEF(A) = (a + b + c + d)/4. \tag{IV.6.11}$$

6.3   Automatic Clustering Techniques

In this section, we briefly review the automatic clustering techniques presented by Yeh and Chen.[29]

6.3.1   An Automatic Clustering Algorithm

In the following, we briefly review the automatic clustering algorithm of Yeh and Chen.[29] The automatic clustering algorithm is shown as follows:

```
Main() // main program //
{
// NOD: Number of Data, Data[]: Data Array //
    sorting Data[] using quick sort [15] in an ascending sequence;
    // assume that the data in ascending sequence is x₁, x₂, ..., xₙ //
    Dif_i[0] = xₙ;
    Dif_i[n] = xₙ;
    Dif_Dif_i[1] = 0;
    Dif_Dif_i[n] = 0;
// compute the difference between two neighboring elements in Data[] //
    for i = 1 to NOD-1 do
    Dif[i] = Data[i + 1] − Data[i];
// compute the difference between two neighboring elements in Dif[i] which
    is not equal to 0, where p: point to the left one, and q: point to the right one //
    for i = 1 to NOD-2 do
  {
    p = i;
    while (Dif[p] = 0 and p > 1) do
      p = p − 1;
      q = i + 1;
    while (Dif[q] = 0 and q < NOD) do
      q = q + 1;
      Dif_Dif[i] = abs(Dif[p] − Dif[q]);
  }
// start clustering //
Clster(1, NOD)
} // end of main() //
Clster(L, R) // subroutine //
{
// set the boundary //
    let Dif[L−1] = Dif[R] = Data[R];
    let Dif_Dif[L−1] = Dif_Dif[R−1] = 0;
// if the two ends of Dif[] are equal to 0, then adjust Dif_Dif[] //
```

temp $= R - 1$;

while (Dif[temp]) $= 0$ **do**

   temp $=$ temp $- 1$;

for $i =$ temp to $R-2$ **do**

   Dif_Dif[i] $=$ Dif[temp];

temp $= L$;

while (Dif[temp] $= 0$) **do**

   temp $=$ temp $+ 1$;

for $i = L$ to temp-1 **do**

   Dif_Dif[i] $=$ Dif[temp];

// Dif_Sum: summation from Dif[L] to Dif[$R-1$];

S: number of elements from Dif[L] to Dif[$R - 1$] which is not equal to 0 //

Dif_Mean $=$ Dif_Sum/S;

let $X = Y = L$;

// Dif_Dif_Max: maximal value from Dif_Dif[L] to Dif_Dif[$R - 2$]//

if $(R = L)$ or (Dif_Mean $\geq$ Dif_Dif_Max) **then**

  Data[L] to Data[R] forms a cluster

else

for $i = L$ to $R - 1$ **do**

{

   if $(i = R-1)$or ((Dif[i] $>$ Dif_Mean) and (Dif[i] $\geq$ Dif_Dif[$i - 1$]) and

   (Dif[i] $\geq$ Dif_Dif[$i - 1$])) **then**

   $Y = i$;

   Cluster(X, Y);

   $Y = Y + 1$;

  }

} // end of Cluster() //

6.3.2   An Example

In the following, we use an example to illustrate the automatic clustering process of the automatic clustering algorithm.

Example 1: Assume that the input data are as follows:

$$2, 8, 3, 21, 36, 27, 28, 28, 4, 10, 85, 35, 21, 15, 23, 36$$

After applying the quick sort, the data can be sorted in an ascending sequence shown as follows:

$$2, 3, 4, 8, 10, 15, 21, 21, 23, 27, 28, 28, 35, 36, 36, 85.$$

[Cluster(1, 16)]

	1	2	3	4	5	6	7	8	9	10	11	12	13	14	15	16
Data[]	2	3	4	8	10	15	21	21	23	27	28	28	35	36	36	85
Dif[] 85	1	1	4	2	5	6	0	2	4	1	0	7	1	0	49	85
Dif_Dif[]	0	0	3	2	3	1	4	4	2	3	6	6	6	48	48	0

$\text{Dif\_Mean} = (1+1+4+2+5+6+2+4+1+7+1+49)/12 = 6.917;$

because of Dif[15] = 49 > Dif_Mean and Dif[15] > Dif_Dif[14]
   and Dif[15] > Dif_Dif[15],
so that CALL Cluster(1, 15);
    CALL Cluster(16, 16);

[Cluster(1, 15)]

	1	2	3	4	5	6	7	8	9	10	11	12	13	14	15
Data[]	2	3	4	8	10	15	21	21	23	27	28	28	35	36	36
Dif[] 36	1	1	4	2	5	6	0	2	4	1	0	7	1	0	36
Dif_Dif[] 0	0	3	2	3	1	4	4	2	3	6	6	6	1	0	

$\text{Dif\_Mean} = (1+1+4+2+5+6+2+4+1+7+1)/11 = 3.091;$

because of Dif[3] = 4 > Dif_Mean and Dif[3] > Dif_Dif[2]
   and Dif[3] > Dif_Dif[3],
so that CALL Cluster(1, 3);
because of Dif[5] = 5 > Dif_Mean and Dif[5] > Dif_Dif[4]
   and Dif[5] > Dif_Dif[5],
so that CALL Cluster(4, 5);
because of Dif[6] = 6 > Dif_Mean and Dif[6] > Dif_Dif[5]
   and Dif[6] > Dif_Dif[6],
so that CALL Cluster(6, 6);
because of Dif[9] = 4 > Dif_Mean and Dif[9] > Dif_Dif[8]
   and Dif[9] > Dif_Dif[9],
so that CALL Cluster(7, 9);
because of Dif[12] = 7 > Dif_Mean and Dif[12] > Dif_Dif[11]
   and Dif[12] > Dif_Dif[12],
so that CALL Cluster(10, 12);
    CALL Cluster(13, 15);

[Cluster(1, 3)]

	1	2	3
Data[]	2	3	4
Dif[]	4	1	1	4
Dif_Dif[]	0	1	0

$\text{Dif\_Mean} = (1+1)/2 = 1;$

because of Dif_Dif_Max = 1 and Dif_Dif_Max = Dif_Mean,
so that {2, 3, 4} forms a cluster.

[Cluster(4, 5)]

	4	5	
Data[]		8	10
Dif[]	10	2	10
Dif_Dif[]	0		

  Dif_Mean $= 2/2 = 1$;
  because of Dif_Dif_Max $= 0 <$ Dif_Mean,
  so that $\{8, 10\}$ forms a cluster.

[Cluster(6, 6)]

		6	
Data[]		15	
Dif[]		15	15
Dif_Dif[]	0		

  because of $L = R$ (i.e., $6 = 6$),
  so that $\{15\}$ forms a cluster.

[Cluster(7, 9)]

	7	8	9	
Data[]		21	21	23
Dif[]	23	0	2	23
Dif_Dif[]		0	2	0

  Dif_Mean $= 2/1 = 2$;
  because of Dif_Mean $=$ Dif_Dif_Max (i.e., $2 = 2$),
  so that $\{21, 21, 23\}$ forms a cluster.

[Cluster(10, 12)]

	10	11	12	
Data[]		27	28	28
Dif[]	28	1	0	28
Dif_Dif[]		0	1	0

  Dif_Mean $= 1$;
  because of Dif_Mean $=$ Dif_Dif_Max (i.e., $1 = 1$),
  so that $\{27, 28, 28\}$ forms a cluster.

[Cluster(13, 15)]

	13	14	15

Data[]		35	36	36
Dif[]	36	1	0	36
Dif_Dif[]	0	1	0	

Dif_Mean = 1;

because of Dif_Mean = Dif_Dif_Max (i.e., 1 = 1),

so that {35, 36, 36} forms a cluster.

[Cluster(16, 16)]

Data[]		85
Dif[]	85	85
Dif_Dif[]	0	

because of $L = R$ (i.e., 16 = 16),

so that {85} forms a cluster.

Therefore, the result of clustering is shown as follows:

Cluster[1] $\rightarrow$ {2, 3, 4},
Cluster[2] $\rightarrow$ {8, 10},
Cluster[3] $\rightarrow$ {15},
Cluster[4] $\rightarrow$ {21, 21, 23},
Cluster[5] $\rightarrow$ {27, 28, 28},
Cluster[6] $\rightarrow$ {35, 36, 36},
Cluster[7] $\rightarrow$ {85}.

Let K_i be a cluster containing n data items $\{d_a, d_{a+1}, \ldots, d_b\}$, where $b - a + 1 = n$. If we can get m clusters after applying the automatic clustering algorithm, then for each cluster K_i we can compute the center C_i and the normalized center T_i of the cluster K_i, respectively, where $1 \leq i \leq m$. The definitions of C_i and T_i are as follows:

$$C_i = (d_a + d_{a+1} + \cdots + d_b)/(b - a + 1),$$

$$T_i = (C_i - S)/(L - S),$$

where $1 \leq i \leq m, 0 \leq T_i \leq 1$, and L and S denote the maximum value and the minimum value of the input data items, respectively. For more details, refer to Yeh and Chen.[29]

6.4 Using Automatic Clustering Techniques for Fuzzy Query Processing in Relational Database Systems

In this section, we present a method for fuzzy query processing in relational database systems based on the automatic clustering algorithm and weighting concepts.

6.4.1 Linguistic Terms in Relational Database Systems

Let us consider the relation EMP in a relational database system shown in Table IV.6.1, where there are five attributes in the relation EMP, i.e., Emp-ID, Name, Degree, Experience, and Salary. The values of the attribute Degree are Ph.D., Master, and Bachelor. The domain of the attribute Experience is between 0 and 10, and the domain of the attribute Salary ranges from 20000 to 70000.

We can use several linguistic terms to describe the values of attributes and to specify a range for each linguistic term as shown in Table IV.6.2.

Table IV.6.3 shows the linguistic terms and their corresponding normalized trapezoidal fuzzy numbers of the attribute Experience; Table IV.6.4 shows the linguistic terms and their corresponding normalized trapezoidal fuzzy numbers of the attribute Salary.

Table IV.6.5 shows the linguistic terms of the attribute Experience and their corresponding trapezoidal fuzzy numbers; Table IV.6.6 shows the linguistic terms of the attribute Salary and their corresponding trapezoidal fuzzy numbers.

Let the weights of the attributes of Experience and Salary given by the user be $W_{Experience}$ and W_{Salary}, respectively, where the weights can be described in linguistic terms or crisp values between zero and one. If the weight W_X of an attribute X is a crisp value c (i.e., $W_X = c$), where $c \in [0, 1]$, then we can convert the crisp value c into the trapezoidal fuzzy number representation (c, c, c, c). Table IV.6.7 shows the linguistic weights and their corresponding trapezoidal fuzzy numbers.

TABLE IV.6.1   Relation EMP in a Relational Database System

Emp-ID	Name	Degree	Experience	Salary
E1	Angel	Ph.D.	7.2	63000
E2	Ann	Master	2.0	37000
E3	Bob	Bachelor	7.0	40000
E4	Carol	Ph.D.	1.2	47000
E5	David	Master	7.5	53000
E6	Gene	Bachelor	1.5	26000
E7	Jack	Bachelor	2.3	29000
E8	Jerome	Ph.D.	2.0	50000
E9	Joan	Ph.D.	3.8	54000
E10	Lee	Bachelor	3.5	35000
E11	Mark	Master	3.5	40000
E12	Mary	Master	3.6	41000
E13	Michael	Master	10	68000
E14	Paul	Ph.D.	5.0	57000
E15	Ray	Bachelor	5.0	36000
E16	Rose	Master	6.2	50000
E17	Sofia	Bachelor	0.5	23000
E18	Sting	Master	7.2	55000
E19	Sue	Master	6.5	51000
E20	Tina	Ph.D.	7.8	65000
E21	Tom	Master	8.1	64000
E22	Vick	Ph.D.	8.5	70000

TABLE IV.6.2   Linguistic Terms Table

Linguistic Terms	Ranges
very very high	[0.9091, 1.0]
very high	[0.8182, 0.9090]
high	[0.7273, 0.8181]
fairly high	[0.6364, 0.7272]
somewhat high	[0.5455, 0.6363]
medium	[0.4546, 0.5454]
somewhat low	[0.3637, 0.4545]
fairly low	[0.2728, 0.3636]
low	[0.1819, 0.2727]
very low	[0.0910, 0.1818]
very very low	[0, 0.0909]

TABLE IV.6.3   Normalized Trapezoidal Fuzzy Numbers of the Linguistic Terms of the Attribute Experience

Linguistic Terms	Normalized Trapezoidal Fuzzy Numbers
very very high	(0.9017, 0.9493, 1.0, 1.0)
very high	(0.8065, 0.8541, 0.9017, 0.9493)
high	(0.7113, 0.7589, 0.8065, 0.8541)
fairly high	(0.6161, 0.6637, 0.7113, 0.7589)
somewhat high	(0.5209, 0.5685, 0.6161, 0.6637)
medium	(0.4257, 0.4733, 0.5209, 0.5685)
somewhat low	(0.3305, 0.3781, 0.4257, 0.4733)
fairly low	(0.2353, 0.2829, 0.3305, 0.3781)
low	(0.1401, 0.1877, 0.2353, 0.2829)
very low	(0.0476, 0.0925, 0.1401, 0.1877)
very very low	(0, 0, 0.0476, 0.0925)

6.4.2   Calculating the Matching Degree of a Record with Respect to the User's Fuzzy Query

In Yeh and Chen,[29] the results of the user's query are not sorted in a descending sequence according to the degree of matching of the records with respect to the user's fuzzy query. It is obvious that if the results of the user's fuzzy query are a large amount of records, the user does not know which record is his best choice among them. In this chapter, we present a method for sorting the results of the user's fuzzy query. It can sort the results of the user's fuzzy query according to the matching degrees of the records with respect to the user's query.

Assume that X_1, X_2, ..., and X_n are linguistic terms of an attribute, and assume that they are represented by trapezoidal fuzzy numbers as shown in Figure IV.6.3. From Figure IV.6.3, we can get two membership grades $\mu_{X_i}(u)$ and $\mu_{X_{i+1}}(u)$ corresponding to different linguistic terms for each $u \in U$, where $\mu_{X_1}(u) + \mu_{X_{i+1}}(u) = 1$, $\mu_{X_i}(u) \in [0, 1]$, $\mu_{X_{i+1}}(u) \in [0, 1]$, and $1 \leq i \leq n - 1$.

Let L and S be the maximum value and the minimum value of the data appearing in the kth cluster (i.e., cluster_k) of an attribute A_j retrieved from a database, where cluster_$k = \{e_1, e_2, \ldots, e_n\}$, and let e_i be an element in the cluster, and let R_i be a record whose value of the attribute A_j is e_i,

TABLE IV.6.4 Normalized Trapezoidal Fuzzy Numbers of the Linguistic Terms of the Attribute Salary

Linguistic Terms	Normalized Trapezoidal Fuzzy Numbers
very very high	(0.9017, 0.9493, 1.0, 1.0)
very high	(0.8065, 0.8541, 0.9017, 0.9493)
high	(0.7113, 0.7552, 0.7884, 0.8064)
fairly high	(0.6638, 0.6752, 0.6978, 0.7112)
somewhat high	(0.5209, 0.5685, 0.6161, 0.6637)
medium	(0.4416, 0.4718, 0.5017, 0.5208)
somewhat low	(0.3782, 0.3991, 0.4212, 0.4415)
fairly low	(0.2353, 0.2829, 0.3305, 0.3781)
low	(0.0562, 0.0982, 0.1452, 0.2352)
very low	(0.0225, 0.0337, 0.0449, 0.0561)
very very low	(0, 0, 0.0112, 0.0224)

TABLE IV.6.5 Linguistic Terms of the Attribute Experience and Their Corresponding Trapezoidal Fuzzy Numbers

Linguistic Terms	Trapezoidal Fuzzy Numbers
absolutely high	(10, 10, 10, 10)
very very high	(9.0476, 9.5238, 10, 10)
very high	(8.0952, 8.5714, 9.0476, 9.5238)
high	(7.1429, 7.6190, 8.0952, 8.5714)
fairly high	(6.1905, 6.6666, 7.1429, 7.6190)
somewhat high	(5.2381, 5.7143, 6.1905, 6.6667)
medium	(4.2857, 4.7619, 5.2381, 5.7143)
somewhat low	(3.3333, 3.8095, 4.2857, 4.7619)
fairly low	(2.3810, 2.8571, 3.3333, 3.8095)
low	(1.4286, 1.9048, 2.3810, 2.8571)
very low	(0.4762, 0.9524, 1.4286, 1.9048)
very very low	(0, 0, 0.4762, 0.9524)
absolutely low	(0, 0, 0, 0)

where $1 \leq i \leq n$, then the membership grade in which the record R_i matches the user's fuzzy query condition for the attribute A_j is calculated as follows:

Case 1: If the record R_i whose attribute value of the attribute A_j is e_i, which falls in the membership function "close to γ" shown in Figure IV.6.1, then

$$\mu_{\text{close to } \gamma}(A_j(R_i)) = \mu_{\text{close to } \gamma}(e_i) = \frac{1}{1 + (\frac{e_i - \gamma}{\beta})^2}, \qquad \text{(IV.6.12)}$$

where $\mu_{\text{close to } \gamma}(A_j(R_i)) \in [0, 1]$ and the parameter β is the half-width of the curve at the crossover point as shown in Figure IV.6.1.

TABLE IV.6.6   Linguistic Terms of the Attribute Salary and
Their Corresponding Trapezoidal Fuzzy Numbers

Linguistic Terms	Trapezoidal Fuzzy Numbers
absolutely high	(70000, 70000, 70000, 70000)
very very high	(65238, 67619, 70000, 70000)
very high	(60476, 62857, 65238, 67619)
high	(55914, 58095, 60476, 62857)
fairly high	(50952, 53333, 55914, 58095)
somewhat high	(46190, 48571, 50952, 53333)
medium	(41429, 43810, 46190, 48571)
somewhat low	(36667, 39048, 41429, 43810)
fairly low	(31905, 34286, 36667, 39048)
low	(27143, 29524, 31905, 34286)
very low	(22381, 24762, 27143, 29524)
very very low	(0, 0, 22381, 24762)
absolutely low	(0, 0, 0, 0)

TABLE IV.6.7   Linguistic Weights and Their Corresponding
Trapezoidal Fuzzy Numbers

Linguistic Weights	Trapezoidal Fuzzy Numbers
absolutely high	(1.0, 1.0, 1.0, 1.0)
very very high	(0.9017, 0.9493, 1.0, 1.0)
very high	(0.8065, 0.8541, 0.9017, 0.9493)
high	(0.7113, 0.7589, 0.8065, 0.8541)
fairly high	(0.6161, 0.6637, 0.7113, 0.7589)
somewhat high	(0.5209, 0.5685, 0.6161, 0.6637)
medium	(0.4257, 0.4733, 0.5209, 0.5685)
somewhat low	(0.3305, 0.3781, 0.4257, 0.4733)
fairly low	(0.2353, 0.2829, 0.3305, 0.3781)
low	(0.1401, 0.1877, 0.2353, 0.2829)
very low	(0.0476, 0.0925, 0.1401, 0.1877)
very very low	(0, 0, 0.0476, 0.0925)
absolutely low	(0.0, 0.0, 0.0, 0.0)

Case 2: If the record R_i whose attribute value of the attribute A_j is e_i, which falls in the
membership function of the linguistic term X_1 shown in Figure IV.6.3, then

$$\mu_{X_1}(A_j(R_i))$$

$$= \begin{cases} 1 - |e_i - \text{Center of the } k\text{th Cluster}|/(L - S), & \text{if } e_i \geq \text{Center of the } k\text{th Cluster} \\ 1, & \text{if } e_i < \text{Center of the } k\text{th Cluster} \end{cases}$$

$$(IV.6.13)$$

where $\mu_{X_1}(A_j(R_i)) \in [0, 1]$, and L and S denote the maximum value and the minimum
value of the kth cluster (i.e., cluster_k), respectively.

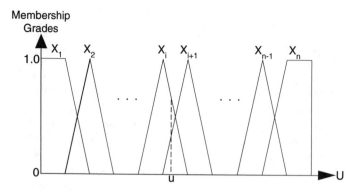

FIGURE IV.6.3   Membership functions of linguistic terms.

Case 3: If the record R_i whose attribute value of the attribute A_j is e_i, which falls in the membership function of the linguistic term X_p, where $2 \leq p \leq n - 1$, shown in Figure IV.6.3, then

$$\mu_{X_p}(A_j(R_i)) = 1 - |e_i - \text{Center of the } k\text{th Cluster}|/(L - S), \qquad \text{(IV.6.14)}$$

where $\mu_{X_p}(A_j(R_i)) \in [0, 1]$, $2 \leq p \leq n - 1$, and L and S denote the maximum value and the minimum value of the kth cluster (i.e., cluster_k), respectively.

Case 4: If the record R_i whose attribute value of the attribute A_j is e_i, which falls in the membership function of the linguistic term X_n shown in Figure IV.6.3, then

$$\mu_{X_n}(A_j(R_i))$$

$$= \begin{cases} 1 - |e_i - \text{Center of the } k\text{th Cluster}|/(L - S), & \text{if } e_i < \text{Center of the } k\text{th Cluster} \\ 1, & \text{if } e_i \geq \text{Center of the } k\text{th Cluster} \end{cases}$$

$$\text{(IV.6.15)}$$

where $\mu_{X_n}(A_j(R_i)) \in [0, 1]$ and L and S denote the maximum value and the minimum value of the kth cluster (i.e., cluster_k), respectively.

6.4.3   Fuzzy Query Processing in Relational Database Systems

Assume that there are two fuzzy sets A and B defined in the universe of discourse $U, U = \{u_1, u_2, \ldots, u_n\}$, with the membership functions μ_A and μ_B, respectively, and $\mu_A(u_i)$ and $\mu_B(u_i)$ indicate the membership values of u_i belonging to the fuzzy sets A and B, respectively, where $u_i \in U, \mu_A(u_i) \in [0, 1], \mu_B(u_i) \in [0, 1]$, and $1 \leq i \leq n$. Pappis et al.[25] introduced the following similarity measure between the fuzzy sets A and B:

$$S(A, B) = \frac{\sum_{i=1}^{n} \min(\mu_A(\mu_i), \mu_B(\mu_i))}{\sum_{i=1}^{n} \max(\mu_A(\mu_i), \mu_B(\mu_i))}, \qquad \text{(IV.6.16)}$$

where $S(A, B) \in [0, 1]$. The larger the value of $S(A, B)$, the more the similarity between the fuzzy sets A and B.

In the following, we present a method for fuzzy query processing in relational database systems, where the retrieval conditions and the weights of query items of the users' fuzzy SQL queries can be described by linguistic terms represented by fuzzy numbers. Let us consider the relation of a relational database system shown in Table IV.6.1. First, we apply the automatic clustering algorithm previously presented[29] to the attributes Experience and Salary, respectively. The clustering results are shown as follows:

1. The clustering results for the attribute Experience are shown as follows:

Cluster 1: 0.5	Center $= 0.5$
Cluster 2: 1.2, 1.5, 2.0, 2.0, 2.3	Center $= 1.8$
Cluster 3: 3.5, 3.5, 3.6, 3.8	Center $= 3.6$
Cluster 4: 5.0, 5.0	Center $= 5.0$
Cluster 5: 6.2, 6.5, 7.0, 7.2, 7.2, 7.5, 7.8, 8.1, 8.5	Center $= 7.3$
Cluster 6: 10	Center $= 10$

   Then, we can get the normalized center N_i(Experience), $1 \leq i \leq 6$, for each cluster i of the attribute Experience shown as follows:

   $$N_1(\text{Experience}) = (0.5 - 0.5)/(10 - 0.5) = 0,$$
   $$N_2(\text{Experience}) = (1.8 - 0.5)/(10 - 0.5) = 0.1368,$$
   $$N_3(\text{Experience}) = (3.6 - 0.5)/(10 - 0.5) = 0.3263,$$
   $$N_4(\text{Experience}) = (5.0 - 0.5)/(10 - 0.5) = 0.4737,$$
   $$N_5(\text{Experience}) = (7.3 - 0.5)/(10 - 0.5) = 0.7189,$$
   $$N_6(\text{Experience}) = (10 - 0.5)/(10 - 0.5) = 1.$$

2. The clustering results for the attribute Salary are as follows:

Cluster 1: 23000	Center $= 23000$
Cluster 2: 26000	Center $= 26000$
Cluster 3: 29000	Center $= 29000$
Cluster 4: 35000, 36000, 37000	Center $= 36000$
Cluster 5: 40000, 40000, 41000	Center $= 40333$
Cluster 6: 47000	Center $= 47000$
Cluster 7: 50000, 50000, 51000, 53000, 54000, 55000, 57000	Center $= 52857$
Cluster 8: 63000, 64000, 65000	Center $= 64000$
Cluster 9: 68000, 70000	Center $= 69000$

   Then, we can get the normalized center N_i(Salary), $1 \leq i \leq 9$, for each cluster i of the attribute Salary shown as follows:

   $$N_1(\text{Salary}) = (23000 - 23000)/(70000 - 23000) = 0,$$
   $$N_2(\text{Salary}) = (26000 - 23000)/(70000 - 23000) = 0.0638,$$
   $$N_3(\text{Salary}) = (29000 - 23000)/(70000 - 23000) = 0.1277,$$
   $$N_4(\text{Salary}) = (36000 - 23000)/(70000 - 23000) = 0.2766,$$
   $$N_5(\text{Salary}) = (40333 - 23000)/(70000 - 23000) = 0.3681,$$
   $$N_6(\text{Salary}) = (47000 - 23000)/(70000 - 23000) = 0.5106,$$
   $$N_7(\text{Salary}) = (52857 - 23000)/(70000 - 23000) = 0.6353,$$
   $$N_8(\text{Salary}) = (64000 - 23000)/(70000 - 23000) = 0.8723,$$
   $$N_9(\text{Salary}) = (69000 - 23000)/(70000 - 23000) = 0.9787.$$

6.4.3.1 Type-1 Query

Assume that A and A_1 are attributes in a relation T of a relational database system. Assume that the user's query is as follows:

$$
\begin{array}{ll}
\text{SELECT} & A \\
\text{FROM} & T \\
\text{WHERE} & A_1 \text{ is close to } c \\
\text{WITH} & \alpha
\end{array}
$$

where c is a crisp value, α is a retrieval threshold value given by the user, and $\alpha \in [0, 1]$. Then:

Step 1: Find the corresponding linguistic term of the fuzzy number "close to c" for the attribute A_1 based on Table IV.6.5, Table IV.6.6, and Equation IV.6.16 According to the attribute A_1, we choose the linguistic term C from Table IV.6.5 or Table IV.6.6 which has the largest similarity degree $S(\text{close to } c, C)$ as the corresponding linguistic term of the fuzzy number "close to c", where $S(\text{close to } c, C) \in [0, 1]$.

Step 2: Based on Table IV.6.3 or Table IV.6.4, find the corresponding normalized trapezoidal fuzzy numbers $T(C)$ of the linguistic term C shown as follows:

$$T(C) = (a, b, c, d)$$

where $0 \le a \le b \le c \le d \le 1$.

Step 3: Based on Equation IV.6.11, perform the defuzzified operation on the trapezoidal fuzzy number $T(C)$ to get the defuzzified value $DEF(T(C))$ of the trapezoidal fuzzy number $T(C)$ shown as follows:

$$DEF(T(C)) = (a + b + c + d)/4,$$

where $DEF(T(C)) \in [0, 1]$.

Step 4: Based on the linguistic terms table shown in Table IV.6.2, we can find that the defuzzified value $DEF(T(C))$ of the linguistic term C of the attribute A_1 is between t_1 and t_2 (i.e., $[t_1, t_2]$). Based on the clustering results of the attribute A_1, we also can find that the normalized center $N_k(A_1)$ of the kth cluster of the attribute A_1 falls in this range.

Step 5: Get the records whose values of the attribute A_1 fall in the kth cluster of the attribute A_1 (i.e., cluster_k), where cluster_$k = \{e_1, e_2, \ldots, e_n\}$. Let e_i be a data value appearing in the kth cluster of the attribute A_1, where $1 \le i \le n$. Based on Equation IV.6.12, compute the membership grade $\mu_{\text{close to } c}(A_1(R_i))$ for each record R_i whose values of the attribute A_1 fall in the kth cluster of the attribute A_1, where $\mu_{\text{close to } c}(A_1(R_i)) \in [0, 1]$.

Step 6: Compute the degree of matching $\text{Mat}(R_i)$ for each record R_i with respect to the user's query. The degree of matching $\text{Mat}(R_i)$ for each record R_i in the kth cluster of the attribute A_1 which matches the user's fuzzy query is calculated as follows:

$$\text{Mat}(R_i) = \mu_{\text{close to } c}(A_1(R_i))$$

The record whose matching degree with respect to the user's query is larger than or equal to the retrieval threshold value α will be retrieved from the database, where $\alpha \in [0, 1]$.

Perform the α-level-cuts $(\text{Mat}(R_i))_\alpha$ on the matching degrees $\text{Mat}(R_i)$, respectively, where $1 \le i \le n$. That is,

$$(\text{Mat}(R_i)) = \begin{cases} \text{Mat}(R_i), & \text{if } \text{Mat}(R_i) \ge \alpha \\ 0 & \text{otherwise,} \end{cases}$$

where $\alpha \in [0, 1]$. Sort the records whose values of the attribute A_1 fall in the cluster_k according to the value of $(\text{Mat}(R_i))_\alpha$ in a descending sequence. The record R_i whose attribute A_1 has the value with the largest matching degree $(\text{Mat}(R_i))_\alpha$ with respect to the user's query will be the best choice of the user, where $1 \le i \le n$.

Example 2: Let us consider the relation of a relational database system shown in Table IV.6.1. Assume the user's fuzzy query is as follows:

> SELECT   Name, Experience, Salary
> FROM   EMP
> WHERE   Experience is close to 7.2
> WITH   0.9

where the retrieval threshold value given by the user is 0.9.

Step 1: We can see that the WHERE-clause of the user's fuzzy query is "Experience is close to 7.2", where the membership function of the fuzzy number "close to 7.2" is shown as follows:

$$
\begin{aligned}
\text{close to } 7.2 = {}& 0.1076/0 + 0.1109/0.12 + 0.1143/0.24 + 0.1178/0.36 \\
& + 0.1216/0.48 + 0.1255/0.6 + 0.1296/0.72 + 0.1338/0.84 \\
& + 0.1379/0.95 + 0.1426/1.07 + 0.1475/1.19 + 0.1527/1.31 \\
& + 0.1581/1.43 + 0.1637/1.55 + 0.1697/1.67 + 0.1760/1.79 \\
& + 0.1826/1.91 + 0.1895/2.03 + 0.1968/2.15 + 0.2045/2.27 \\
& + 0.2120/2.38 + 0.2205/2.5 + 0.2296/2.62 + 0.2391/2.74 \\
& + 0.2491/2.86 + 0.2598/2.98 + 0.2710/3.1 + 0.2829/3.22 \\
& + 0.2944/3.33 + 0.3077/3.45 + 0.3217/3.57 + 0.3366/3.69 \\
& + 0.3523/3.81 + 0.3689/3.93 + 0.3865/4.05 + 0.4050/4.17 \\
& + 0.4246/4.29 + 0.4453/4.41 + 0.4672/4.53 + 0.4901/4.65 \\
& + 0.5121/4.76 + 0.5373/4.88 + 0.5636/5 + 0.5909/5.12 \\
& + 0.6193/5.24 + 0.6486/5.36 + 0.6787/5.48 + 0.7094/5.60 \\
& + 0.7379/5.71 + 0.7691/5.83 + 0.8/5.95 + 0.5304/6.07 \\
& + 0.8597/6.19 + 0.8875/6.31 + 0.9134/6.43 + 0.9367/6.55 \\
& + 0.9570/6.67 + 0.9738/6.79 + 0.9867/6.91 + 0.9954/7.03 \\
& + 0.9996/7.15 + 1.0/7.2 + 0.9992/7.27 + 0.9943/7.39 \\
& + 0.9849/7.51 + 0.9726/7.62 + 0.9554/7.74 + 0.9348/7.86 \\
& + 0.9113/7.98 + 0.8853/8.1 + 0.8573/8.22 + 0.8279/8.34 \\
& + 0.7974/8.46 + 0.7691/8.57 + 0.7379/8.69 + 0.7068/8.81 \\
& + 0.6762/8.93 + 0.6462/9.05 + 0.6169/9.17 + 0.5886/9.29 \\
& + 0.6242/9.41 + 0.5373/9.52 + 0.5121/9.64 + 0.4881/9.76 \\
& + 0.4653/9.88 + 0.4436/10.
\end{aligned}
$$

In order to find the corresponding linguistic term of the fuzzy number "close to 7.2," based on Equation IV.6.16 and Table IV.6.5, we can see that

$$S(\text{close to 7.2, very very high}) = 0.5026,$$
$$S(\text{close to 7.2, very high}) = 0.5889,$$
$$S(\text{close to 7.2, high}) = 0.6263,$$
$$S(\text{close to 7.2, fairly high}) = 0.6306,$$
$$S(\text{close to 7.2, somewhat high}) = 0.6214,$$
$$S(\text{close to 7.2, medium}) = 0.5431,$$
$$S(\text{close to 7.2, somewhat low}) = 0.4417,$$
$$S(\text{close to 7.2, fairly low}) = 0.3382,$$
$$S(\text{close to 7.2, low}) = 0.2598,$$
$$S(\text{close to 7.2, very low}) = 0.1971,$$
$$S(\text{close to 7.2, very very low}) = 0.1448.$$

Thus, we choose the linguistic term "fairly high" as the corresponding linguistic term of the fuzzy number "close to 7.2" due to the fact that it has the largest similarity degree among the linguistic terms shown in Table IV.6.5.

Step 2: From Table IV.6.3, we can see that the corresponding normalized trapezoidal fuzzy number representation of the linguistic term "fairly high" is $T(\text{fairly high}) = (0.6161, 0.6637, 0.7113, 0.7589)$.

Step 3: Based on Equation IV.6.11, we can see that the defuzzified value $DEF(T(\text{fairly high}))$ of the normalized trapezoidal fuzzy number $T(\text{fairly high})$ is equal to

$$DEF(T(\text{fairly high})) \quad = (0.6161 + 0.6637 + 0.7113 + 0.7589)/4 = 0.6875.$$

Step 4: Based on Table IV.6.2, we can see that the defuzzified value $DEF(T(\text{fairly high}))$ of the linguistic term "fairly high" of the attribute Experience is between 0.6364 and 0.7272 (i.e., [0.6364, 0.7272]). Based on the clustering results of the attribute Experience presented previously, we also can see that the normalized center $N_5(\text{Experience})$ of the 5th cluster of the attribute Experience is 0.6875 which falls in this range.

Step 5: Because the query condition "Experience is close to 7.2" falls in the 5th cluster of the attribute Experience, the records whose values of the attribute Experience fall in the 5th cluster of the attribute Experience will be retrieved from the database. Thus, we can get the following results with respect to the user's query condition "Experience is close to 7.2":

Emp-ID	Name	Degree	Experience	Salary
E1	Angel	Ph.D.	7.2	63000
E3	Bob	Bachelor	7.0	40000
E5	David	Master	7.5	53000
E16	Rose	Master	6.2	50000
E18	Sting	Master	7.2	55000
E19	Sue	Master	6.5	51000
E20	Tina	Ph.D.	7.8	65000
E21	Tom	Master	8.1	64000
E22	Vick	Ph.D.	8.5	70000

Then, based on Equation IV.6.12, where we assume that β is equal to 2.5 for the attribute Experience, the membership grade of each record R_i whose value of the attribute Experience falls in the 5th cluster of the attribute Experience is calculated as follows:

$$\mu_{\text{close to } 7.2}(\text{Experience}(R_1)) = 1/(1 + ((7.2 - 7.2)/2.5)^2 = 1,$$

$$\mu_{\text{close to } 7.2}(\text{Experience}(R_3)) = 1/(1 + ((7.0 - 7.2)/2.5)^2 = 0.9936,$$

$$\mu_{\text{close to } 7.2}(\text{Experience}(R_5)) = 1/(1 + ((7.5 - 7.2)/2.5)^2 = 0.9858,$$

$$\mu_{\text{close to } 7.2}(\text{Experience}(R_{16})) = 1/(1 + ((6.2 - 7.2)/2.5)^2 = 0.8621,$$

$$\mu_{\text{close to } 7.2}(\text{Experience}(R_{18})) = 1/(1 + ((7.2 - 7.2)/2.5)^2 = 1,$$

$$\mu_{\text{close to } 7.2}(\text{Experience}(R_{19})) = 1/(1 + ((6.5 - 7.2)/2.5)^2 = 0.9273,$$

$$\mu_{\text{close to } 7.2}(\text{Experience}(R_{20})) = 1/(1 + ((7.8 - 7.2)/2.5)^2 = 0.9455,$$

$$\mu_{\text{close to } 7.2}(\text{Experience}(R_{21})) = 1/(1 + ((8.1 - 7.2)/2.5)^2 = 0.8853,$$

$$\mu_{\text{close to } 7.2}(\text{Experience}(R_{22})) = 1/(1 + ((8.5 - 7.2)/2.5)^2 = 0.7872.$$

Step 6: The degree of matching for each record with respect to the user's fuzzy query is calculated as follows:

1. The degree of matching for the record R_1 with respect to the user's fuzzy query is equal to:

$$\text{Mat}(R_1) = \mu_{\text{close to } 7.2}(\text{Experience}(R_1))$$
$$= 1.0,$$

2. The degree of matching for the record R_3 with respect to the user's fuzzy query is equal to:

$$\text{Mat}(R_3) = \mu_{\text{close to } 7.2}(\text{Experience}(R_3))$$
$$= 0.9936,$$

3. The degree of matching for the record R_5 with respect to the user's fuzzy query is equal to:

$$\text{Mat}(R_5) = \mu_{\text{close to } 7.2}(\text{Experience}(R_5))$$
$$= 0.9858,$$

4. The degree of matching for the record R_{16} with respect to the user's fuzzy query is equal to:

$$\text{Mat}(R_{16}) = \mu_{\text{close to } 7.2}(\text{Experience}(R_{16}))$$
$$= 0.8621,$$

5. The degree of matching for the record R_{18} with respect to the user's fuzzy query is equal to:

$$\text{Mat}(R_{18}) = \mu_{\text{close to } 7.2}(\text{Experience}(R_{18}))$$
$$= 1.0,$$

6. The degree of matching for the record R_{19} with respect to the user's fuzzy query is equal to:

$$\text{Mat}(R_{19}) = \mu_{\text{close to } 7.2}(\text{Experience}(R_{19}))$$
$$= 0.9273,$$

7. The degree of matching for the record R_{20} with respect to the user's fuzzy query is equal to:

$$\text{Mat}(R_{20}) = \mu_{\text{close to } 7.2}(\text{Experience}(R_{20}))$$
$$= 0.9455,$$

8. The degree of matching for the record R_{21} with respect to the user's fuzzy query is equal to:

$$\text{Mat}(R_{21}) = \mu_{\text{close to } 7.2}(\text{Experience}(R_{21}))$$
$$= 0.8853,$$

9. The degree of matching for the record R_{22} with respect to the user's fuzzy query is equal to:

$$\text{Mat}(R_{22}) = \mu_{\text{close to } 7.2}(\text{Experience}(R_{22}))$$
$$= 0.7872.$$

The degrees of matching for the records in the database system with respect to the user's fuzzy query can be represented by a fuzzy set shown as follows:

$$1.0/R_1 + 0.9936/R_3 + 0.9858/R_5 + 0.8621/R_{16} + 1.0/R_{18} + 0.9273/R_{19}$$
$$+ 0.9455/R_{20} + 0.8853/R_{21} + 0.7872/R_{22}.$$

Because the retrieval threshold value given by the user is 0.9, by performing the 0.9-level-cut operation based on Definition 1, we can get the following results:

$$\text{Mat}(R_1)_{0.9\text{-level}} = 1.0,$$
$$\text{Mat}(R_3)_{0.9\text{-level}} = 0.9936,$$
$$\text{Mat}(R_5)_{0.9\text{-level}} = 0.9858,$$
$$\text{Mat}(R_{16})_{0.9\text{-level}} = 0,$$
$$\text{Mat}(R_{18})_{0.9\text{-level}} = 1.0,$$
$$\text{Mat}(R_{19})_{0.9\text{-level}} = 0.9273,$$
$$\text{Mat}(R_{20})_{0.9\text{-level}} = 0.9455,$$
$$\text{Mat}(R_{21})_{0.9\text{-level}} = 0,$$
$$\text{Mat}(R_{22})_{0.9\text{-level}} = 0.$$

The results of the user's fuzzy query are represented by a fuzzy set shown as follows:

$$1.0/R_1 + 0.9936/R_3 + 0.9858/R_5 + 1.0/R_{18} + 0.9273/R_{19} + 0.9455/R_{20}.$$

By sorting the records according to the matching degrees in a descending sequence, the results of the user's fuzzy query are as follows:

Name	Experience	Salary	Degree of Matching
Angel	7.2	63000	1.0
Sting	7.2	55000	1.0
Bob	7.0	40000	0.9936
David	7.5	53000	0.9858
Tina	7.8	65000	0.9455
Sue	6.5	51000	0.9273

6.4.3.2　Type-2 Query

Assume that A and A_1 are attributes in a relation T of a relational database system. Assume that the user's query is as follows:

$$\begin{aligned} &\text{SELECT} &&A \\ &\text{FROM} &&T \\ &\text{WHERE} &&A_1 \text{ is } C \\ &\text{WITH} &&\alpha \end{aligned}$$

where C is a crisp value or a linguistic term, α is a retrieval threshold value given by the user, and $\alpha \in [0, 1]$.

If C is a crisp value, then the records whose attribute A_1 has the value C are returned to the users, where the matching degrees of the records that match the user's query are set to 1.0.

If C is a linguistic term, then:

Step 1:　Based on Table IV.6.3 or Table IV.6.4, find the corresponding normalized trapezoidal fuzzy numbers $T(C)$ of the linguistic term C of the attribute A_1 as follows:

$$T(C) = (a, b, c, d),$$

where $0 \leq a \leq b \leq c \leq d \leq 1$.

Step 2:　Based on Equation IV.6.11, perform the defuzzified operation on the trapezoidal fuzzy number $T(C)$ to get the defuzzified value $DEF(T(C))$ of the trapezoidal fuzzy number $T(C)$ shown as follows:

$$DEF(T(C)) = (a + b + c + d)/4,$$

where $DEF(T(C)) \in [0, 1]$.

Step 3:　Based on the linguistic terms data shown in Table IV.6.2, we can find that the defuzzified value $DEF(T(C))$ of the linguistic term C of the attribute A_1 is between t_1 and t_2 (i.e., $[t_1, t_2]$). Based on the clustering results of the attribute A_1, we also can find that the normalized center $N_k(A_1)$ of the kth cluster of the attribute A_1 falls in this range.

Step 4:　Get the records R_i whose values of the attribute A_1 fall in the kth cluster of the attribute A_1 (i.e., cluster_k), where cluster_$k = \{e_1, e_2, \ldots, e_n\}$. Let e_i be a data value appearing in the kth cluster of the attribute A_1, where $1 \leq i \leq n$. Based on Equations IV.6.13–IV.6.15, compute the membership grade $\mu_c(A_1(R_i))$ for each record R_i whose values of the attribute A_1 falls in the kth cluster of the attribute A_1.

Step 5: Compute the degree of matching $\text{Mat}(R_i)$ for each record R_i with respect to the user's query. The degree of matching for each record R_i in the kth cluster of the attribute A_1 which matches the user's query is calculated as follows:

$$\text{Mat}(R_i) = \mu_c(A_1(R_i)).$$

The record whose matching degree with respect to the user's query is larger than or equal to the retrieval threshold value α will be retrieved from the database, where $\alpha \in [0, 1]$. Perform the α-level-cuts $(\text{Mat}(R_i))_\alpha$ on the matching degrees $\text{Mat}(R_i)$, respectively, where $1 \le i \le n$. That is,

$$(\text{Mat}(R_i))_\alpha = \begin{cases} \text{Mat}(R_i), & \text{if Mat } (R_i) \ge \alpha \\ 0 & \text{otherwise,} \end{cases}$$

where $\alpha \in [0, 1]$. Sort the records whose values of the attribute A_1 fall in the cluster_k according to the value of $(\text{Mat}(R_i))_\alpha$ in a descending sequence. The record R_i whose attribute A_1 has the value with the largest matching degree $(\text{Mat}(R_i))_\alpha$ with respect to the user's query will be the best choice of the user, where $1 \le i \le n$.

Example 3: Let us consider the relation of a relational database system shown in Table IV.6.1. Assume that the user's fuzzy query is as follows:

> SELECT   Name, Experience, Salary
> FROM   EMP
> WHERE   Experience is fairly low
> WITH   0.98

where the retrieval threshold value given by the user is 0.98.

Step 1: We can see that the WHERE-clause of the user's fuzzy query is "Experience is fairly low." From Table IV.6.3, we can see that the normalized trapezoidal fuzzy number representation of the linguistic term "fairly low" is $T(\text{fairly low}) = (0.2353, 0.2829, 0.3305, 0.3781)$.

Step 2: Based on Equation IV.6.11, we can see that the defuzzified value of the normalized trapezoidal fuzzy number $T(\text{fairly low})$ is equal to

$$DEF(T(\text{fairly low})) = (0.2353 + 0.2829 + 0.3305 + 0.3781)/4$$
$$= 0.3067.$$

Step 3: Based on Table IV.6.2, we can see that the defuzzified value of the linguistic term "fairly low" of the attribute Experience is between 0.2728 and 0.3636 (i.e., [0.2728, 0.3636]). Based on the clustering results of the attribute Experience presented previously, we also can see that the normalized center $N_3(\text{Experience})$ of the 3rd cluster of the attribute Experience is 0.3263 which falls in this range.

Step 4: Because the query condition "Experience is fairly low" falls in the 3rd cluster of the attribute Experience, the records whose values of the attribute Experience fall in the 3rd cluster of the attribute Experience will be retrieved from the database. Thus, we can get the following result with respect to the user's query condition "Experience is fairly low":

Emp-ID	Name	Degree	Experience	Salary
E9	Joan	Ph.D.	3.8	54000
E10	Lee	Bachelor	3.5	35000
E11	Mark	Master	3.5	40000
E12	Mary	Master	3.6	41000

Then, based on Equation IV.6.14, the membership grade of each record R_i whose value of the attribute Experience falls in the 3rd cluster of the attribute Experience is calculated as follows:

$$\mu_{\text{fairly low}}\ (\text{Experience}(R_9)) = 1 - |3.8 - 3.6|/(10 - 0.5) = 0.9789,$$
$$\mu_{\text{fairly low}}\ (\text{Experience}(R_{10})) = 1 - |3.5 - 3.6|/(10 - 0.5) = 0.9895,$$
$$\mu_{\text{fairly low}}\ (\text{Experience}(R_{11})) = 1 - |3.5 - 3.6|/(10 - 0.5) = 0.9895,$$
$$\mu_{\text{fairly low}}\ (\text{Experience}(R_{12})) = 1 - |3.6 - 3.6|/(10 - 0.5) = 1.0.$$

Step 5: The degree of matching for each record with respect to the user's fuzzy query is calculated as follows:

1. The degree of matching for the record R_9 with respect to the user's fuzzy query is equal to:

$$\text{Mat}(R_9) = \mu_{\text{fairly low}}(\text{Experience}(R_9))$$
$$= 0.9789.$$

2. The degree of matching for the record R_{10} with respect to the user's fuzzy query is equal to:

$$\text{Mat}(R_{10}) = \mu_{\text{fairly low}}(\text{Experience}(R_{10}))$$
$$= 0.9895.$$

3. The degree of matching for the record R_{11} with respect to the user's fuzzy query is equal to:

$$\text{Mat}(R_{11}) = \mu_{\text{fairly low}}(\text{Experience}(R_{11}))$$
$$= 0.9895.$$

4. The degree of matching for the record R_{12} with respect to the user's fuzzy query is equal to:

$$\text{Mat}(R_{12}) = \mu_{\text{fairly low}}(\text{Experience}(R_{12}))$$
$$= 1.0.$$

The degrees of matching for the records in the database system with respect to the user's fuzzy query can be represented by a fuzzy set shown as follows:

$$0.9789/R_9 + 0.9895/R_{10} + 0.9895/R_{11} + 1.0/R_{12}.$$

Because the retrieval threshold value given by the user is 0.98, by performing the 0.98-level-cut operation based on Definition 1, we can get the following results:

$$\text{Mat}(R_9)_{0.98\text{-level}} = 0,$$
$$\text{Mat}(R_{10})_{0.98\text{-level}} = 0.9895,$$
$$\text{Mat}(R_{11})_{0.98\text{-level}} = 0.9895,$$
$$\text{Mat}(R_{12})_{0.98\text{-level}} = 1.0.$$

The results of the user's fuzzy query represented by a fuzzy set is as follows:

$$0.9895/R_{10} + 0.9895/R_{11} + 1.0/R_{12}.$$

By sorting the records according to the matching degrees in a descending sequence, the results of the user's fuzzy query are as follows:

Name	Experience	Salary	Degree of Matching
Mary	3.6	41000	1.0
Lee	3.5	35000	0.9895
Mark	3.5	40000	0.9895

6.4.3.3 Type-3 Query

Assume that A, A_1, and A_2 are attributes in a relation T of a relational database system. Assume that the user's fuzzy query is as follows:

SELECT	A
FROM	T
WHERE	A_1 is C_1 AND A_2 is C_2
WEIGHT	A_1 is W_{A_1}; A_2 is W_{A_2}
WITH	α

where C_1 and C_2 are linguistic terms; W_{A_1} and W_{A_2} are linguistic weights of the attributes A_1 and A_2, respectively, represented by fuzzy numbers or crisp values between zero and one; α is a retrieval threshold value given by the user; and $\alpha \in [0, 1]$. Then:

Step 1: Based on Table IV.6.3 and Table IV.6.4, find the corresponding normalized trapezoidal fuzzy numbers $T(C_1)$ and $T(C_2)$ of the linguistic terms C_1 and C_2 of the attributes A_1 and A_2, respectively, where

$$T(C_1) = (a_1, b_1, c_1, d_1), \text{ and } 0 \le a_1 \le b_1 \le c_1 \le d_1 \le 1,$$
$$T(C_2) = (a_2, b_2, c_2, d_2), \text{ and } 0 \le a_2 \le b_2 \le c_2 \le d_2 \le 1.$$

Step 2: Based on Equation IV.6.11, perform the defuzzified operation on $T(C_1)$ and $T(C_2)$ to get the defuzzified value $DEF(T(C_1))$ and $DEF(T(C_2))$, respectively, shown as follows:

$$DEF(T(C_1)) = (a_1 + b_1 + c_1 + d_1)/4,$$
$$DEF(T(C_2)) = (a_2 + b_2 + c_2 + d_2)/4,$$

where $DEF(T(C_1)) \in [0, 1]$ and DEF $(T(C_2)) \in [0, 1]$.

Step 3: Based on the linguistic terms table shown in Table IV.6.2, we can find that the defuzzified value of the linguistic term C_1 of the attribute A_1 is between s_1 and s_2 (i.e., $[s_1, s_2]$). Based on the clustering results of the attribute A_1, we can also see that the normalized center $N_p(A_1)$ of the pth cluster of the attribute A_1 falls in this range. In the same way, we can find that the defuzzified value of the linguistic term C_2 of the attribute A_2 is between t_1 and t_2 (i.e., $[t_1, t_2]$). Based on the clustering results of the attribute A_2, we can also see that the normalized center $N_q(A_2)$ of the qth cluster of the attribute A_2 falls in this range.

Step 4: Get the records whose values of the attributes A_1 and A_2 fall in the pth cluster (i.e., cluster$\_p$) and the qth cluster (i.e., cluster$\_q$), respectively, where cluster$\_p = \{e_{p1}, e_{p2}, \ldots, e_{pn}\}$ and cluster$\_q = \{e_{q1}, e_{q2}, \ldots, e_{qm}\}$. Let e_{pi} and e_{qj} be data values appearing in the pth cluster and the qth cluster of the attributes A_1 and A_2, respectively, where $1 \leq i \leq n$ and $1 \leq j \leq m$. Based on Equations IV.6.13–IV.6.15, compute the membership grades of the records R_{pi} and R_{qj} in the pth cluster of the attribute A_1 and the qth cluster of the attribute A_2, respectively, where $1 \leq i \leq n$ and $1 \leq j \leq n$.

Step 5: If the weight of an attribute is a crisp value c, where $c \in [0, 1]$, then we convert the crisp value c into the trapezoidal fuzzy number representation (c, c, c, c). If the weights of the attributes A_1 and A_2 are linguistic terms W_{A_1} and W_{A_2}, respectively, then based on Table IV.6.7, we can find the corresponding trapezoidal fuzzy numbers $T(W_{A_1})$ and $T(W_{A_2})$ of the linguistic weights W_{A_1} and W_{A_2}, respectively, where

$$T(W_{A_1}) = (w_{a1}, w_{b1}, w_{c1}, w_{d1}), 0 \leq w_{a1} \leq w_{b1} \leq w_{c1} \leq w_{d1} \leq 1,$$
$$T(W_{A_2}) = (w_{a2}, w_{b2}, w_{c2}, w_{d2}), \text{ and } 0 \leq w_{a2} \leq w_{b2} \leq w_{c2} \leq w_{d2} \leq 1.$$

Then, calculate the normalized weights $T(\overline{W}_{A_1})$ and $T(\overline{W}_{A_2})$ of $T(W_{A_1})$ and $T(W_{A_2})$, respectively, based on the arithmetic operations of fuzzy numbers shown as follows:

$$T(\overline{W}_{A_1}) = T(W_{A_1}) \oslash (T(W_{A_1}) \oplus T(W_{A_2})),$$
$$T(\overline{W}_{A_2}) = T(W_{A_2}) \oslash (T(W_{A_1}) \oplus T(W_{A_2})).$$

Based on Equation IV.6.11, perform the defuzzified operations on $T(\overline{W}_{A_1})$ and $T(\overline{W}_{A_2})$ to get the defuzzified values $DEF(T(\overline{W}_{A_1}))$ and $DEF(T(\overline{W}_{A_2}))$ of $T(\overline{W}_{A_1})$ and $T(\overline{W}_{A_2})$, respectively, where $DEF(T(\overline{W}_{A_1})) \in [0, 1]$ and $DEF(T(\overline{W}_{A_2})) \in [0, 1]$.

Step 6: Because the logical operator in the WHERE-clause is AND, based on Equations IV.6.13–IV.6.15, the degree of matching Mat(R_i) for each record R_i which matches the user's weighted fuzzy query is calculated as follows:

$$\text{Mat}(R_i) = \text{Min}(1, \mu_{c_1}(A_1(R_{pi})) \times DEF(T(\overline{W}_{A_1}))$$
$$+ \mu_{c2}(A_2(R_{qj})) \times DEF(T(\overline{W}_{A_2}))$$

where $R_i \in \{$ the records whose values of the attribute A_1 fall in the pth cluster of the attribute $A_1\} \cup \{$ the records whose values of the attribute A fall in the qth cluster of the attribute $A_2\}$; R_{pi} and R_{qj} are the records whose values of the attribute A_1 fall in the pth cluster of the attribute A_1 and whose values of the attribute A_2 fall in the qth cluster of the attribute A_2, respectively. The record whose matching degree with respect to the user's weighted fuzzy query is larger than or equal to the retrieval threshold value α will be retrieved from the database, where $\alpha \in [0, 1]$.

Step 7: Perform the α-level-cut $(\text{Mat}(R_i))_\alpha$ on the matching degree $\text{Mat}(R_i)$, respectively, where $1 \leq i \leq n$. That is,

$$(\text{Mat}(R_i))_\alpha = \begin{cases} \text{Mat}(R_i), & \text{if Mat}(R_i) \geq \alpha \\ 0 & \text{otherwise}, \end{cases}$$

where $\alpha \in [0, 1]$. Sort the records according to the value of $(\text{Mat}(R_i))_\alpha$ in a descending sequence. The record R_i with the largest matching degree $(\text{Mat}(R_i))_\alpha$ with respect to the user's query will be the best choice of the user, where $1 \leq i \leq n$.

Example 4: Let us consider the relation of a relational database system shown in Table IV.6.1 Assume that the user's fuzzy query is as follows:

SELECT	Name, Experience, Salary
FROM	EMP
WHERE	Experience is 5.0 AND Salary is somewhat high
WEIGHT	Experience is high; Salary is low
WITH	0.80

where the retrieval threshold value given by the user is 0.80.

Step 1: We can see that the WHERE-clause of the user's fuzzy query is "Experience is 5.0 AND Salary is somewhat high". Because the user's query is focused on Experience = 5.0, where 5.0 is a crisp value, we can get the following result with respect to the user's query condition, "Experience is 5.0":

Emp-ID	Name	Degree	Experience	Salary
E14	Paul	Ph.D.	5.0	57000
E15	Ray	Bachelor	5.0	36000

Thus, we set the matching degrees of the above two records to one, respectively. From Table IV.6.4, we can see that the corresponding normalized trapezoidal fuzzy number representation of the linguistic term "somewhat high" is T(somewhat high) = (0.5209, 0.5685, 0.6161, 0.6637).

Step 2: Based on Equation IV.6.11, we can see that the defuzzified value of the normalized trapezoidal fuzzy number T(somewhat high) is equal to

$$DEF(T(\text{somewhat high})) = (0.5209 + 0.5685 + 0.6161 + 0.6637)/4$$
$$= 0.5923.$$

Step 3: Based on Table IV.6.2, we can see that the defuzzified value of the linguistic term "somewhat high" of the attribute Salary is between 0.5455 and 0.6363 (i.e., [0.5455, 0.6363]). Based on the clustering results of the attribute Salary presented previously, we also can see that the normalized center N_7(Salary) of the 7th cluster of the attribute Salary is 0.6353 which falls in this range.

Step 4: Because the query condition "Salary is somewhat high" falls in the 7th cluster of the attribute Salary, the records whose values of the attribute Salary fall in the 7th cluster of the attribute Salary will be retrieved from the database. Thus, we can get the following result with respect to the user's query condition "Salary is somewhat high":

Emp-ID	Name	Degree	Experience	Salary
E5	David	Master	7.5	53000
E8	Jerome	Ph.D.	2.0	50000
E9	Joan	Ph.D.	3.8	54000
E14	Paul	Ph.D.	5.0	57000
E16	Rose	Master	6.2	50000
E18	Sting	Master	7.2	55000
E19	Sue	Master	6.5	51000

Then, based on Equation IV.6.13, the membership grade $\mu_{\text{somewhat high}}(\text{Salary}(R_i))$ of each record R_i whose value of the attribute Salary falls in the 7th cluster of the attribute Salary is calculated as follows:

$$\mu_{\text{somewhat high}}(\text{Salary}(R_5)) = 1 - |53000 - 52857|/(70000 - 23000) = 0.9970,$$

$$\mu_{\text{somewhat high}}(\text{Salary}(R_8)) = 1 - |50000 - 52857|/(70000 - 23000) = 0.9392,$$

$$\mu_{\text{somewhat high}}(\text{Salary}(R_9)) = 1 - |54000 - 52857|/(70000 - 23000) = 0.9757,$$

$$\mu_{\text{somewhat high}}(\text{Salary}(R_{14})) = 1 - |57000 - 52857|/(70000 - 23000) = 0.9119,$$

$$\mu_{\text{somewhat high}}(\text{Salary}(R_{16})) = 1 - |50000 - 52857|/(70000 - 23000) = 0.9392,$$

$$\mu_{\text{somewhat high}}(\text{Salary}(R_{18})) = 1 - |55000 - 52857|/(70000 - 23000) = 0.9544,$$

$$\mu_{\text{somewhat high}}(\text{Salary}(R_{19})) = 1 - |51000 - 52857|/(70000 - 23000) = 0.9605.$$

Step 5: Because the fuzzy weights of the attributes Experience and Salary are high and low, respectively, based on Table IV.6.7, we can see that the fuzzy number representations of the fuzzy weights of the attributes Experience and Salary are $T(W_{\text{Experience}})$ and $T(W_{\text{Salary}})$, respectively, where $T(W_{\text{Experience}}) = (0.7113, 0.7589, 0.8065, 0.8541)$ and $T(W_{\text{Salary}}) = (0.1401, 0.1877, 0.2353, 0.2829)$. Then, the normalized fuzzy weights $T(\overline{W}_{\text{Experience}})$ and $T(\overline{W}_{\text{Salary}})$ of $T(W_{\text{Experience}})$ and $T(W_{\text{Salary}})$ can be calculated, respectively, shown as follows:

$$T(\overline{W}_{\text{Experience}}) = T(W_{\text{Experience}}) \oslash (T(W_{\text{Experience}}) \oplus T(W_{\text{Salary}}))$$

$$= (0.7113, 0.7589, 0.8065, 0.8541) \oslash ((0.7113, 0.7589,$$

$$0.8065, 0.8541) \oplus (0.1401, 0.1877, 0.2353, 0.2829))$$

$$= (0.7113, 0.7589, 0.8065, 0.8541) \oslash (0.8514, 0.9466, 1.0418, 1.1370)$$

$$= (0.6256, 0.7285, 0.8520, 1.0032),$$

$$T(\overline{W}_{\text{Salary}}) = T(W_{\text{Salary}}) \oslash (T(W_{\text{Experience}}) \oplus T(W_{\text{Salary}}))$$

$$= (0.1401, 0.1877, 0.2353, 0.2829) \oslash ((0.7113, 0.7589, 0.8065,$$

$$0.8541) \oplus (0.1401, 0.1877, 0.2353, 0.2829))$$

$$= (0.1401, 0.1877, 0.2353, 0.2829) \oslash (0.8514, 0.9466, 1.0418, 1.1370)$$

$$= (0.1232, 0.1802, 0.2486, 0.3323).$$

Based on Equation IV.6.11, we can see that the defuzzified value $DEF(T(\overline{W}_{\text{Experience}}))$ of $T(\overline{W}_{\text{Experience}})$ is equal to

$$DEF(T(\overline{W}_{\text{Experience}})) = (0.6256 + 0.7285 + 0.8520 + 1.0032)/4$$
$$= 0.8023,$$

and the defuzzified value $DEF(T(\overline{W}_{\text{Salary}}))$ of $T(\overline{W}_{\text{Salary}})$ is equal to

$$DEF(T(\overline{W}_{\text{Salary}})) = (0.1232 + 0.1802 + 0.2486 + 0.3323)/4$$
$$= 0.2211.$$

Step 6: Because the logical operator in the WHERE-clause is AND, the degree of matching for each record with respect to the user's fuzzy query is calculated as follows:

1. The degree of matching for the record R_5 with respect to the user's weighted fuzzy query is equal to:

$$\text{Mat}(R_5) = \text{Min}(1, \mu_{5.0}(\text{Experience}(R_5)) \times DEF(T(\overline{W}_{\text{Experience}}))$$
$$+ \mu_{\text{somewhat high}}(\text{Salary}(R_5)) \times DEF(T(\overline{W}_{\text{Salary}}))$$
$$= \text{Min}(1, 0 \times 0.8023 + 0.9970 \times 0.2211)$$
$$= \text{Min}(1, 0 + 0.2204)$$
$$= 0.2204.$$

2. The degree of matching for the record R_8 with respect to the user's weighted fuzzy query is equal to:

$$\text{Mat}(R_8) = \text{Min}(1, \mu_{5.0}(\text{Experience}(R_8)) \times DEF(T(\overline{W}_{\text{Experience}}))$$
$$+ \mu_{\text{somewhat high}}(\text{Salary}(R_8)) \times DEF(T(\overline{W}_{\text{Salary}}))$$
$$= \text{Min}(1, 0 \times 0.8023 + 0.9392 \times 0.2211)$$
$$= \text{Min}(1, 0 + 0.2077)$$
$$= 0.2077.$$

3. The degree of matching for the record R_9 with respect to the user's weighted fuzzy query is equal to:

$$\text{Mat}(R_9) = \text{Min}(1, \mu_{5.0}(\text{Experience}(R_9)) \times DEF(T(\overline{W}_{\text{Experience}}))$$
$$+ \mu_{\text{somewhat high}}(\text{Salary}(R_9)) \times DEF(T(\overline{W}_{\text{Salary}}))$$
$$= \text{Min}(1, 0 \times 0.8023 + 0.9757 \times 0.2211)$$
$$= \text{Min}(1, 0 + 0.2157)$$
$$= 0.2157.$$

4. The degree of matching for the record R_{14} with respect to the user's weighted fuzzy query is equal to:

$$\text{Mat}(R_{14}) = \text{Min}(1, \mu_{5.0}(\text{Experience}(R_{14})) \times DEF(T(\overline{W}_{\text{Experience}}))$$
$$+ \mu_{\text{somewhat high}}(\text{Salary}(R_{14})) \times DEF(T(\overline{W}_{\text{Salary}}))$$
$$= \text{Min}(1, 1 \times 0.8023 + 0.9119 \times 0.2211)$$
$$= \text{Min}(1, 0.8023 + 0.2016)$$
$$= 1.0.$$

5. The degree of matching for the record R_{15} with respect to the user's weighted fuzzy query is equal to:

$$\text{Mat}(R_{15}) = \text{Min}(1, \mu_{5.0}(\text{Experience}(R_{15})) \times \text{DEF}(T(\overline{W}_{\text{Experience}}))$$

$$+ \mu_{\text{somewhat high}}(\text{Salary}(R_{15})) \times \text{DEF}(T(\overline{W}_{\text{Salary}}))$$

$$= \text{Min}(1, 1 \times 0.8023 + 0 \times 0.2211)$$

$$= \text{Min}(1, 0.8023 + 0)$$

$$= 0.8023.$$

6. The degree of matching for the record R_{16} with respect to the user's weighted fuzzy query is equal to:

$$\text{Mat}(R_{16}) = \text{Min}(1, \mu_{5.0}(\text{Experience}(R_{16})) \times \text{DEF}(T(\overline{W}_{\text{Experience}}))$$

$$\mu_{\text{somewhat high}}(\text{Salary}(R_{16})) \times \text{DEF}(T(\overline{W}_{\text{Salary}}))$$

$$= \text{Min}(1, 0 \times 0.8023 + 0.9392 \times 0.2211)$$

$$= \text{Min}(1, 0 + 0.2077)$$

$$= 0.2077.$$

7. The degree of matching for the record R_{18} with respect to the user's weighted fuzzy query is equal to:

$$\text{Mat}(R_{18}) = \text{Min}(1, \mu_{5.0}(\text{Experience}(R_{18})) \times \text{DEF}(T(\overline{W}_{\text{Experience}}))$$

$$+ \mu_{\text{somewhat high}}(\text{Salary}(R_{18})) \times \text{DEF}(T(\overline{W}_{\text{Salary}}))$$

$$= \text{Min}(1, 0 \times 0.8023 + 0.9544 \times 0.2211)$$

$$= \text{Min}(1, 0 + 0.2110)$$

$$= 0.2110.$$

8. The degree of matching for the record R_{19} with respect to the user's weighted fuzzy query is equal to:

$$\text{Mat}(R_{19}) = \text{Min}(1, \mu_{5.0}(\text{Experience}(R_{19})) \times \text{DEF}(T(\overline{W}_{\text{Experience}}))$$

$$+ \mu_{\text{somewhat high}}(\text{Salary}(R_{19})) \times \text{DEF}(T(\overline{W}_{\text{Salary}}))$$

$$= \text{Min}(1, 0 \times 0.8023 + 0.9605 \times 0.2211)$$

$$= \text{Min}(1, 0 + 0.2124)$$

$$= 0.2124.$$

Step 7: The degrees of matching for the records in the database system with respect to the user's fuzzy query can be represented by a fuzzy set shown as follows:

$$0.2204/R_5 + 0.2077/R_8 + 0.2157/R_9 + 1.0/R_{14} + 0.8023/R_{15}$$

$$+ 0.2077/R_{16} + 0.2110/R_{18} + 0.2124/R_{19}.$$

Because the retrieval threshold value given by the user is 0.8, by performing the 0.8-level-cut operation, we can get the following results:

$$\text{Mat}(R_5)_{0.8\text{-level}} = 0,$$
$$\text{Mat}(R_8)_{0.8\text{-level}} = 0,$$
$$\text{Mat}(R_9)_{0.8\text{-level}} = 0,$$
$$\text{Mat}(R_{14})_{0.8\text{-level}} = 1.0,$$
$$\text{Mat}(R_{15})_{0.8\text{-level}} = 0.8023,$$
$$\text{Mat}(R_{16})_{0.8\text{-level}} = 0,$$
$$\text{Mat}(R_{18})_{0.8\text{-level}} = 0,$$
$$\text{Mat}(R_{19})_{0.8\text{-level}} = 0.$$

The results of the user's fuzzy query are represented by a fuzzy set shown as follows:

$$1.0/R_{14} + 0.8023/R_{15}.$$

By sorting the records according to the matching degrees in a descending sequence, the results of the user's fuzzy query are as follows:

Name	Experience	Salary	Degree of Matching
Paul	5.0	57000	1.0
Ray	5.0	36000	0.8023

6.4.3.4   (D) Type-4 Query

Assume that A, A_1, and A_2 are attributes in a relation T of a relational database system. Assume that the user's query is as follows:

SELECT A
FROM T
WHERE A_1 is C_1 OR A_2 is C_2
WITH α

where C_1 and C_2 are linguistic terms, α is a retrieval threshold value given by the user, and $\alpha \in [0, 1]$. Then:

Step 1:   Based on Tables IV.6.3 and IV.6.4, find the corresponding normalized trapezoidal fuzzy numbers $T(C_1)$ and $T(C_2)$ of the linguistic terms C_1 and C_2 of the attributes A_1 and A_2, respectively:

$$T(C_1) = (a_1, b_1, c_1, d_1), \text{ where } 0 \leq a_1 \leq b_1 \leq c_1 \leq d_1 \leq 1,$$
$$T(C_2) = (a_2, b_2, c_2, d_2), \text{ where } 0 \leq a_2 \leq b_2 \leq c_2 \leq d_2 \leq 1.$$

Step 2: Based on formula Equation IV.6.11, perform the defuzzified operation on the $T(C_1)$ and $T(C_2)$ to get the defuzzified value $DEF(T(C_1))$ and $DEF(T(C_2))$, respectively, shown as follows:

$$DEF(T(C_1)) = (a_1 + b_1 + c_1 + d_1)/4,$$
$$DEF(T(C_2)) = (a_2 + b_2 + c_2 + d_2)/4,$$

where $DEF(T(C_1)) \in [0, 1]$ and $DEF(T(C_2)) \in [0, 1]$.

Step 3: Based on the linguistic terms table shown in Table IV.6.2, we can find that the defuzzified value of the linguistic term C_1 of the attribute A_1 is between s_1 and s_2 (i.e., $[s_1, s_2]$). Based on the clustering results of the attribute A_1, we also can see that the normalized center $N_p(A_1)$ of the pth cluster of the attribute A_1 falls in this range. In the same way, we can find that the defuzzified value of the linguistic term C_2 of the attribute A_2 is between t_1 and t_2 (i.e., $[t_1, t_2]$). Based on the clustering results of the attribute A_2, we also can see that the normalized center $N_q(A_2)$ of the qth cluster of the attribute A_2 falls in this range.

Step 4: Get the records whose values of the attribute A_1 and A_2 fall in the pth cluster (i.e., cluster_p) and the qth cluster (i.e., cluster_q), respectively, where cluster_p = $\{e_{p1}, e_{p2}, \ldots, e_{pn}\}$ and cluster_q = $\{e_{q1}, e_{q2}, \ldots, e_{qm}\}$. Let e_{pi} and e_{qj} be data values appearing in the pth cluster and qth cluster, respectively, where $1 \le i \le n$ and $1 \le j \le m$. Based on Equations IV.6.13–IV.6.15, compute the membership grades of the records R_{pi} and R_{qj} in the pth cluster of the attribute A_1 and the qth cluster of the attribute A_2, respectively, where $1 \le i \le n$ and $1 \le j \le n$.

Step 5: Because the logical operator in the WHERE-clause is OR, based on Equations IV.6.13–IV.6.15, the degree of matching Mat(R_i) for each record R_i which matches the user's fuzzy query is calculated as follows:

$$\text{Mat}(R_i) = \text{Max}(\mu_{c_1}(A_1(R_{pi})), \mu_{c_2}(A_2(R_{qj})))$$

where $R_i \in \{$ the records whose values of the attribute A_1 fall in the pth cluster of the attribute $A_1\} \cup \{$ the records whose values of the attribute A_2 fall in the qth cluster of the attribute $A_2\}$; R_{pi} and R_{qj} are the records whose values of the attribute A_1 fall in the pth cluster of the attribute A_1 and whose values of the attribute A_2 fall in the qth cluster of the attribute A_2.

Step 6: Perform the α-level-cut $(\text{Mat}(R_i))_\alpha$ on the matching degree Mat(R_i), respectively, where $1 \le i \le n$. That is,

$$(\text{Mat}(R_i))_\alpha = \begin{cases} \text{Mat}(R_i), & \text{if Mat}(R_i) \ge \alpha \\ 0 & \text{otherwise,} \end{cases}$$

where $\alpha \in [0, 1]$. Sort the records according to the value of $(\text{Mat}(R_i))_\alpha$ in a descending sequence. The record R_i with the largest matching degree $(\text{Mat}(R_i))_\alpha$ with respect to the user's query will be the best choice of the user, where $1 \le i \le n$.

6.5 Conclusions

We have presented a method for fuzzy query processing in relational database systems, where weighted fuzzy queries are allowed to users. The proposed method allows the query conditions and the weights of query items of the user's fuzzy SQL queries to be described by linguistic terms represented by fuzzy numbers. We have used Borland C^{++} Builder Version 4.0 to implement a system for fuzzy query processing. Because the proposed fuzzy query processing method allows the users to construct their fuzzy queries more conveniently, the existing relational database systems will be more intelligent and more flexible to the users.

References

1. Bhatia, S.K. and Deogun, J.S., Conceptual clustering in information retrieval, *IEEE Trans. Syst. Man Cybernetics*, 28(3), 427, 1998.
2. Bosc, P. et al., Fuzzy databases, in *Fuzzy Sets in Approximate Reasoning and Information Systems*, Bezdek, J.C., Dubois, D., and Prade, H., Eds., Kluwer Academic Publishers, Boston, MA, 403, 1999.
3. Bosc, P., Galibourg, M., and Hamon, G., Fuzzy querying with SQL: extensions and implementation aspects, *Fuzzy Sets Syst.*, 28, 333, 1988.
4. Bosc, P. and Lietard, L., Fuzzy integrals and database flexible querying, *Proc. Fifth IEEE Int. Conf. Fuzzy Syst.*, 1, 100, 1996.
5. Bosc, P. and Pivert, O., Some approaches for relational databases flexible querying, *Int. J. Intelligent Inf. Syst.*, 1, 323, 1992.
6. Bosc, P. and Pivert, O., SQLF: a relational database language for fuzzy querying, *IEEE Trans. Fuzzy Syst.*, 3(1), 1, 1995.
7. Bosc, P., Pivert, O., and Farquhar, K., Integrating fuzzy queries into an existing database management system: an example, *Int. J. Intelligent Syst.*, 9(5), 475, 1994.
8. Bosc, P. and Prade, H., An introduction to the fuzzy set and probability theory-based treatment of flexible queries and uncertain or imprecise databases, in *Uncertainty Management in Information Systems—From Needs to Solutions*, Motro, A. and Smets, P., Eds. Kluwer Academic Publishers, Boston, MA, 285, 1997.
9. Chang, S.K. and Ke, S.E., Database skeleton and its application to fuzzy query translation, *IEEE Trans. Software Eng.*, SE-4(1), 31, 1978.
10. Chang, S.K. and Ke, J.S., Translation of fuzzy queries for relational database systems, *IEEE Trans. Pattern Anal. Machine Intelligence*, PAMI-1, 281, 1979.
11. Chen, S.M., Ke, J.S., and Chang, J.F., Techniques of fuzzy query translation for database systems, Proceedings of the 1986 International Computer Symposium, Tainan, Taiwan, Republic of China, 1281, 1986.
12. Chen, S.M. and Jong, W.T., Fuzzy query translation for relational database systems, *IEEE Trans. Syst. Man Cybernetics*, 27(4), 714, 1997.
13. Chen, S.M., Using fuzzy reasoning techniques for fault diagnosis of the J-85 jet engines, Proceedings of the Third National Conference on Science and Technology of National Defense, Taoyuan, Taiwan, Republic of China, 29, 1995.
14. Date, C.J., *An Introduction to Database Systems*, Addison-Wesley, Reading, MA, 1990.
15. Horowitz, E. and Sahni, S., *Fundamentals of Data Structures*, Computer Science Press, New York, 1982.
16. Ichikawa, T. and Hirakawa, M., ARES: a relational database with the capability of performing flexible interpretation of queries, *IEEE Trans. Software Eng.*, 12(5), 624, 1986.
17. Kacprzyk, J., Fuzzy logic in DBMSs and querying, Proceedings of the Second New Zealand International Two-Stream Conference on Artificial Neural Networks and Expert Systems, 106, 1995.
18. Kacprzyk, J., Zadrozny, S., and Ziolkowski, A., FQUERY III +: a "human-consistent" database querying system based on fuzzy logic with linguistic quantifiers, *Inf. Syst.*, 14(6), 443, 1989.
19. Kacprzyk, J. and Ziolkowski, A., Database queries with fuzzy linguistic quantifiers, *IEEE Trans. Syst. Man Cybernetics*, SMC-16(3), 474, 1989.
20. Kamel, M., Hadfield, B., and Ismail, M., Fuzzy query using clustering techniques, *Inf. Process. Manage.*, 26(2), 279, 1990.

21. Kandel, A., *Fuzzy Mathematical Techniques with Applications*, Addison-Wesley, Reading, MA, 1986.
22. Kaufmann, A. and Gupta, M.M., *Introduction to Fuzzy Arithmetic: Theory and Applications*, Van Nostrand Reinhold, New York, 1985.
23. Kaufmann, A. and Gupta, M.M., *Fuzzy Mathematical Models in Engineering and Management Science*, North-Holland, Amsterdam, 1988.
24. Lin, Y.S. and Chen, S.M., Using automatic clustering techniques for fuzzy query processing in relational database systems, Proceedings of the 11th National Conference on Information Management, Kaohsiung, Taiwan, Republic of China, 2000.
25. Pappis, C.P. and Karacapilidis, N.I., A comparative assessment of measures of similarity of fuzzy values, *Fuzzy Sets Syst.*, 56(2), 171, 1993.
26. Tahani, V., A conceptual framework for fuzzy query processing—a step toward very intelligent database systems, *Inf. Process. Manage.*, 13, 289, 1977.
27. Waller, W.G. and Kraft, D.H., A mathematical model of a weighted Boolean retrieval system, *Inf. Process. Manage.*, 15(5), 235, 1979.
28. Wong, M.H. and Leung, K.S., A fuzzy database-query language, *Inf. Syst.*, 15(5), 583, 1990.
29. Yeh, M.S. and Chen, S.M., A new method for fuzzy query processing using automatic clustering techniques, *J. Comput.*, 6(1), 1, 1994.
30. Zadeh, L.A., Fuzzy sets, *Inf. Control*, 8, 338, 1965.
31. Zadeh, L.A., The concepts of a linguistic variable and its application to approximate reasoning (I), *Inf. Sci.*, 8, 199, 1975.

7

A Method for Generating Fuzzy Rules from Relational Database Systems for Estimating Null Values

Shyi-Ming Chen
National Taiwan University of Science and Technology

Ming-Shiou Yeh
National Chiao Tung University

Abstract. Traditionally, the knowledge base of a rule-based system is constructed by the knowledge acquisition process. In recent years, some researchers focused on the research topic of rules generation by learning from examples. In this chapter, we present a method to construct fuzzy decision trees and to generate fuzzy rules from relational database systems for estimating null values. The proposed method can get a good forecasting result for estimating null values in relational database systems.

7.1 Introduction

During the last decade, the research fields of applied computer science have established the need for formulation of models of imprecise information systems that could simulate human approximate reasoning.[10] In 1965, Zadeh proposed the fuzzy set theory.[21] It has been widely used for representing and reasoning with imprecise and uncertain information. Traditionally, the knowledge base of a rule-based system is constructed by the knowledge acquisition process. In recent years, some researchers focused on the research topic of rule generation by learning from examples.[4-9,11,13,16-20] Chen, Lee, and Lee[5] presented a method for generating fuzzy rules from numerical data for handling fuzzy classification problems. Chen and Chen[6] presented a method to generate fuzzy rules for fuzzy classification systems. Chang and Chen[7] presented a method to generate fuzzy rules from numerical data based on the exclusion of attribute terms. Hart[8] applied the induction of decision trees to knowledge acquisition for expert systems, where experts supply various examples to construct decision trees, and the resulting decision trees are used to generate fuzzy rules. CLS[9] and ID3[13] are the well-known inductive learning algorithms used to induce classification rules in the form of decision trees. Kao and Chen[11] presented a method to generate fuzzy rules from training data containing noise for handling classification problems. Wang and Mendel[17] used the fuzzy associative memory to construct a fuzzy rule base from numerical data and linguistic information and apply it to truck "backer-upper" control. Sudkamp and Hammell[16] presented a method to learn fuzzy rules using the concepts of interpolation and learning. Yasdi[19] used the inductive learning techniques to learn classification rules from databases.

In this article, we present a method to construct fuzzy decision trees and generate fuzzy rules from relational database systems for estimating null values. The proposed method can get a good forecasting result for estimating null values in relational database systems.

The rest of this article is organized as follows. In Section 7.2, we briefly review basic concepts of fuzzy set theory.[21] In Section 7.3, we introduce the concepts of generating fuzzy rules from relational database systems. In Section 7.4, we present an algorithm called fuzzy concept learning system (FCLS)[20] to construct fuzzy decision trees from relational databases and to generate fuzzy rules from the constructed fuzzy decision trees. We also present a method to deal with the completeness of the generated fuzzy decision tree and present a method to forecast null values in a relational database system. The conclusions are discussed in Section 7.5.

7.2 Fuzzy Set Theory

Imprecision and uncertainty in the knowledge of human beings are difficult to properly express in real-world applications via conventional mathematical models. Since Professor Zadeh proposed the theory of fuzzy sets,[21] it has been widely utilized to represent and manipulate the uncertain and imprecise knowledge and has recently been extensively and successfully utilized in several different kinds of fields, such as decision making, medical diagnosis, automatic control, psychology, economics, . . . , etc. In the following, we briefly review the definition of fuzzy sets.[21]

7.2.1 Basic Concepts of Fuzzy Set Theory

Definition 1: A fuzzy set **A** of the universe of discourse **U** is defined by

$$\mathbf{A} = \{(u, \mu_A(u)) | u \in \mathbf{U}\}, \tag{IV.7.1}$$

where μ_A is the membership function of the fuzzy set **A** to describe the characteristics of the fuzzy set **A**, $\mu_A : \mathbf{U} \rightarrow [0, 1]$; $\mu_A(u)$ indicates the membership grade in which element u belongs to the fuzzy set **A**.

If $\mathbf{U} = \{u_1, u_2, \ldots, u_n\}$ is a finite set, then the fuzzy set **A** can be represented as follows:

$$\mathbf{A} = \mu_A(u_1)/u_1 + \mu_A(u_2)/u_2 + \cdots + \mu_A(u_n)/u_n, \tag{IV.7.2}$$

where the symbol + means union and the symbol / means the separator. For convenience, the fuzzy set **A** also can be represented by

$$\mathbf{A} = \{(u_i, \mu_A(u_i)) \mid u_i \in \mathbf{U}\}.$$

If **U** is an infinite set, then the fuzzy set **A** can be represented as follows:

$$\mathbf{A} = \int_u \mu_A(u)/u, \quad u \in \mathbf{U}, \tag{IV.7.3}$$

where the symbol $\int$ means union rather than the integration symbol in calculus.

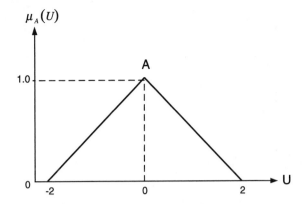

FIGURE IV.7.1   Continuous triangular membership function.

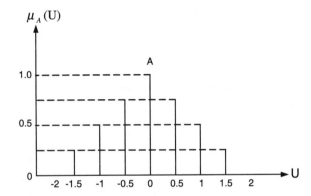

FIGURE IV.7.2   Discrete triangular membership function.

Many kinds of membership functions can be used to represent fuzzy sets. In the following, we describe some of the most widely used membership functions shown as follows.

7.2.1.1   Triangular Membership Function

1. **Continuous triangular membership function:** Let **A** be a fuzzy set of the universe of discourse **U**,

$$\mathbf{A} = \int_{-2}^{0} \left(\frac{2+u}{2} \right) \bigg/ u + \int_{0}^{2} \left(\frac{2-u}{2} \right) \bigg/ u, u \in \mathbf{U}. \tag{IV.7.4}$$

The continuous triangular membership function curve of the fuzzy set **A** is shown in Figure IV.7.1.

2. **Discrete triangular membership function:** Let **A** be a fuzzy set of the universe of discourse **U**,

$$\mathbf{U} = \{-2, -1.5, -1, -0.5, 0, 0.5, 1, 1.5, 2\},$$
$$\mathbf{A} = 0.25/-1.5 + 0.5/-1 + 0.75/-0.5 + 1.0/0$$
$$+ 0.75/0.5 + 0.5/1 + 0.25/1.5.$$

The discrete triangular membership function curve of the fuzzy set **A** is shown in Figure IV.7.2.

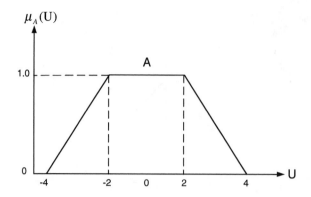

FIGURE IV.7.3   Trapezoidal membership function.

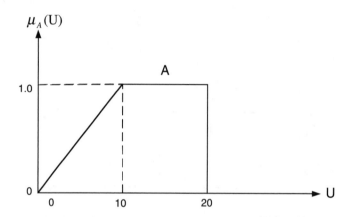

FIGURE IV.7.4   Linear membership function.

7.2.1.2   Trapezoidal Membership Function

Let **A** be a fuzzy set of the universe of discourse **U**,

$$\mathbf{A} = \int_{-4}^{-2} \frac{4+u}{4} \bigg/ u + \int_{-2}^{2} 1/u + \int_{2}^{4} \frac{4-u}{4} \bigg/ u, \quad u \in \mathbf{U}. \tag{IV.7.5}$$

The trapezoidal membership function curve of the fuzzy set **A** is shown in Figure IV.7.3.

7.2.1.3   Linear Membership Function

Let **A** be a fuzzy set of the universe of discourse **U**,

$$\mathbf{A} = \int_{0}^{10} 0.1u/u + \int_{10}^{20} 1/u, \quad u \in [0, 20]. \tag{IV.7.6}$$

The membership function curve of the fuzzy set **A** is shown in Figure IV.7.4.

7.2.1.4   Bell Membership Function

Let **A** be a fuzzy set of the universe of discourse **U**,

$$\mathbf{A} = \int_{u} e^{-0.5(x-5)^2}/x \tag{IV.7.7}$$

The membership function curve of the fuzzy set **A** is shown in Figure IV.7.5.

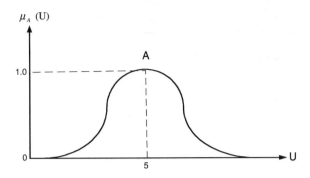

FIGURE IV.7.5   Bell membership function.

7.2.2 Fuzzy Logic

Let A and B be two fuzzy sets of the universe of discourse $\mathbf{U}$, where $\mathbf{U} = \{u_1, u_2, \ldots, u_n\}$, and let μ_A and μ_B be the membership functions of the fuzzy sets A and B, respectively, where

$$A = \{(u_i, \mu_A(u_i)) | u_i \in \mathbf{U}\},$$

$$B = \{(u_i, \mu_B(u_i)) | u_i \in \mathbf{U}\},$$

$\mu_A(u_i) \in [0, 1]$, $\mu_B(u_i) \in [0, 1]$, and $1 \leq i \leq n$. The union of the fuzzy sets A and B, denoted as $A \cup B$, is defined by

$$A \cup B = \{(u_i, \mu_{A \cup B}(u_i)) | \mu_{A \cup B}(u_i) = Max(\mu_A(u_i), \mu_B(u_i)), u_i \in U\}. \qquad \text{(IV.7.8)}$$

The intersection of the fuzzy sets A and B, denoted as $A \cap B$, is defined by

$$A \cap B = \{(u_i, \mu_{A \cap B}(u_i)) | \mu_{A \cap B}(u_i) = Min(\mu_A(u_i), \mu_B(u_i)), u_i \in U\}. \qquad \text{(IV.7.9)}$$

The complement of the fuzzy set A, denoted as A, is defined by

$$\overline{A} = \{(u_i, f_{\overline{A}}(u_i)) | f_{\overline{A}}(u_i) = 1 - f_A(u_i), u_i \in U\}. \qquad \text{(IV.7.10)}$$

7.2.3 Linguistic Variables and Linguistic Terms

One of the major applications of fuzzy sets is in the field of computational linguistics. The goal of computational linguistics is to let the calculation of natural language statements be analogous to the way that logic calculates with logic statements. A linguistic variable is a variable whose values are linguistic terms[22] (also called linguistic values) represented by fuzzy sets. A linguistic term is an expression represented by a word, phrase, or sentence in a natural language term. Table IV.7.1 shows some linguistic variables and typical values that might be assigned to them.

7.2.4 Summary

In this section, we have introduced basic concepts of the fuzzy set theory, including various shapes of membership functions, linguistic variables, and linguistic terms. The next section will introduce some basic concepts for generating fuzzy rules from relational database systems.

TABLE IV.7.1   Linguistic Variables and Linguistic Terms

Linguistic Variables	Linguistic Terms
Height	Short, Average, Tall, Giant
Age	Young , Middle-Age , Old
Number	Almost None, Several, Few, Many
Color	Dark, Bright

7.3   Basic Concepts of Generating Fuzzy Rules from Relational Database Systems

7.3.1   Fuzzification of Relations

In order to generate fuzzy rules from a relational database, the theory of fuzzy sets[21] and the concepts of fuzzification[17] are used. The goal of the fuzzification process is to transfer the input crisp numerical data into fuzzy data and incorporate the imprecision. The attributes of a relation in a relational database system can be considered as linguistic variables.[22] First, we must partition the input domains (i.e., the domains of the attributes) into several fuzzy regions (linguistic terms). This means to define the membership functions of the linguistic terms (fuzzy regions) for each linguistic variable (attribute). Let X be an attribute in a domain V and let $X_1, X_2, \ldots,$ and X_n be its possible linguistic terms, where the membership function curves of the linguistic terms $X_1, X_2, \ldots,$ and X_n are shown in Figure IV.7.6. From Figure IV.7.6, we can see that for every $v \in V$, we can get one or two membership grades corresponding to different linguistic terms. For instance, from Figure IV.7.6, we can get $\mu_{Xi}(v)$ and $\mu_{Xi+1}(v)$, where $\mu_{Xi}(v) + \mu_{Xi+1}(v) = 1$, $\mu_{Xi}(v) \in [0, 1]$, $\mu_{Xi+1}(v) \in [0, 1]$, and $1 \leq i \leq n-1$. We must ensure that this partition satisfies the ε-completeness property,[10,17] where $\varepsilon \in [0, 1]$. In this chapter, we let the value of ε be equal to 0.5. This means that there exists at least one X_i, such that $\mu_{Xi}(v) \geq 0.5$ for every $v \in V$. In the process of constructing fuzzy decision trees, we must transform a crisp value v into a singleton fuzzy set $\{\mu_{Xi}(v)/X_i\}$, where X_i is a linguistic term and $\mu_{Xi}(v) \geq 0.5$. If $\mu_{Xi}(v) = \mu_{Xi+1}(v) = 0.5$, then we transform v into $\{\mu_{Xi}(v)/X_i\}$.

If attribute X is an unfuzzifiable attribute, then we let each domain value of the attribute X be a singleton fuzzy set. For example, Figure IV.7.7 shows a relation in a relational database system, where A, B, and C are attributes. Figure IV.7.8 shows the membership functions of the linguistic terms Low, Medium, and High. Based on Figure IV.7.8, the relation shown in Figure IV.7.7 can be fuzzified into Figure IV.7.9.

7.3.2   Fuzzy Decision Trees

In the following, we introduce the concepts of deriving a fuzzy decision tree from a fuzzy relation. Quinlan[15] pointed out that a decision tree is a recursive structure for the expression of classification rules. He also pointed out that such a tree may be a leaf associated with one class or may consist of a test that has a set of mutually exclusive possible outcomes together with a subsidiary decision tree for each such outcome, where each test is commonly restricted to be a function of an attribute. The concept of fuzzy decision trees is an extension of Quinlan's decision trees.[15] In a fuzzy decision tree, a nonterminal node is also called a decision node. There are two kinds of terminal nodes in a fuzzy decision tree, i.e., certainty factor nodes (CF) denoted by O, and hypothetical certainty factor nodes (HCF) denoted by ⌈⌉. The certainty factor nodes and the hypothetical certainty factor nodes are associated with values between zero and one. The path from the root node to each terminal node

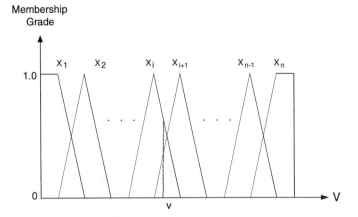

FIGURE IV.7.6   Membership functions of linguistic terms.

A	B	...	C
7	10	...	7
7	2	...	1.4
.	.	.	.
.	.	.	.
.	.	.	.
5	3	...	1.5

FIGURE IV.7.7   A relation in a relational database system.

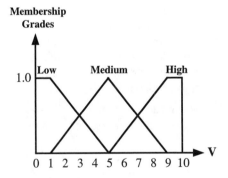

FIGURE IV.7.8   Membership functions of the linguistic terms low, medium, and high.

A	B	...	C
{0.5/Medium}	{1.0/High}	...	{0.5/High}
{0.5/Medium}	{0.75/Low}	...	{0.9/Low}
.	.	.	.
.	.	.	.
.	.	.	.
{1.0/Medium}	{0.5/Medium}	...	{0.88/Low}

FIGURE IV.7.9   Fuzzified relation of Figure IV.7.7.

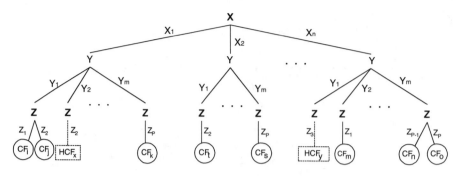

FIGURE IV.7.10   A fuzzy decision tree.

X	**Y**	...	**Z**
$\{c_{11}/X_1\}$	$\{c_{12}/Y_1\}$	⋮	$\{c_{1n}/Z_1\}$
$\{c_{21}/X_1\}$	$\{c_{22}/Y_1\}$	...	$\{c_{2n}/Z_2\}$
⋮	⋮	...	⋮
$\{c_{m1}/X_1\}$	$\{c_{m2}/Y_m\}$	...	$\{c_{mn}/Z_p\}$
⋮	⋮	⋮	⋮

FIGURE IV.7.11   An example of a fuzzy relation.

(i.e., certainty factor node or hypothetical certainty factor node) forms a fuzzy rule. For example, Figure IV.7.10 shows an example of a fuzzy decision tree, where **X**, **Y**, and **Z** are attributes in a relational database; X_i, Y_j, and $Z_k (1 \leq i \leq n, 1 \leq j \leq m,$ and $1 \leq k \leq p$) are linguistic terms, respectively. From Figure IV.7.10, we can see that the value of attribute **Z** is determined by the values of attributes **X** and **Y**. In Figure IV.7.10, the certainty factor node CF_i indicates that there are some tuples in the database shown in Figure IV.7.11 which satisfy the classification in a degree denoted by CF_i, where $CF_i \in [0, 1]$. A hypothetical certainty factor node HCF_x exists due to the fact that there are no tuples which satisfy the classification and which are generated by the tree growing process to be described in Section 7.4. Let us consider the path $\mathbf{X} \xrightarrow{X_1} \mathbf{Y} \xrightarrow{Y_1} \mathbf{Y} \xrightarrow{Z_2} CF_j$ in the fuzzy decision tree shown in Figure IV.7.10. This indicates that there is a fuzzy rule.

$$\text{IF } \mathbf{X} \text{ is } X_1 \text{ and } \mathbf{Y} \text{ is } Y_1 \text{ THEN } \mathbf{Z} \text{ is } Z_2 \quad (CF = CF_j) \tag{IV.7.11}$$

in the fuzzy rule base. A null path is a path whose terminal node is a hypothetical certainty factor node. For instance, from Figure IV.7.10, we can see that the path $\mathbf{X} \xrightarrow{X_1} \mathbf{Y} \xrightarrow{Y_2} \mathbf{Z} \xrightarrow{Z_2} HCF_x$ is a null path. A non-null path is a path whose terminal node is a certainty factor node. For instance, from Figure IV.7.10, we can see that the path $\mathbf{X} \xrightarrow{X_1} \mathbf{Y} \xrightarrow{Y_1} \mathbf{Z} \xrightarrow{Z_1} CF_i$ is a non-null path.

7.3.3   Fuzzy Rules

In the following paragraphs, we introduce the concepts of generating fuzzy rules from a constructed fuzzy decision tree. A fuzzy rule has the following form:

$$\text{IF } \mathbf{X} \text{ is } X_i \text{ and } \mathbf{Y} \text{ is } Y_j \text{ THEN } \mathbf{Z} \text{ is } Z_k \quad (CF = c) \tag{IV.7.12}$$

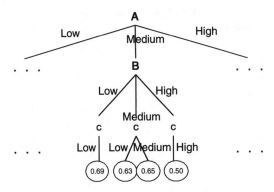

FIGURE IV.7.12   A subtree of a fuzzy decision tree.

where

1. **X**, **Y**, and **Z** are linguistic variables; X_i, Y_j, Z_k are linguistic terms represented by fuzzy sets.
2. CF is the certainty factor of the fuzzy rule indicating the degree of belief of the rule, where $c \in [0, 1]$. The larger the value of c, the more the rule is believed.

After constructing a fuzzy decision tree, we can generate fuzzy rules from the constructed fuzzy decision tree. In a fuzzy decision tree, the path from the root node to each terminal node (certainty factor node or a hypothetical certainty factor node) forms a fuzzy rule. Thus, if we have a fuzzy decision tree as shown in Figure IV.7.12, we can generate the following fuzzy rules:

$$\vdots$$

IF A is Medium and **B** is High	**THEN C** is High	(CF = 0.50)
IF A is Medium and **B** is Medium	**THEN C** is Medium	(CF = 0.65)
IF A is Medium and **B** is Medium	**THEN C** is Low	(CF = 0.63)
IF A is Medium and **B** is Low	**THEN C** is Low	(CF = 0.69)

$$\vdots$$

where the attributes **A** and **B** are called antecedent attributes, respectively, and the attribute **C** is called a consequent attribute.

7.3.4 Summary

In this section, we have briefly introduced the basic concepts of the generation of fuzzy rules from relational database systems. In the next section, we will present an algorithm to construct a fuzzy decision tree from a relation of a relational database system and to generate fuzzy rules from the constructed fuzzy decision tree.

7.4   Generating Fuzzy Rules from Relational Database Systems for Estimating Null Values

In 1966, Hunt, Marin, and Stone[9] presented a learning algorithm called the concept learning system (CLS).[9] It is intended to solve single-concept learning tasks and to use the learned concepts to classify new instances. ID3 is a more recent version of the CLS developed by Quinlan[14] and can efficiently construct a decision tree. In this section, we present an algorithm called fuzzy concept learning system algorithm (FCLS)[20] to construct a fuzzy decision tree from a relation of a relational database system and to generate fuzzy rules from the constructed fuzzy decision tree for estimating null values in a relational database system.

The proposed FCLS algorithm uses a set of training instances (i.e., tuples of a relation) to learn or generate some fuzzy rules and uses these generated fuzzy rules to estimate the null values in a relational database system. If the training instances to be learned do not contain all kinds of conditions, then several null paths will be generated in the constructed fuzzy decision tree. Sudkamp and Hammell[16] presented a region growing method and a weighted average method to complete the entries of a fuzzy associative memory and make an experiment for the approximation of real-value functions. In this section, we present a method to complete the null paths in a constructed fuzzy decision tree for generating fuzzy rules and to estimate null values in a relational database system based on the generated fuzzy rules.

7.4.1   A Fuzzy Concept Learning System Algorithm (FCLS)

Let S be a set of attributes (i.e., $S = \{X, Y, \ldots, W\}$) determining attribute Z, and let Z be a consequent attribute. To select different elements in the set S as a decision node may construct different kinds of fuzzy decision trees. In the ID3 learning algorithm,[13] Quinlan used an entropy function of information theory to choose decision nodes. In this article, the concept of fuzziness of attribute (FA)[20] is used to select an attribute in the set S as a decision node which has the smallest value of FA.

Definition 2: Let X be an attribute of S and let $t_j(X)$ be the value of the attribute X of the jth training instance (i.e., the jth tuple of a relation) in a relational database. The degree of fuzziness $FA(X)$ of the attribute X is defined as follows:

$$FA(X) = \frac{\sum_{j=1}^{c}(1 - \mu_{Xi}(t_j(X)))}{c}, \tag{IV.7.13}$$

where c is the number of training instances.

Definition 3: Each certainty factor node in a path of a fuzzy decision tree is associated with a certainty factor value (CF), where the certainty factor value CF is defined as follows:

$$CF = \min\{Avg(F_1), Avg(F_2), Avg(F_3)\} \tag{IV.7.14}$$

where $Avg(F_1)$, $Avg(F_2)$, and $Avg(F_3)$ are the average values of the linguistic terms F_1, F_2, and F_3, respectively, and F_1, F_2, and F_3 are on the path

$$D_1 \xrightarrow{F_1} D_2 \xrightarrow{F_2} D_3 \xrightarrow{F_3} \text{(CF)}$$

in the fuzzy decision tree, and

$$Avg(F_i) = \frac{\sum_{j=1}^{s} \mu_{Fi}(t_j(D_i))}{s}, \tag{IV.7.15}$$

where $t_j(D_i)$ denotes the attribute D_i of the jth tuple of a relation, s denotes the number of training instances (i.e., the number of tuples in the relation) in which the value of attribute D_i is the linguistic term F_i, and $1 \leq i \leq 3$.

First, we fuzzify a relation in a relational database system into a fuzzy relation as described in Section 7.3. Let S be a set of antecedent attributes and let Z be a consequent attribute of a relation in a relational database system. Furthermore, let T be a set of training instances (i.e., a set of tuples in a relation of a relational database system). In a relational database system, a tuple in a relation forms a training instance. The FCLS algorithm[20] is presented as follows:

Step 1: Fuzzify the relation into the fuzzy relation.

Step 2: Select an attribute among the set **S** of antecedent attributes which has the smallest FA. Assume attribute **X** with the smallest FA, then partition the set **T** of the training instances into subsets $T_1, T_2, \ldots, T_n$ according to the linguistic terms $X_1, X_2, \ldots, X_n$ of the attribute **X**, respectively. Compute the average value $\text{Avg}(X_i)$ of X_i based on Equation IV.7.15, where $1 \leq i \leq n$.

Step 3: Let the attribute **X** be the decision node, and sprout the tree according to the linguistic terms of the attribute **X** shown as follows:

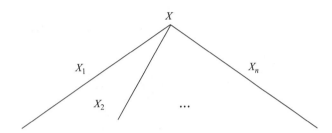

where $X_1, X_2, \ldots, X_n$ are linguistic terms of the attribute **X** represented by fuzzy sets.

Step 4: Let $\mathbf{S} = \mathbf{S} - \{\mathbf{X}\}$, where "$-$" is the set difference operator.

Step 5: For $i = 1$ to n do

    {

        Let $\mathbf{T} \leftarrow T_i$;

        If $\mathbf{S} = \phi$ then

          {

              create a decision node for consequent attribute **Z**;

              partition the training instances **T** into $T_1, T_2, \ldots, T_k$ according to

                  the linguistic terms $Z_1, Z_2, \ldots, Z_k$ of the attribute **Z**;

              compute the average value $\text{Avg}(Z_i)$ of Z_i, where $1 \leq i \leq k$;

              create a terminal node for every T_i with $\text{Avg}(Z_i) \neq 0$ and compute

                  the value CF_i associated with the created certainty factor node

                  for every non-null path in the tree

          }

        else

          go to Step 2

    }.

In the following, we use an example to illustrate the fuzzy rules generation process.

Example 1: Assume that there is a relation in a relational database as shown in Table IV.7.2.

From Table IV.7.2, we can see that the attribute Salary is determined by the attributes Degree and Experience. In this case, we can see that $\mathbf{S} = \{\text{Degree, Experience}\}$ and $\mathbf{Z} = \text{Salary}$, where the attributes Degree and Experience are called the antecedent attributes, and the attribute Salary is

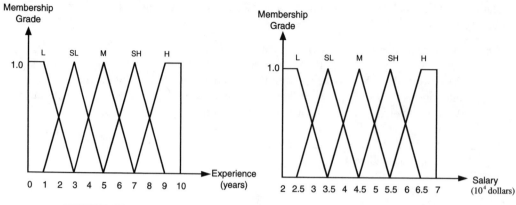

FIGURE IV.7.13 Membership functions of the attributes Experience and Salary.

TABLE IV.7.2 A Relation in a Relational Database System

Emp-ID	Degree	Experience	Salary
E1	Ph.D.	8.5	70000
E2	Master	2.0	37000
E3	Bachelor	7.0	40000
E4	Ph.D.	1.2	47000
E5	Master	7.5	53000
E6	Bachelor	1.5	26000
E7	Bachelor	2.3	29000
E8	Ph.D.	2.0	50000
E9	Ph.D.	3.8	54000
E10	Bachelor	3.5	35000
E11	Master	3.5	40000
E12	Master	3.6	41000
E13	Master	10	68000
E14	Ph.D.	5.0	57000
E15	Bachelor	5.0	36000
E16	Master	6.2	50000
E17	Bachelor	0.5	23000
E18	Master	7.2	55000
E19	Master	6.5	51000
E20	Ph.D.	7.8	65000
E21	Master	8.1	64000
E22	Ph.D.	7.2	63000

called the consequent attribute. From Table IV.7.2, we can see that the values of the attribute Degree are Ph.D., Master, and Bachelor, and the domains of the attributes Experience and Salary are from 0 to 10 and from 20000 to 70000, respectively. In this example, we let the fuzzy domain of the attribute Degree be Ph.D., Master, and Bachelor, and we let the fuzzy domain of the attributes Experience and Salary be high (H), somewhat-high (SH), medium (M), somewhat-low (SL), and low (L), respectively, where the membership functions of these linguistic terms are shown in Figure IV.7.13.

First, we fuzzify the relation shown in Table IV.7.2, where the result of fuzzification of Table IV.7.2 is a fuzzy relation shown in Table IV.7.3.

TABLE IV.7.3   A Fuzzy Relation

Emp-ID	Degree	Experience	Salary
E1	{1.0/Ph.D.}	{0.75/H}	{1.0/H}
E2	{1.0/Master}	{0.5/SL}	{0.8/SL}
E3	{1.0/Bachelor}	{1.0/SH}	{0.5/M}
E4	{1.0/Ph.D.}	{0.9/L}	{0.8/M}
E5	{1.0/Master}	{0.75/SH}	{0.8/SH}
E6	{1.0/Bachelor}	{0.75/L}	{0.9/L}
E7	{1.0/Bachelor}	{0.65/SL}	{0.6/L}
E8	{1.0/Ph.D.}	{0.5/L}	{0.5/SH}
E9	{1.0/Ph.D.}	{0.6/SL}	{0.9/SH}
E10	{1.0/Bachelor}	{0.75/SL}	{1.0/SL}
E11	{1.0/Master}	{0.75/SL}	{0.5/SL}
E12	{1.0/Master}	{0.7/SL}	{0.6/M}
E13	{1.0/Master}	{1.0/H}	{1.0/H}
E14	{1.0/Ph.D.}	{1.0/M}	{0.8/SH}
E15	{1.0/Bachelor}	{1.0/M}	{0.9/SL}
E16	{1.0/Master}	{0.6/SH}	{0.5/M}
E17	{1.0/Bachelor}	{1.0/L}	{1.0/L}
E18	{1.0/Master}	{0.9/SH}	{1.0/SH}
E19	{1.0/Master}	{0.75/SH}	{0.6/SH}
E20	{1.0/Ph.D.}	{0.6/SH}	{1.0/H}
E21	{1.0/Master}	{0.55/H}	{0.9/H}
E22	{1.0/Ph.D.}	{0.9/SH}	{0.8/H}

Then, based on Equation IV.7.13, we can compute the fuzziness of each attribute in the set S, S = {Degree, Experience}, shown as follows:

$$FA(Degree) = 0,$$

$$FA(Experience) = [(1 - 0.75) + (1 - 0.5) + (1 - 1.0) + (1 - 0.9) + (1 - 0.75) + (1 - 0.75)$$
$$+ (1 - 0.65) + (1 - 0.5) + (1 - 0.6) + (1 - 0.75) + (1 - 0.75) + (1 - 0.7)$$
$$+ (1 - 1.0) + (1 - 1.0) + (1 - 1.0) + (1 - 0.6) + (1 - 1.0) + (1 - 0.9)$$
$$+ (1 - 0.75) + (1 - 0.6) + (1 - 0.55) + (1 - 0.9)]/22$$
$$= 0.23.$$

After applying the FCLS algorithm, a fuzzy decision tree is constructed as shown in Figure IV.7.14. Finally, from Figure IV.7.14, we can get 16 fuzzy rules, shown as follows:

Rule 1: **IF** Degree is Ph.D. **AND** Experience is L **THEN** Salary is M (CF = 0.70)
Rule 2: **IF** Degree is Ph.D. **AND** Experience is SL **THEN** Salary is SH (CF = 0.60)
Rule 3: **IF** Degree is Ph.D. **AND** Experience is M **THEN** Salary is SH (CF = 0.80)
Rule 4: **IF** Degree is Ph.D. **AND** Experience is SH **THEN** Salary is H (CF = 0.75)
Rule 5: **IF** Degree is Ph.D. **AND** Experience is H **THEN** Salary is H (CF = 0.75)
Rule 6: **IF** Degree is Master **AND** Experience is L **THEN** Salary is SL (CF = 0.50)
Rule 7: **IF** Degree is Master **AND** Experience is SL **THEN** Salary is SL (CF = 0.50)
Rule 8: **IF** Degree is Master **AND** Experience is SL **THEN** Salary is M (CF = 0.60)

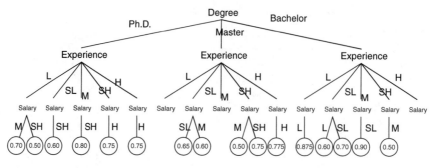

FIGURE IV.7.14   Fuzzy decision tree of Example IV.7.1.

Rule 9: **IF** Degree is Master **AND** Experience is SH **THEN** Salary is M (CF = 0.50)
Rule 10: **IF** Degree is Master **AND** Experience is SH **THEN** Salary is SH (CF = 0.75)
Rule 11: **IF** Degree is Master **AND** Experience is H **THEN** Salary is H (CF = 0.775)
Rule 12: **IF** Degree is Bachelor **AND** Experience is L **THEN** Salary is L (CF = 0.875)
Rule 13: **IF** Degree is Bachelor **AND** Experience is SL **THEN** Salary is L (CF = 0.60)
Rule 14: **IF** Degree is Bachelor **AND** Experience is SL **THEN** Salary is SL (CF = 0.70)
Rule 15: **IF** Degree is Bachelor **AND** Experience is M **THEN** Salary is SL (CF = 0.90)
Rule 16: **IF** Degree is Bachelor **AND** Experience is SH **THEN** Salary is SL (CF = 0.50)

7.4.2   Completeness of Fuzzy Decision Trees

In the following, we present a method to deal with the completeness of the constructed fuzzy decision tree. Let α be a mapping function from fuzzy terms to ordinary numbers, and let β be a mapping function from ordinary numbers to linguistic terms. For example, in Figure IV.7.13, we let

$$\alpha(L) = 1, \quad \beta(1) = L,$$
$$\alpha(SL) = 2, \quad \beta(2) = SL,$$
$$\alpha(M) = 3, \quad \beta(3) = M,$$
$$\alpha(SH) = 4, \quad \beta(4) = SH,$$
$$\alpha(H) = 5, \quad \beta(5) = H.$$

For the fuzzy decision tree constructed by the proposed FCLS algorithm, if there are some null paths, then a hypothetical certainty factor node HCF is created for each null path. For example, Figure IV.7.15 shows a subtree of a fuzzy decision tree, where Y_{i-1}, Y_i, and Y_{i+1} are fuzzy values in the fuzzy domain of **Y**; Y_{i-1} and Y_{i+1} are on the non-null path with the rightmost value Z_L and the leftmost value Z_R of **Z**, where Z_L and Z_R are linguistic terms, where Y_i is on the null path of the tree. Assume that we want to sprout the branch Z_U denoted by the dotted line, as shown in Figure IV.7.15, where Z_U is a linguistic term. Then,

- **Case 1:** Assume that we want to sprout the branch Z_U denoted by the dotted line shown in Figure IV.7.16, then Z_U can be evaluated as follows:

$$Z_U = \begin{cases} \beta(\lfloor \frac{\alpha(Z_L)+\alpha(Z_R)-1}{2} \rfloor), & \text{if } |\alpha(Z_K) - \alpha(Z_R)| \text{ is an odd number and } CF_L \geq CF_R \\ \beta(\lfloor \frac{\alpha(Z_L)+\alpha(Z_R)+1}{2} \rfloor), & \text{if } |\alpha(Z_K) - \alpha(Z_R)| \text{ is an odd number and } CF_L \geq CF_R \\ \beta(\lfloor \frac{\alpha(Z_L)+\alpha(Z_R)}{2} \rfloor), & \text{otherwise} \end{cases}$$

(IV.7.16)

and let the associated value of the hypothetical certainty node HCF$_U$ be equal to 0.5.

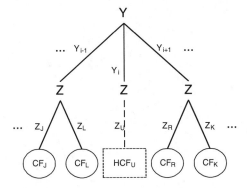

FIGURE IV.7.15   A subtree of a fuzzy decision tree.

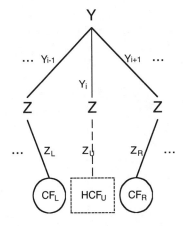

FIGURE IV.7.16   A subtree of a fuzzy decision tree (Case 1).

- **Case 2:** Assume that we want to sprout the branch Z_U denoted by the dotted line shown in Figure IV.7.17, where the node **Z** of Y_1 to Y_{i-1} cannot sprout out any branches, and assume that we have the following metaknowledge: *the smaller the values of the attribute* **Y**, *the smaller the values of the attribute* **Z**. Then, Z_U can be evaluated as follows:

$$Z_U = \beta \left(\left\lfloor \frac{1 + \alpha(Z_R)}{2} \right\rfloor \right). \tag{IV.7.17}$$

Assume that we have the following metaknowledge: *the smaller the values of the attribute* **Y**, *the greater the values of the attribute* **Z**. Then, Z_U can be evaluated as follows:

$$Z_U = \beta \left(\left\lfloor \frac{\alpha(Z_k) + \alpha(Z_K)}{2} \right\rfloor \right), \tag{IV.7.18}$$

where $\alpha(Z_k)$ has the largest ordinary number in the fuzzy domain of **Z**, and let the value of the hypothetical certainty node HCF_U be equal to 0.5.

- **Case 3:** Assume that we want to sprout the branch Z_U denoted by the dotted line shown in Figure IV.7.18, where the decision node Z of Y_{i+1} to Y_m cannot sprout out any branches, and

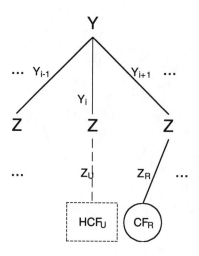

FIGURE IV.7.17   A subtree of a fuzzy decision tree (Case 2).

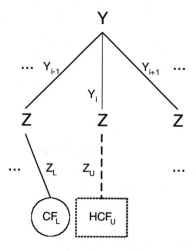

FIGURE IV.7.18   A subtree of a fuzzy decision tree (Case 3).

assume that we have the following metaknowledge: *the greater the values of the attribute Y, the smaller the values of the attribute Z.* Then, Z_U can be evaluated as follows:

$$Z_U = \beta \left(\left\lfloor \frac{1 + \alpha(Z_L)}{2} \right\rfloor \right).$$
                                                                (IV.7.19)

Assume that we have the following metaknowledge: *the greater the values of the attribute Y, the greater the values of the attribute Z.* Then, Z_U can be evaluated as follows:

$$Z_U = \beta \left(\left\lfloor \frac{\alpha(Z_K) + \alpha(Z_K)}{2} \right\rfloor \right),$$
                                                                (IV.7.20)

where $\alpha(Z_k)$ has the largest ordinary number in the fuzzy domain of Z, and let the value of the hypothetical certainty node HCF_U be equal to 0.5.

This process will go on continuously until there is no null path in the constructed fuzzy decision tree.

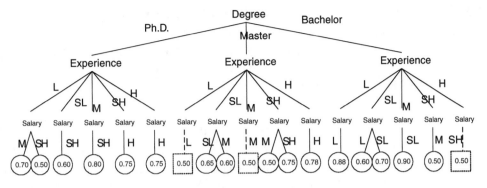

FIGURE IV.7.19   Complete fuzzy decision tree of Example IV.7.2.

Example 2: Assume the same conditions as those shown in Example 1, where Figure IV.7.14 is the constructed fuzzy decision tree of Example 1. After performing the proposed FCLS algorithm, the complete fuzzy decision tree is constructed as shown in Figure IV.7.19. It is obvious that Figure IV.7.19 shows a complete fuzzy decision tree derived from Figure IV.7.14 after performing the proposed method.

For example, for the path Degree $\xrightarrow{\text{Master}}$ Experience $\xrightarrow{M}$ Salary, shown in Figure IV.7.19, based on Case 1 of the proposed method, we can see that

$$Y_i = M, Y_{i-1} = SL, Y_{i+1} = SH, Z_L = M, \text{ and } Z_R = M$$

Because $|\alpha(Z_L) - \alpha(Z_R)| = 0$, we can see that

$$Z_U = \beta\left(\left\lfloor \left| \frac{\alpha(Z_L) + \alpha(Z_R)}{2} \right| \right\rfloor\right)$$

$$= \beta\left(\left\lfloor \left| \frac{\alpha(M) + \alpha(M)}{2} \right| \right\rfloor\right)$$

$$= \beta\left(\left\lfloor \left| \frac{3+3}{2} \right| \right\rfloor\right)$$

$$= \beta(3)$$

$$= M.$$

Thus, we can get a virtual fuzzy rule shown as follows:

$$\text{Degree} \xrightarrow{\text{Master}} \text{Experience} \xrightarrow{M} \text{Salary} \xrightarrow{M} \boxed{0.50}$$

From Figure IV.7.19, we can see that there are three null-paths in the constructed fuzzy decision tree, i.e.,

$$\text{Degree} \xrightarrow{\text{Master}} \text{Experience} \xrightarrow{M} \text{Salary} \xrightarrow{M} \boxed{0.50}$$

$$\text{Degree} \xrightarrow{\text{Bachelor}} \text{Experience} \xrightarrow{H} \text{Salary} \xrightarrow{SH} \boxed{0.50}$$

Thus, there are two virtual fuzzy rules to be generated from the constructed fuzzy decision tree, shown as follows:

Rule 17: **IF** Degree is Master **AND** Experience is M **THEN** Salary is M (CF = 0.50).
Rule 18: **IF** Degree is Bachelor **AND** Experience is H **THEN** Salary is SH (CF = 0.50).

7.4.3 Estimating Null Values in Relational Database Systems

In the following, we present a method to estimate null values in relational database systems based on the generated fuzzy rules. First, we briefly review the defuzzification method from Chen[2] and Kandel.[10] Chen[2] presented a defuzzification technique of trapezoidal fuzzy sets based on Kandel,[10] as shown in Figure IV.7.20, where e is the defuzzification value of the trapezoidal fuzzy set A and

$$e = (a + b + c + d)/4 \qquad\qquad (IV.7.21)$$

It is obvious that a triangular fuzzy set also can be considered as a special case of a trapezoidal fuzzy set, where $b = c$. Thus, the defuzzification value DEF(A) of the triangular fuzzy set A shown in Figure IV.7.21 is

$$DEF(A) = e = (a + b + b + d)/4. \qquad\qquad (IV.7.22)$$

If v is a crisp value of attribute **V** in some tuples of a relational database, then we let $\{\mu_{v_s}(v)/V_s, \mu_{v_t}(v)/V_t\}$ be the fuzzified value of v, where V_s and V_t are linguistic terms represented by fuzzy sets, $\mu_{v_s}(v) \geq \mu_{v_t}(v)$ and $\mu_{v_s}(v) + \mu_{v_t}(v) = 1.0$. If v is a crisp value of an unfuzzifiable attribute V in a relational database, then we let $\{1.0/v\}$ be the fuzzified value of v. In order to forecast

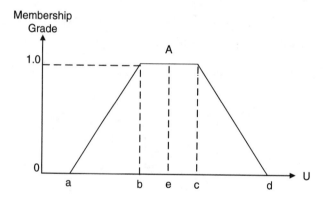

FIGURE IV.7.20   Defuzzification of a trapezoidal fuzzy set.

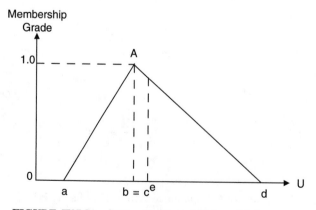

FIGURE IV.7.21   Defuzzification of a triangular fuzzy set.

a null value in a relational database system, first, we must modify the fuzzified value of v into the form $F_v = \{\mu_{v_m}(v)/V_m, \mu_{v_n}(v)/V_n\}$, where V_m and V_n are linguistic terms represented by fuzzy sets.

- **Case 1:** If **V** is a fuzzifiable attribute (linguistic variable) and $\mu_{v_s}(v) \neq 1.0$, then we let

$$V_m = V_s,$$
$$V_n = V_t,$$
$$\mu_{v_m}(v) = \mu_{v_s}(v),$$
$$\mu_{v_n}(v) = \mu_{v_t}(v).$$

- **Case 2:** If **V** is a fuzzifiable attribute (linguistic variable) and $\mu_{v_s}(v) = 1.0$, then we let

$$V_m = V_n = V_s,$$
$$\mu_{v_m}(v) = \mu_{v_n}(v) = \mu_{v_s}(v) = 1.0.$$

- **Case 3:** If **V** is an unfuzzifiable attribute, then we let

$$V_m = V_n = v,$$
$$\mu_{v_m}(v) = \mu_{v_n}(v) = 1.0.$$

Assume that x and y are crisp domain values of attributes **X** and **Y** in some tuples of a relational database, respectively, and z is a null value of attribute **Z**, then let

$$F_x = \{\mu_{X_A}(x)/X_a, \mu_{X_B}(x)/X_b\} \quad \text{and} \quad F_y = \{\mu_{Y_C}(y)/Y_C, \mu_{Y_d}(y)/Y_d\}$$

be the modified forms of the fuzzified values of x and y, respectively. Assume that the fuzzy rule base contains the following fuzzy rules generated by the proposed FCLS algorithm:

IF **X** is X_a and **Y** is Y_c THEN **Z** is Z_{M_1} $(CF = C_1)$
IF **X** is X_a and **Y** is Y_c THEN **Z** is Z_{M_2} $(CF = C_2)$
IF **X** is X_b and **Y** is Y_d THEN **Z** is Z_{N_1} $(CF = D_1)$
IF **X** is X_b and **Y** is Y_d THEN **Z** is Z_{N_2} $(CF = D_2)$

where **X** and **Y** are antecedent attributes; **Z** is a consequent attribute; X_a, X_b, Y_c, Y_d, Z_{M_1}, Z_{M_2}, Z_{N_1}, and Z_{N_2} are linguistic terms represented by fuzzy sets. Then, the null value z can be evaluated as follows:

$$z = \frac{\mu_{X_a}(x) \times \mu_{Y_c}(y) \times \frac{\sum_{i=1}^{2} C_i * DEF(Z_{M_i})}{\sum_{i=1}^{2} C_i} + \mu_{X_b}(x) \times \mu_{Y_d}(y) \times \frac{\sum_{i=1}^{2} D_i * DEF(Z_{N_i})}{\sum_{i=1}^{2} D_i}}{\mu_{X_a}(x) \times \mu_{Y_c}(y) + \mu_{X_b}(x) \times \mu_{Y_d}(y)}$$

(IV.7.23)

where $DEF(Z_{M_i})$ and $DEF(Z_{N_i})$ are the defuzzified values of the fuzzy sets Z_{M_i} and Z_{N_i}, respectively.

Example 3: Assume that there is a relation in a relational database system containing a null value in the attribute Salary as shown in Table IV.7.4, and assume that we want to forecast the null value of the attribute Salary.

TABLE IV.7.4   A Relation Contains Null Values

Emp-ID	Degree	Experience	Salary
E1	Ph.D.	8.5	70000
E2	Master	2.0	37000
E3	Bachelor	7.0	40000
E4	Ph.D.	1.2	47000
E5	Master	7.5	53000
E6	Bachelor	1.5	26000
E7	Bachelor	2.3	29000
E8	Ph.D.	2.0	50000
E9	Ph.D.	3.8	54000
E10	Bachelor	3.5	35000
E11	Master	3.5	40000
E12	Master	3.6	41000
E13	Master	10	68000
E14	Ph.D.	5.0	57000
E15	Bachelor	5.0	36000
E16	Master	6.2	50000
E17	Bachelor	0.5	23000
E18	Master	7.2	55000
E19	Master	6.5	51000
E20	Ph.D.	7.8	65000
E21	Master	8.1	64000
E22	Ph.D.	7.2	63000
E23	Master	4.5	Null

From Table IV.7.4, we can see that the tuple with Emp-ID = E23 has a null value in the attribute Salary. Based on the membership functions shown in Figure IV.7.13 and after performing the fuzzification process, we can see that Table IV.7.4 can be fuzzified into Table IV.7.5.

Thus, we can see that $F_{Master} = \{1.0/Master, 1.0/Master\}$ and $F_{4.5} = \{0.75/M, 0.25/SL\}$. After executing the proposed FCLS algorithm and according to the generated fuzzy rules, Rule 8 and Rule 17 shown in Examples 1 and 2, the null value of the attribute Salary can be estimated, where Rule 8 and Rule 17 are shown as follows:

Rule 8: **IF** Degree is Master **AND** Experience is SL **THEN** Salary is M ($CF = 0.60$).
Rule 17: **IF** Degree is Master **AND** Experience is M **THEN** Salary is M ($CF = 0.50$).

Based on Equation IV.7.23, Table IV.7.5, Rule 8, and Rule 17, the null value z of the attribute Salary of the employee whose Emp-ID is E23 can be estimated as follows:

$$z = \frac{1 \times 0.75 \times \frac{0.5*DEF(M)}{0.5} + 1 \times 0.25 \times \frac{0.6*DEF(M)}{0.6}}{1 \times 0.75 + 1 \times 0.25}$$

$$= \frac{1 \times 0.75 \times 45000 + 1 \times 0.25 \times \frac{0.6*45000}{0.6}}{1 \times 0.75 + 1 \times 0.25}$$

$$= 45000.$$

That is, the salary of the employee whose Emp-ID = E23 is about 45000.

TABLE IV.7.5   A Fuzzified Relation Contains Null Values

Emp-ID	Degree	Experience	Salary
E1	{1.0/Ph.D.}	{0.9/SH}	{0.8/H}
E2	{1.0/Master}	{0.5/L}	{0.8/SL}
E3	{1.0/Bachelor}	{1.0/SH}	{0.5/SL}
E4	{1.0/Ph.D.}	{0.9/L}	{0.8/M}
E5	{1.0/Master}	{0.75/SH}	{0.8/SH}
E6	{1.0/Bachelor}	{0.75/L}	{0.9/L}
E7	{1.0/Bachelor}	{0.65/SL}	{0.6/L}
E8	{1.0/Ph.D.}	{0.5/L}	{0.5/M}
E9	{1.0/Ph.D.}	{0.6/SL}	{0.9/SH}
E10	{1.0/Bachelor}	{0.75/SL}	{1.0/SL}
E11	{1.0/Master}	{0.75/SL}	{0.5/SL}
E12	{1.0/Master}	{0.7/SL}	{0.6/M}
E13	{1.0/Master}	{1.0/H}	{1.0/H}
E14	{1.0/Ph.D.}	{1.0/M}	{0.8/SH}
E15	{1.0/Bachelor}	{1.0/M}	{0.9/SL}
E16	{1.0/Master}	{0.6/SH}	{0.5/M}
E17	{1.0/Bachelor}	{1.0/L}	{1.0/L}
E18	{1.0/Master}	{0.9/SH}	{1.0/SH}
E19	{1.0/Master}	{0.75/SH}	{0.6/SH}
E20	{1.0/Ph.D.}	{0.6/SH}	{1.0/H}
E21	{1.0/Master}	{0.55/H}	{0.9/H}
E22	{1.0/Ph.D.}	{0.75/H}	{1.0/H}
E23	{1.0/Master, 1.0/Master}	{0.75/M, 0.25/SL}	Null

In the same way, after applying the proposed FCLS algorithm to the relation shown in Table IV.7.2, the estimated salaries of the employees and the estimated errors are shown in Table IV.7.6, where the estimated error is calculated as follows:

$$\text{Estimated Error} = \frac{\text{Estimated Value} - \text{Original Value}}{\text{Original Value}}. \tag{IV.7.24}$$

From Table IV.7.6, we can see that the average estimated error of the example is 0.0508. That is, the average forecasting accuracy rate of the example is 94.92%.

7.5   Conclusions

In this chapter, we have presented a method for constructing fuzzy decision trees from relational databases and generating fuzzy rules from the constructed decision trees for estimating null values. The proposed method can properly deal with the completeness of the constructed fuzzy decision trees. It can get a good forecasting result for estimating null values in relational database systems.

TABLE IV.7.6 An Estimated Relation Compared to the Original Relation shown in Table IV.7.2

Emp-ID	Degree	Experience	Salary (Original)	Salary (Estimated)	Estimated Error
E1	Ph.D.	8.5	70000	65000	−0.0714
E2	Master	2.0	37000	30704	−0.1702
E3	Bachelor	7.0	40000	35000	−0.1250
E4	Ph.D.	1.2	47000	46000	−0.0213
E5	Master	7.5	53000	54500	0.0283
E6	Bachelor	1.5	26000	26346	0.0133
E7	Bachelor	2.3	29000	28500	−0.0172
E8	Ph.D.	2.0	50000	50000	0.0000
E9	Ph.D.	3.8	54000	55000	0.0185
E10	Bachelor	3.5	35000	31538	−0.0989
E11	Master	3.5	40000	41590	0.0398
E12	Master	3.6	41000	45159	0.1014
E13	Master	10	68000	65000	−0.0441
E14	Ph.D.	5.0	57000	55000	−0.0351
E15	Bachelor	5.0	36000	35000	−0.0278
E16	Master	6.2	50000	48600	−0.0280
E17	Bachelor	0.5	23000	25000	0.0870
E18	Master	7.2	55000	52400	−0.0473
E19	Master	6.5	51000	49500	−0.0294
E20	Ph.D.	7.8	65000	65000	0.0000
E21	Master	8.1	64000	58700	−0.0828
E22	Ph.D.	7.2	63000	65000	0.0317
Average Estimated Error					0.0508

References

1. Berenji, H.R. and Khedkar, P., Learning and tuning fuzzy logic controllers through reinforcements, *IEEE Trans. Neural Networks*, 3(5), 724, 1992.
2. Chen, S.M., Using fuzzy reasoning techniques for fault diagnosis of the J-85 jet engines, Proceedings of the Third National Conference on Science and Technology of National Defence, Taoyuan, Taiwan, Republic of China, October, 29, 1994.
3. Chen, S.M. and Yeh, M.S., Generating fuzzy rules from relational database systems for estimating null values, *Cybernetics Syst. Int. J.*, 28(8), 695, 1997.
4. Chen, S.M. and Lin, S.Y., A new method for constructing fuzzy decision trees and generating fuzzy classification rules from training examples, *Cybernetics Syst. Int. J.*, 31(7), 763, 2000.
5. Chen, S.M., Lee, S.H., and Lee, C.H., Generating fuzzy rules from numerical data for handling fuzzy classification problems, Proceedings of the 1999 National Computer Symposium, Taipei, Taiwan, Republic of China, December, 336, 1999.
6. Chen, Y.J. and Chen, S.M., A new method to generate fuzzy rules for fuzzy classification systems, Proceedings of the 2000 Eighth National Conference on Fuzzy Theory and Its Applications, Taipei, Taiwan, Republic of China, December 2000.

7. Chang, C.H. and Chen, S.M., A new method to generate fuzzy rules from numerical data based on the exclusion of attribute terms, Proceedings of the 2000 International Computer Symposium: Workshop on Artificial Intelligence, Chiayi, Taiwan, Republic of China, December, 57, 2000.

8. Hart, A., The role of induction in knowledge elicitation, *Expert Syst.*, 2(1), 24, 1985.

9. Hunt, E.B., Marin, J., and Stone, P.J., *Experience in Induction*, Academic Press, New York, 1966.

10. Kandel, A., *Fuzzy Mathematical Techniques with Applications*, Addison-Wesley, Reading, MA, 1986.

11. Kao, C.H. and Chen, S.M., A new method to generate fuzzy rules from training data containing noise for handling classification problems, Proceedings of the Fifth Conference on Artificial Intelligence and Applications, Taipei, Taiwan, Republic of China, November, 324, 2000.

12. Kaufmann, A. and Gupta, M.M., *Fuzzy Mathematical Models in Engineering and Management Science*, North-Holland, Amsterdam, 1988.

13. Quinlan, J.R., Discovering rules by induction from large collection of examples, in *Expert Systems in the Micro Electronic Age*, Michie, D., Ed., Edinburgh University Press, Edinburgh, 1979.

14. Quinlan, J.R., Induction of decision trees, *Machine Learning*, 1(1) 81, 1986.

15. Quinlan, J.R., Decision trees and decision making, *IEEE Trans. Syst. Man Cybernetics*, 20(2), 339, 1990.

16. Sudkamp, T. and Hammell, R.J., II, Interpolation, completion, and learning fuzzy rules, *IEEE Trans. Syst. Man Cybernetics*, 24(2), 332, 1994.

17. Wang, L.X. and Mendel, J.M., Generating fuzzy rules by learning from examples, *IEEE Trans. Syst. Man Cybernetics*, 22(6), 1414, 1992.

18. Wu, T.P. and Chen, S.M., A new method for constructing membership functions and fuzzy rules from training examples, *IEEE Trans. Syst. Man Cybernetics Part B: Cybernetics*, 29(1), 25, 1999.

19. Yasdi, R., Learning classification rules from database in the context of knowledge acquisition and representation, *IEEE Trans. Knowledge Data Eng.*, 3(3), 293, 1991.

20. Yeh, M.S. and Chen, S.M., An algorithm for generating fuzzy rules from relational database systems, Proceedings of the 6th International Conference on Information Management, Taipei, Taiwan, Republic of China, May, 219, 1995.

21. Zadeh, L.A., Fuzzy sets, *Inf. Control*, 8, 338, 1965.

22. Zadeh, L.A., The concepts of a linguistic variable and its application to approximate reasoning (1), *Inf. Sci.*, 8, 199, 1975.

8

Adaptation of Cluster Sizes in Objective Function Based Fuzzy Clustering

Annette Keller
German Aerospace Center

Frank Klawonn
University of Applied Sciences
Wolfenbüttel

Abstract. This paper discusses new approaches in objective-function based fuzzy clustering. Some well-known approaches are extended by a supplementary component. The resulting new clustering techniques are able to adapt single clusters to the expansion of the corresponding group of data in an iterative optimization procedure. Another new approach based on volume centers as cluster representatives with varying radii for individual groups is also described. The corresponding objective functions are presented, and alternating optimization schemes are derived. Experimental results demonstrate the significance of the presented techniques.

Key words: Objective-function based fuzzy clustering, size-adaptable clustering, alternating optimization.

8.1 Introduction

Standard fuzzy clustering methods like the fuzzy c-means algorithm are based on the idea of optimizing an objective function. This objective function depends on the distances of the data to the cluster centers weighted by the membership degrees. By taking the first derivative of the objective function with respect to the cluster parameters, one obtains necessary conditions for the objective function to have an optimum. These conditions are then applied in an iteration procedure and define a clustering algorithm. Numerous approaches have been developed to detect different forms of cluster shapes in data sets. The more flexible the clustering algorithms are in general, the more they depend on a suitable cluster initialization. Also, with the flexibility of cluster structures, the complexity of the proposed algorithms highly increases. In this chapter, we present an extension that can be applied to well-known simple and fast clustering techniques enabling these to adapt to the cluster sizes without highly increasing the computational effort.

In Section 8.2, we review the necessary background on objective-function based fuzzy clustering techniques. Well-known and often applied clustering techniques such as probabilistic, possibilistic, and noise clustering are discussed as well as several possible distance measures. The restrictions imposed by the algorithms lead us to the idea of our new approaches. In Section 8.3, we show that only slight modifications of the basic ideas are necessary to enable better adaptations by clustering. Nevertheless, our new techniques are not based on special cluster shapes that could be restrictive for the underlying models. The results of the proposed methods compared to the basic algorithms presented in Section 8.2 are illustrated in Section 8.4. Some artificial data sets are used to demonstrate the possibilities of the presented algorithms. The last example consists of real-world data and demonstrates the applicability of the new approaches. Section 8.5 finally summarizes our experiences.

8.2   Objective-Function Based Fuzzy Clustering

In this section, we present a short introduction to objective-function based fuzzy clustering and describe some well-known and often applied techniques. For a detailed overview on fuzzy clustering, see Höppner et al.[8] Most objective-function based fuzzy clustering algorithms aim at minimizing an objective-function that evaluates the partition of data into a given number of clusters.

8.2.1   Basic Objective Functions

Before discussing several special clustering techniques, general forms of objective functions for fuzzy clustering are introduced that still depend on the choice of a suitable distance measure. Two very common basic clustering techniques are probabilistic and possibilistic clustering. Both depend on a distance or dissimilarity measure weighted by the membership degrees. Probabilistic clustering[1] uses a constraint ensuring that all data points totally belong to the partition, whereas possibilistic clustering[15] considers outliers with small membership degrees to all groups of data. A third approach related to possibilistic clustering is called noise clustering.[3] The idea of this approach is to assign outliers to a special group of data called a noise cluster and reduce the influence of this group on the whole partition. Selim and Ismail[18] introduced other approaches to avoid the drawback of probabilistic clustering. They suggest letting a datum belong to a maximum number of clusters, to set the membership degrees to zero if a predefined maximal distance is exceeded or to define a minimum threshold for the membership degrees.

8.2.1.1   Probabilistic Clustering

In the case of probabilistic fuzzy clustering, the objective function has the form

$$J^{prob}(X, U, v) = \sum_{k=1}^{n} \sum_{i=1}^{c} u_{ik}^{m} \cdot d^2(v_i, x_k) \tag{IV.8.1}$$

$X = \{x_1, \ldots, x_n\} \in \mathbb{R}^p$ is the data set, n is the number of data points, c denotes the number of fuzzy clusters, $u_{ik} \in [0, 1]$ is the membership degree of datum x_k to cluster i, v_i is the prototype or the vector of parameters for cluster i, and $d(v_i, x_k)$ is a distance between prototype v_i and datum x_k.

The parameter $m > 1$ is called the fuzziness index. For $m \to 1$, the clusters tend to be crisp, i.e., either $u_{ik} \to 1$ or $u_{ik} \to 0$. For $m \to \infty$, we have $u_{ik} \to 1/c$. Usually, $m = 2$ is chosen.

To avoid the trivial solution that all membership degrees u_{ik} are 0, constraints have to be taken into account. In this case, the constraints are

$$\sum_{k=1}^{n} u_{ik} > 0 \qquad \text{for all } i \in \{1, \ldots, c\} \tag{IV.8.2}$$

and

$$\sum_{i=1}^{c} u_{ik} = 1 \qquad \text{for all } k \in \{1, \ldots, n\}. \tag{IV.8.3}$$

Equation IV.8.2 guarantees that only non-empty clusters are admitted in the partition. Equation IV.8.3 ensures that the sum of all membership degrees for one datum equals 1.0. This can be interpreted as "each datum is fully divided among the clusters and belongs totally to the partition of the data set."

Differentiating Equation IV.8.1 and taking the constraints into account by a Lagrange function leads to the necessary condition

$$u_{ik} = \frac{1}{\sum_{j=1}^{c} \left(\frac{d^2(v_i, x_k)}{d^2(v_j, x_k)} \right)^{\frac{1}{m-1}}} \tag{IV.8.4}$$

for Equation IV.8.1 to have a (local) minimum. Therefore, Equation IV.8.4 is used in an iteration procedure for updating the membership degrees u_{ik}. If a suitable distance function and parameter form is chosen, equations for the prototypes can be derived analogously, assuming the membership degrees are fixed. The alternating optimization scheme starts with a random initialization and applies the equations for the u_{ik} and the prototypes until the difference between the membership matrices (u_{ik}^{old}) and (u_{ik}^{new}) in two succeeding iterations is smaller than a given bound ε.

8.2.1.2 Possibilistic Clustering

In probabilistic clustering, the strong constraint Equation IV.8.3 possibly leads to undesirable membership degrees of some data. Assume a data point in great distance to all clusters exists in the data set. This noise point would be assigned the same high membership degree $\frac{1}{c}$ to all c clusters and, therefore, would have a greater influence on the partition than desired. To avoid such a drawback, the approach of possibilistic clustering was introduced with remaining constraint Equation IV.8.2 but modified constraint Equation IV.8.3:

$$\sum_{i=1}^{c} u_{ik} > 0 \qquad \text{for all } k \in \{1, \ldots, n\}. \tag{IV.8.5}$$

With these constraints, the membership degree u_{ik} could be interpreted as a degree of representativeness of datum x_k for cluster i. To avoid the trivial solution all $u_{ik} \to 0$ by minimizing Equation IV.8.1 considering Equation IV.8.5, the objective function has to be modified as well:

$$J^{poss}(X, U, v) = \sum_{k=1}^{n} \sum_{i=1}^{c} u_{ik}^m \cdot d^2(v_i, x_k) + \sum_{i=1}^{c} \eta_i \sum_{k=1}^{n} (1 - u_{ik}^m) \tag{IV.8.6}$$

The additional parameter η_i determines the permissible extension of cluster i. Differentiating Equation IV.8.6 considering Equation IV.8.2 and IV.8.5 leads to

$$u_{ik} = \frac{1}{1 + (\frac{d^2(v_i, x_k)}{\eta_i})^{\frac{1}{m-1}}} \qquad \text{(IV.8.7)}$$

To illustrate the influence of η_i, assume $\eta_i = d^2(v_i, x_k)$. The resulting membership degrees are $u_{ik} = (1 + 1^{\frac{1}{m-1}})^{-1} = \frac{1}{2}$. Defining a membership degree of $\frac{1}{2}$ as a lower bound for assigning a data point x_k to cluster i gives parameter η_i the mentioned meaning. The permissible extension of cluster i is in some way defined by η_i. If the cluster shapes are known in advance, η_i could be estimated for all $i = 1, \ldots, c$ easily. Otherwise, additional assumptions have to be made. One possible approach is to assume clusters containing about the same number of data points and estimate

$$\eta_i = \frac{\sum_{k=1}^{n} u_{ik}^m \cdot d^2(v_i, x_k)}{\sum_{k=1}^{n} u_{ik}^m}. \qquad \text{(IV.8.8)}$$

Krishnapuram and Keller[15] have also proposed other methods to estimate the parameters η_i.

8.2.1.3 Noise Clustering

Possibilistic clustering is one approach to deal with noisy data. Another related technique is called noise clustering; see, e.g., Davé[3] or Davé and Krishnapuram[4] and the references therein. The principle idea is to add one noise cluster to the set of clusters. Since the objective function considers only the distance function and the membership degrees, the noise cluster could be represented by the weighted membership degrees of the data to this cluster. The second term in Equation IV.8.9 expresses the noise cluster:

$$J^{noise}(X, U, v) = \sum_{k=1}^{n} \sum_{i=1}^{c} u_{ik}^m \cdot d^2(v_i, x_k) + \sum_{k=1}^{n} \delta^2 \left(1 - \sum_{i=1}^{c} u_{ik}\right)^m \qquad \text{(IV.8.9)}$$

Parameter δ has to be chosen in advance and is supposed to be the (large) constant distance of each datum to the noise cluster. In this case, Equations IV.8.2 and IV.8.5 have to be considered as in possibilistic clustering in order to derive equations for the membership degrees:

$$u_{ik} = \frac{1}{\sum_{j=1}^{c}(\frac{d^2(v_i, x_k)}{d^2(v_j, x_k)})^{\frac{1}{m-1}} + (\frac{d^2(v_i, x_k)}{\delta^2})^{\frac{1}{m-1}}} \qquad \text{(IV.8.10)}$$

as necessary conditions for Equation IV.8.9 to have a minimum. An interesting result is that

$$\sum_{i=1}^{c} u_{ik} < 1 \qquad \text{for all } k \in \{1, \ldots, n\} \qquad \text{(IV.8.11)}$$

for every datum with non-zero distance to all clusters. This illustrates that each datum belongs, at least with a small membership degree, to the noise cluster. In 1984, Ohashi[16] made an attempt to consider noise in data. Davé and Krishnapuram[4] showed that the minimization of Ohashi's objective function is equivalent to the approach introduced by Davé.[3]

8.2.2 Distance Measures

In the previous section, several general clustering concepts have been described. All techniques rely on the definition of suitable distance measures. Choosing a certain dissimilarity measure defines the structure which is searched for in the sample data. Different distance measures that are able to describe varying forms or shapes of clusters are possible.

8.2.2.1 The Fuzzy c-Means Algorithm

One simple fuzzy clustering technique is the fuzzy c-means algorithm (FCM); see, e.g., Bezdek,[1] where the distance $d(v_i, x_k)$ is simply the Euclidean distance

$$\mathfrak{D}_{FCM} = d^2(v_i, x_k) = \| x_k - v_i \|^2 = \sum_{\nu=1}^{p}(x_k^{(\nu)} - v_i^{(\nu)})^2, \tag{IV.8.12}$$

and the prototypes are vectors $v_i \in \mathbb{R}^p$, where p is the dimensionality of the data. $x_k^{(\nu)}$ ($v_i^{(\nu)}$) denotes the νth coordinate of the data vector (cluster center representative). Due to the Euclidean distance measure, this technique searches for spherical clusters of approximately the same size. By differentiating Equations IV.8.1, IV.8.6, or IV.8.9, we obtain the necessary condition

$$v_i = \frac{\sum_{k=1}^{n} u_{ik}^m \cdot x_k}{\sum_{k=1}^{n} u_{ik}^m} \tag{IV.8.13}$$

as prototype calculation instruction for the objective functions to have a (local) minimum using $\mathfrak{D}_{FCM}$. Since the first summand is identical in the three mentioned objective functions (see Section 8.2.1), and the second term in Equations IV.8.6 and IV.8.9 does not depend on a certain distance measure, the derived prototype equation holds in all three cases. These prototypes could be used alternately with Equation IV.8.4 in the iteration procedure. The update equation for the membership degrees depends on the chosen basic objective function, as described in the previous section.

8.2.2.2 The Algorithm by Gustafson and Kessel

Gustafson and Kessel[7] designed a fuzzy clustering method that is able to adapt to hyper-ellipsoidal forms. The prototypes consist of the cluster centers v_i as in FCM and (positive definite) covariance matrices C_i. The Gustafson and Kessel algorithm (GK) replaces the Euclidean distance by the transformed Euclidean distance:

$$\mathfrak{D}_{GK} = d^2(v_i, x_k) = (\rho_i \det C_i)^{1/p} \cdot (x_k - v_i)^{\top} C_i^{-1}(x_k - v_i). \tag{IV.8.14}$$

The factor $(\rho_i \det C_i)^{1/p}$ in $\mathfrak{D}_{GK}$ guarantees the volume for all clusters to be constant. Factor ρ_i could be used to determine the size of cluster i and is not changed during the alternating optimization. If the sizes cannot be estimated in advance, the parameters ρ_i might be set to one. The covariance matrices C_i are computed using Equation IV.8.15:

$$C_i = \sum_{k=1}^{n} u_{ik}^m \cdot (x_k - v_i)(x_k - v_i)^{\top}. \tag{IV.8.15}$$

The prototype calculation instruction of Equation IV.8.13 does not depend on the norm used in the distance measure, so again we obtain

$$v_i = \frac{\sum_{k=1}^{n} u_{ik}^m \cdot x_k}{\sum_{k=1}^{n} u_{ik}^m}$$

as a necessary condition for the objective functions Equations IV.8.1, IV.8.6, or IV.8.9 to have a (local) minimum. With these equations, the alternating iteration procedure for the Gustafson–Kessel algorithm is defined. In the update equation for the membership degrees, $\mathfrak{D}_{GK}$, see Equation IV.8.14, has to be inserted as a distance measure. This general form searches for hyper-ellipsoidal forms in the domain of interest.

Considering, for example, the task of rule learning, where fuzzy clusters are projected to single dimensions, non-axes parallel ellipsoids would lead, in general, to a serious loss of information caused by the construction of the fuzzy sets for the single domains. One approach to avoid this drawback is to restrict the covariance matrices C_i to diagonal matrices resulting in axes-parallel hyper-ellipsoids.[11,14] The distance measure in this case can be rewritten as

$$\mathfrak{D}_{AGK} = d^2(v_i, x_k) = \left(\rho_i \prod_{v=1}^{p} c_i^{(v)} \right)^{1/p} \cdot \left(\sum_{v=1}^{p} (x_k^{(v)} - v_i^{(v)})^2 \cdot \frac{1}{c_i^{(v)}} \right). \tag{IV.8.16}$$

Here, p is the dimensionality of the data vectors, and $x_k^{(v)}$ and $v_i^{(v)}$ denote the vth component of the kth data point and ith cluster center, respectively. For the alternating optimization, the calculation instruction for the covariance matrices can be simplified in the following way:

$$c_i^{(\gamma)} = \sum_{k=1}^{n} u_{ik}^m \cdot (x_k^{(\gamma)} - v_i^{(\gamma)})^2, \tag{IV.8.17}$$

where $c_i^{(\gamma)}$ denotes the γth diagonal element of the covariance matrix. $\mathfrak{D}_{AGK}$ could be used as the distance measure in the membership update equations from Section 8.2.1.

8.2.2.3   The Algorithm by Gath and Geva

Another clustering technique (GG) was designed by Gath and Geva.[5] This extension of the Gustafson–Kessel algorithm is, in some way, able to adapt the cluster size and, like the GK, adapts to hyper-ellipsoidal forms. Actually, this approach is not based on an objective function optimizer. Instead, the GG is a heuristic method derived from the fuzzification of a maximum likelihood estimator. Here, the distance measure is of the following form:

$$\mathfrak{D}_{GG} = d^2(v_i, x_k) = \frac{1}{\pi_i} \cdot \sqrt{\det(A_i)} \cdot \exp^{(\frac{1}{2} \cdot (x_k - v_i)^\top A_i^{-1} (x_k - v_i))}. \tag{IV.8.18}$$

The parameter π_i denotes the *a priori* probability for a datum to belong to the ith normal distribution. π_i is estimated as described by Equation IV.8.19, i.e., "number of data belonging to cluster i in relation to total number of data."

$$\pi_i = \frac{\sum_{k=1}^{n} u_{ik}^m}{\sum_{j=1}^{c} \sum_{k=1}^{n} u_{jk}^m} \tag{IV.8.19}$$

The covariance matrix of the ith normal distribution is denoted by A_i, where Equation IV.8.20 is the calculation instruction for estimating matrix A_i.

$$A_i = \frac{\sum_{k=1}^{n} u_{ik}^m \cdot (x_k - v_i)(x_k - v_i)^\top}{\sum_{k=1}^{n} u_{ik}^m} \tag{IV.8.20}$$

The prototype coordinates are the estimated expected values of the assumed normal distribution for cluster i. Again, the calculation of the prototypes can be done using Equation IV.8.13 as in the FCM or GK, respectively:

$$v_i = \frac{\sum_{k=1}^{n} u_{ik}^m \cdot x_k}{\sum_{k=1}^{n} u_{ik}^m}.$$

Now the equations for the alternating iteration procedure of the Gath–Geva algorithm are complete. To obtain equations for the prototypes as a necessary condition for the optimization function having a local minimum, the objective functions described in Section 8.2.1 would have to be differentiated.

Using $\mathfrak{D}_{GG}$ as a distance measure would lead to equations for which no analytical solution exists. Therefore, the estimation analogous to probability theory provides a good heuristic method.

Since GG is able to adapt to hyper-ellipsoidal forms as well as to different cluster sizes, the same problems as with the Gustafson–Kessel algorithm arise in the task of rule learning. As with GK, it is possible to restrict this approach to detect axes-parallel hyper-ellipsoids.[11,14] Nevertheless, the axes-parallel version of the algorithm introduced by Gath and Geva is able to adapt to different cluster sizes. In this case, the distance measure can be rewritten as

$$\mathfrak{D}_{AGG} = d^2(v_i, x_k) = \frac{1}{\pi_i} \sqrt{\prod_{v=1}^{p} a_i^{(v)}} \cdot exp^{\frac{1}{2} \cdot (\sum_{v=1}^{p} (x_k^{(v)} - v_i^{(v)})^2 \cdot \frac{1}{a_i^{(v)}})} . \tag{IV.8.21}$$

Again, p denotes the dimensionality of the data, $x_k^{(v)}$ and $v_i^{(v)}$ designate the vth component of the kth data point and the ith cluster center, respectively. In the alternating optimization, the covariance matrices could be simplified in the following way:

$$a_i^{(\gamma)} = \frac{\sum_{k=1}^{n} u_{ik}^m \cdot (x_k^{(\gamma)} - v_i^{(\gamma)})^2}{\sum_{k=1}^{n} u_{ik}^m}, \tag{IV.8.22}$$

where $a_i^{(\gamma)}$ denotes the γth diagonal element of the covariance matrix. Parameter π_i is estimated again as denoted in Equation IV.8.19.

Trauwaert, Kaufmann, and Rousseeuw[21] describe other clustering methods based on the maximum likelihood principle.

8.2.3   Alternating Optimization Approaches

In order to increase the influence of the user in extracting functional models from data, Runkler and Bezdek[17] have developed alternative approaches based on the presented basic ideas. They call the general clustering form with interchanging update equations for prototypes and membership degrees as presented above *alternating optimization*. In one approach, the expert has to specify the input space components in the form of prototype parameters. In the case of the fuzzy c-means algorithm (see Section 8.2.2.1), the expert would have to state prototype coordinates for each input domain. For other clustering algorithms, a suitable distance measure would also have to be chosen. The components for the output space are alternately updated during the optimization phase of that algorithm. Runkler and Bezdek call this form of alternating optimization *regular alternating optimization*, or rAO. Since some parameters are defined by the user and do not have to be updated during the alternating optimization, the computational effort is reduced.

By *alternating cluster estimation* (ACE), Runkler and Bezdek denote a clustering method where the expert has to select suitable membership function shapes and thereby defines the update equations for the cluster parameters.

The combination of both approaches where the expert has to specify suitable prototype parameters for the input domain as well as to choose suitable membership function shapes is called *regular alternating cluster estimation*, or rACE. The algorithm generates a partition of the data and then evaluates the projections of the cluster centers into the output space.

Although the resulting functional models are easy to understand and reflect the experts' interpretation of the modeled system, they are not necessarily based on objective functions. Problems may arise if the expert associates a system behavior with the data and assigns suitable parameters for the clustering algorithm but the defined model has a different basis. The greater the influence of the expert, the greater are the restrictions of the associated functional model. Knowledge about unknown dependencies in the data is difficult to extract with these models, but under assumptions about the system behavior these are computationally fast methods resulting in interpretable and easily understandable functional models.

The algorithms from Section 8.2 can be applied to learn fuzzy rules from data for classification problems[6,13] or function approximation.[12,14,20] Fuzzy rules are usually obtained from fuzzy clusters by projecting the clusters to the coordinate spaces leading to a certain loss of information. The more flexible the cluster algorithms are in finding different shape forms, the greater is the resulting loss of information in rule generation. One method to avoid a major part of this information loss is described by Klawonn and Keller.[10] There we start with a partition of the single domains in fuzzy sets and try to find a suitable partition for the data under consideration. Here we propose another approach. We modify Equation IV.8.1 in a way that enables simple fuzzy clustering algorithms like FCM to adapt to the cluster size—meaning to make the algorithm more flexible with respect to the cluster shape—without increasing the existing loss of information in the case of rule learning.

8.3   Adaptation of Cluster Volumes

In this section, we present some approaches to adapt to clusters with different volumes. The first approach was previously presented[9] to reduce the loss of information in rule learning. Therefore, only small modifications of some algorithms displayed in Section 8.2 have to be made. The principle of objective-function based fuzzy clustering with cluster representatives in the form of real-valued prototypes remains unchanged. The second presented method no longer uses multidimensional center-points as representatives, but uses center-volumes instead. As we will see, this approach has a non-negligible drawback. A combination of the presented methods seems to eliminate this lack and is presented in Section 8.3.3. Another approach using volume prototypes to enable the fuzzy c-means algorithm to detect clusters with different densities was presented by Setnes and Kaymak.[19]

8.3.1   Center-Based Clustering Algorithm

For each cluster, we introduce an additional parameter τ_i to the objective function in order to enable the clustering algorithm to adapt the cluster volumes. τ_i can be considered the (relative) radius of the corresponding cluster. The resulting probabilistic objective function is shown in Equation IV.8.23, with constant real-valued parameter $l > 0$.

$$J_{SACB}^{prob}(X, U, v) = \sum_{i=1}^{c} \sum_{k=1}^{n} u_{ik}^m \cdot \frac{1}{\tau_i^l} \cdot d^2(x_k, v_i) \qquad \text{(IV.8.23)}$$

To avoid the trivial solution that all $\tau_i \to \infty$, the constraint

$$\sum_{i=1}^{c} \tau_i = \tau \qquad \text{(IV.8.24)}$$

has to be taken into account, where τ is a predefined constant parameter, e.g., $\tau = c$ or $\tau = 1$.

Since the objective function Equation IV.8.23 does not require special properties of the distance measure $\mathfrak{D}$, most of the described distance measures need only small modifications to use the advantages of the proposed objective function.

Let us define

$$\mathfrak{D}_{SACB,*} = d_\tau^2(x_k, v_i) = \frac{1}{\tau_i^l} \cdot d^2(x_k, v_i) \qquad \text{(IV.8.25)}$$

as a new group of distance measures. Then, the objective function (Equation IV.8.23) can be rewritten as

$$J_{SACB,*}^{prob}(X, U, v) = \sum_{i=1}^{c} \sum_{k=1}^{n} u_{ik}^{m} \cdot d_{\tau}^{2}(x_k, v_i). \tag{IV.8.26}$$

Considering Equations IV.8.2 and IV.8.3 from Section 8.2.1, we obtain the same equations for the membership degrees as in those equations except that we have to replace the old distance $d^2(v_i, x_k)$ by $d_{\tau}^2(v_i, x_k)$, i.e., in the probabilistic case

$$u_{ik} = \frac{1}{\sum_{j=1}^{c}(\frac{d_{\tau}^2(x_k,v_i)}{d_{\tau}^2(x_k,v_j)})^{\frac{1}{m-1}}}.$$

Analogously, the distance measure for the membership equations in the case of possibilistic and noise clustering could be replaced. The modified distance measure for the FCM is shown in Equation IV.8.27:

$$\mathfrak{D}_{SACB,FCM} = d_{\tau}^2(x_k, v_i) = \frac{1}{\tau_i^l} \cdot (x_k - v_i)^T (x_k - v_i) \tag{IV.8.27}$$

Analogously to the objective function from Section 8.2.1 minimizing Equation IV.8.23 leads to the necessary condition (Equation IV.8.13)

$$v_i = \frac{\sum_{k=1}^{n} u_{ik}^m \cdot x_k}{\sum_{k=1}^{n} u_{ik}^m}$$

for the evaluation of the prototype coordinates.

Assuming that the parameters $l > 0$ and $\tau > 0$ are fixed, we have to take Equation IV.8.24 into account, to determine the values τ_i with predefined and, during the iteration procedure, unchanged $l > 0$ and $\tau > 0$. So, we obtain the Lagrange function

$$J_{SACB,\lambda}^{prob}(X, U, v) = \sum_{i=1}^{c} \sum_{k=1}^{n} u_{ik}^m \cdot \frac{1}{\tau_i^l} \cdot d^2(x_k, v_i) + \lambda \left(\sum_{i=1}^{c} \tau_i - \tau \right). \tag{IV.8.28}$$

Note that the last term of Equation IV.8.28 depends on neither u_{ik} nor on v_i so that the formulae for the optimal choices of u_{ik} and v_i remain valid.

Since the distance measure is independent on τ_i, differentiating Equation IV.8.28 gives us

$$\frac{\partial J_{SACB,\lambda}^{prob}(X, U, v)}{\partial \tau_i} = -\frac{l}{\tau_i^{l+1}} \cdot \sum_{k=1}^{n} u_{ik}^m \cdot d^2(x_k, v_i) + \lambda \overset{!}{=} 0 \tag{IV.8.29}$$

and, therefore,

$$\tau_i = \left(\frac{l \cdot \sum_{k=1}^{n} u_{ik}^m \cdot d^2(x_k, v_i)}{\lambda} \right)^{\frac{1}{l+1}}. \tag{IV.8.30}$$

With Equation IV.8.24, λ evaluates to

$$\lambda = \frac{(\sum_{j=1}^{c}(l \cdot \sum_{k=1}^{n} u_{jk}^m \cdot d^2(x_k, v_j))^{\frac{1}{l+1}})^{l+1}}{\tau^{l+1}}. \tag{IV.8.31}$$

After inserting Equation IV.8.31 into Equation IV.8.30, Equation IV.8.32 represents the resulting calculation instruction for the τ_i.

$$\tau_i = \frac{(\sum_{k=1}^{n} u_{ik}^m \cdot d^2(x_k, v_i))^{\frac{1}{l+1}}}{\sum_{j=1}^{c}(\sum_{k=1}^{n} u_{jk}^m \cdot d^2(x_k, v_j))^{\frac{1}{l+1}}} \cdot \tau \qquad \text{(IV.8.32)}$$

The parameter $l > 0$ plays a similar role as the fuzzifier m. When we choose a small value for l, a strong emphasis is put on adapting to the cluster size. Too small values for l can have a negative effect on algorithms as the GK, since the priority is put on the cluster size instead of the cluster shape. For $l \to \infty$, no more adaptation of cluster sizes is carried out, and we obtain the original algorithms.

Equation IV.8.32 can be used alternately with Equations IV.8.4 and IV.8.13 and a suitable distance measure for fuzzy clustering algorithms. We call this group of clustering techniques size-adaptable center-based clustering algorithms (SACB). Applying our results to the described FCM or GK enables these algorithms to detect clusters of different sizes. In the case of the FCM, rule generation only results in a small loss of information. Adapting the sizes of the detected spherical structures has no influence on the precision of the resulting fuzzy rules. Also, the axes-parallel version of the GK, i.e., AGK, does not lead to a significant loss of information in rule learning. Not only considering the task of rule learning, this approach can be, in combination with the GK as well as the AGK, an objective-function based alternative to the GG or AGG, respectively. It is possible to combine our approach with the objective function approaches of possibilistic (Section 8.2.1.2) or noise clustering (Section 8.2.1.3). The difference in these methods compared to the probabilistic objective function of Section 8.2.1.1 does not depend on a special distance measure. In the case of possibilistic clustering, Equations IV.8.7 for the membership degrees u_{ik}, IV.8.32 for the size parameters τ_i, and the necessary conditions derived from the chosen distance measure ($\mathfrak{D}_{FCM}$, $\mathfrak{D}_{GK}$, or $\mathfrak{D}_{AGK}$), e.g., the equations for cluster centers or covariance matrices, could be used in an alternating optimization procedure. Applying noise clustering to the size-adapting clustering approach, Equation IV.8.10 has to be used to calculate the accessory membership degrees. The other parameters are equivalent to probabilistic and possibilistic clustering.

We call the corresponding alternating optimization incorporating cluster size adaptation the *sized algorithm* (FCM-sized, GK-sized, etc.).

8.3.2 Volume-Based Clustering Algorithm

In the previously described fuzzy clustering techniques, the clusters are characterized by a vector, consisting of real-valued attributes, and a distance measure. Only the data points that coincide with a prototype may be assigned to the corresponding cluster with a membership degree of 1.0. Let us imagine dense spherical clusters. Instead of having just one ideal prototype for each cluster, to which we calculate the distances of the data points, we now assume that we have a complete circle or (hyper-) ball as the cluster center. This means that data within this area have distance zero to the cluster. This idea was proposed by Selnes and Kaymak.[19] However, there it was not based on an objective function, but on pure heuristic considerations. Here, we want to derive an alternating optimization scheme for this approach as well.

Taking these considerations into account, we obtain a probabilistic objective function (Equation IV.8.33) reflecting this idea of volume prototypes using Equation IV.8.34 as the distance function.

$$J_{SAVB}^{prob}(X, U, v) = \sum_{i=1}^{c} \sum_{k=1}^{n} u_{ik}^m \cdot \max\{0, \ (x_k - v_i)^T (x_k - v_i) - \tau_i\} \qquad \text{(IV.8.33)}$$

$$\mathfrak{D}_{SAVB} = d^2(v_i, x_k) = \max\{0, \ (x_k - v_i)^T (x_k - v_i) - \tau_i\} \qquad \text{(IV.8.34)}$$

If the clusters' radii τ_i are known in advance, these values should be used directly. Otherwise the τ_i have to be adapted during the alternating optimization, taking Equation IV.8.24 into account, to avoid the trivial solution $\tau_i \to \infty$ for all $i \in \{1, \ldots, c\}$ in minimizing the objective function (Equation IV.8.33). Here, τ is a predefined constant parameter. Assigning 0 to τ (all τ_i are 0) leads to the previously described fuzzy c-means (FCM) clustering technique (see Section 8.2.2.1). To derive equations for prototype coordinates (Equation IV.8.37) and radii values (Equation IV.8.40), respectively, the partial derivatives (Equations IV.8.36, IV.8.38, and IV.8.39) of Equation IV.8.35 have to be computed.

$$J_{SAVB,\lambda}^{prob}(X, U, v) = \sum_{i=1}^{c} \sum_{k=1}^{n} u_{ik}^m \cdot d^2(v_i, x_k) + \lambda \cdot \left(\sum_{i=1}^{c} \tau_i^2 - \tau \right), \tag{IV.8.35}$$

with $d^2(v_i, x_k)$ representing the distance measure from Equation IV.8.34.

$$\frac{\partial J_{SAVB,\lambda}^{prob}}{\partial v_i} = 2 \cdot \sum_{k:(x_k-v_i)^T(x_k-v_i)>\tau_i} u_{ik}^m \cdot (v_i - x_k) \overset{!}{=} 0 \tag{IV.8.36}$$

$$\overset{(IV.8.36)}{\Rightarrow} v_i = \frac{\sum_{k:(x_k-v_i)^T(x_k-v_i)>\tau_i} u_{ik}^m \cdot x_k}{\sum_{k:(x_k-v_i)^T(x_k-v_i)>\tau_i} u_{ik}^m} \tag{IV.8.37}$$

$$\frac{\partial J_{SAVB,\lambda}^{prob}}{\partial \tau_i} = \sum_{k:(x_k-v_i)^T(x_k-v_i)>\tau_i} u_{ik}^m + 2 \cdot \lambda \cdot \tau_i \overset{!}{=} 0 \tag{IV.8.38}$$

$$\frac{\partial J_{SAVB,\lambda}^{prob}}{\partial \lambda} = \sum_{i=1}^{c} \tau_i^2 - \tau \overset{!}{=} 0 \tag{IV.8.39}$$

$$\overset{(IV.8.38), (IV.8.39)}{\Rightarrow} \tau_i = \frac{\sum_{k:(x_k-v_i)^T(x_k-v_i)>\tau_i} u_{ik}^m}{\sqrt{\sum_{i=1}^{c} (\sum_{k:(x_k-v_i)^T(x_k-v_i)>\tau_i} u_{ik}^m)^2}} \cdot \sqrt{\tau} \tag{IV.8.40}$$

In the alternating optimization scheme, the distance measure has to be replaced by Equation IV.8.34. Depending on the clustering technique used (probabilistic, possibilistic, or noise), the corresponding update equations of the membership degrees u_{ik} (Equations IV.8.4, IV.8.7, or IV.8.10) have to be used.

It has to be noted that the objective function is not differentiable everywhere. The necessary conditions Equations IV.8.36 and IV.8.40 lead to a local minimum if no data points leave a volume center or wander into a volume center. This is why, in Equations IV.8.37 and IV.8.40, only data points with Euclidean distance greater than τ_i to the prototypes v_i have influence on the next alternating parameters τ_i^{new} and v_i^{new} for cluster i. Imagine two well-separated spherical clusters are given. In the first alternating optimization steps, the structures are identified correctly. The τ_i are assigned the correct radius values of the circles containing the data points. In the next step, each prototype and each radius is only calculated on the basis of the data points assigned to the opposite cluster. So, the cluster parameters are alternately interchanged. Even if the τ_i are smaller than the correct radius values, convergence is neither guaranteed nor plausible. To ensure convergence of the alternating optimization scheme, we can simply check, before updating the radii and the cluster centers, that the updated values decrease the objective function. If this is not the case, we skip the corresponding single step in the optimization scheme. In the following section, we propose a better way to solve this problem by combining our two different approaches.

8.3.3 Combination of Volume-Based and Center-Based Clustering Algorithm

To avoid restrictions as in the case of the volume-based clustering technique (SAVB) in Section 8.3.2, we have modified the objective function Equation IV.8.33 so as to combine the distance measure (Equation IV.8.34) with the Euclidean distance used for the fuzzy c-means algorithm. The resulting objective function is shown in Equation IV.8.41. Parameter $0 < q < 1$ determines the influence of each summand in the distance function (Equation IV.8.42).

$$
J_{SAVCB}^{prob}(X, U, v)
$$

$$
= \sum_{i=1}^{c} \sum_{k=1}^{n} u_{ik}^{m} \cdot (q \cdot \max\{0, \ (x_k - v_i)^T (x_k - v_i) - \tau_i\}
$$

$$
+ (1 - q) \cdot (x_k - v_i)^T (x_k - v_i)) \tag{IV.8.41}
$$

$$
\mathfrak{D}_{SAVCB} = d^2(v_i, \ x_k) = q \cdot \max\{0, \ (x_k - v_i)^T (x_k - v_i) - \tau_i\}
$$

$$
+ (1 - q) \cdot (x_k - v_i)^T (x_k - v_i) \tag{IV.8.42}
$$

To adapt the cluster radii during the alternating optimization, again, Equation IV.8.24 has to be considered, leading to Equation IV.8.43:

$$
J_{SAVCB, \lambda}^{prob}(X, U, v)
$$

$$
= \sum_{i=1}^{c} \sum_{k=1}^{n} u_{ik}^{m} \cdot (q \cdot \max\{0, \ (x_k - v_i)^T (x_k - v_i) - \tau_i\}
$$

$$
+ (1 - q) \cdot (x_k - v_i)^T (x_k - v_i)) + \lambda \cdot \left(\sum_{i=1}^{c} \tau_i^2 - \tau \right) \tag{IV.8.43}
$$

The alternating optimization scheme needs calculation instructions for the cluster centers v_i, the radii τ_i, and the membership degrees u_{ik}. Equations IV.8.44 and IV.8.45 are the corresponding partial derivatives of our objective function. The partial derivative for λ remains as denoted in Section 8.3.2, Equation IV.8.39.

$$
\frac{\partial J_{SAVCB, \lambda}^{prob}}{\partial \lambda} = \sum_{i=1}^{c} \tau_i^2 - \tau \overset{!}{=} 0
$$

$$
\frac{\partial J_{SAVCB, \lambda}^{prob}}{\partial v_i} = 2 \cdot (1 - q) \cdot \sum_{k=1}^{n} u_{ik}^{m} \cdot (v_i - x_k)
$$

$$
+ 2 \cdot q \cdot \sum_{k:(x_k - v_i)^T (x_k - v_i) > \tau_i} u_{ik}^{m} \cdot (v_i - x_k) \overset{!}{=} 0 \tag{IV.8.44}
$$

$$
\frac{\partial J_{SAVCB, \lambda}^{prob}}{\partial \tau_i} = -q \cdot \sum_{k:(x_k - v_i)^T (x_k - v_i) > \tau_i} u_{ik}^{m} + 2 \cdot \lambda \cdot \tau_i \overset{!}{=} 0 \tag{IV.8.45}
$$

The resulting calculation instructions are shown in Equations IV.8.46 and IV.8.47, respectively. The greater the influence of the Euclidean distance ($q \to 0$), the smaller are the calculated center radii τ_i.

$$\stackrel{(IV.8.44)}{\Rightarrow} \quad v_i = \frac{\sum_{k=1}^{n} u_{ik}^m \cdot x_k \; - \; q \cdot \sum_{k:(x_k-v_i)^T \cdot (x_k-v_i) \leq \tau_i} u_{ik}^m \cdot x_k}{\sum_{k=1}^{n} u_{ik}^m \; - \; q \cdot \sum_{k:(x_k-v_i)^T \cdot (x_k-v_i) \leq \tau_i} u_{ik}^m} \qquad (IV.8.46)$$

$$\stackrel{(IV.8.45)}{\Rightarrow} \quad \tau_i = \frac{\sum_{k:(x_k-v_i)^T (x_k-v_i) > \tau_i} u_{ik}^m}{\sqrt{\sum_{i=1}^{c} (\sum_{k:(x_k-v_i)^T (x_k-v_i) > \tau_i} u_{ik}^m)^2}} \cdot \sqrt{\tau} \cdot q \qquad (IV.8.47)$$

Depending on the chosen basic clustering technique (probabilistic, possibilistic, or noise), the adequate calculation instruction for the membership degrees u_{ik} has to be chosen. Therefore, the distance measure has to be replaced by $\mathfrak{D}_{SAVCB}$ (Equation IV.8.42). Even if the influence of the Euclidean distance is rather small ($q \approx 0.99$), the alternating optimization converges reliably in our experiments.

We refer to the corresponding alternating optimization scheme as the FCM-volume algorithm.

8.4 Examples

To demonstrate the properties of our new approaches, we have designed two artificial test data sets, as can be seen in Figure IV.8.1. Figure IV.8.1a shows two spherical clusters with uniformly distributed data points for both clusters but different radii. Here, the number of data points for each cluster is the same. In Figure IV.8.1b, two ellipsoidal clusters with equally distributed data points but to different extents are displayed. The larger cluster has about twice as many data points as the smaller one.

In Figure IV.8.2, the results for the data set from Figure IV.8.1a with the algorithms using the Euclidean distance measure are compared. The fuzzifier m was, in all cases, set to 2.0. The constraint parameter τ was set to 1.0 in both cases, FCM-sized and FCM-volume. For the size-adaptable version of the fuzzy c-means algorithm, the exponent l was set to 0.5. For the influence of the radii part in the case of the FCM-volume approach, $q = 0.99$ was chosen. The original fuzzy c-means algorithm has difficulty assigning the data to the correct clusters (see Figure IV.8.2a). A datum is assigned to the cluster with the highest membership degree. The approach using the Euclidean distance combined with volume centers is in a position to adapt the volume centers and, therefore, yields slightly better results than the original FCM (see Figure IV.8.2c). Only the size-adaptable approach (Figure IV.8.2b) has the ability to assign most data points correctly.

In Figure IV.8.3, the results for the ellipsoidal test data set from Figure IV.8.1b are shown. As clustering algorithms, the axes-parallel versions of the Gustafson–Kessel algorithm and our new size-adaptable version of that algorithm have been chosen. The fuzzifier m was assigned the value 2.0 in both cases. For the size-adaptable approach, constraint parameter τ has been set to 1 and the exponent l was assigned 0.4. Our new approach is able to adapt to the ellipses' content (Figure IV.8.3b), whereas the result in Figure IV.8.3 shows that the Gustafson–Kessel algorithm searches for groups of about the same size. Our approach can be further improved if a smaller value for parameter l, e.g., $l = 0.3$, is chosen.

As another example, we used the Wisconsin breast cancer database[2,22] to test our new approaches with the probabilistic objective function.

This classified data set originally contains 699 data points with 9 attributes and a classification attribute. 16 data points with missing values have been deleted from the data set for our tests. From the remaining 683 data points, 444 were classified as benign and 239 as malignant. In Figure IV.8.4,

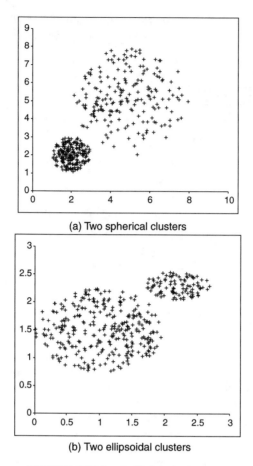

(a) Two spherical clusters

(b) Two ellipsoidal clusters

FIGURE IV.8.1   Artificial test data sets.

the results for the original fuzzy c-means algorithm (FCM) are compared to our size-adaptable version of this algorithm (FCM-sized) and the combination of the FCM with the volume–center based approach (FCM-volume). In Figure IV.8.4, the percentage of wrong classified data for two to ten clusters is displayed. The fuzzifier m was, in all cases, set to 2.0. The values for the other parameters are $\tau = 1.0$ for FCM-sized as well as for FCM-volume, $l = 0.8$ for FCM-sized, and $q = 0.9$ for FCM-volume. In this case, the priority of the FCM-volume rests with the volume-based component. The best results are obtained with our new algorithms. The FCM-sized algorithm yields the best classification, where 2.6% of the data entries are misclassified with four clusters. A similar good result (2.8% misclassified data points) is reached in the case of the FCM-volume at 5 clusters. The best result for the FCM (3.1% wrong classified data) is obtained with 4 clusters. Our approaches both seem to improve the results for the Wisconsin breast cancer database.

As can be seen in Figure IV.8.5, the results for the axes-parallel version of the method designed by Gustafson and Kessel (GK parallel) are compared to our size-adaptable version of this algorithm (GK-parallel sized). As mentioned, the percentage of wrong classified data for two to twelve clusters is shown. The fuzzifier m was again set to 2.0. The values for the other parameters are $\tau = 1$ and $l = 0.5$ for the GK parallel-sized. That both algorithms show similar worse results indicates that the model of ellipsoidal clusters is not suited for this data set and that the GK and its variants, although they are extensions of the FCM variants, are not able to find the spherical clusters of the good results found by the FCM and its variants. Not shown are the results obtained by the non-restricted version

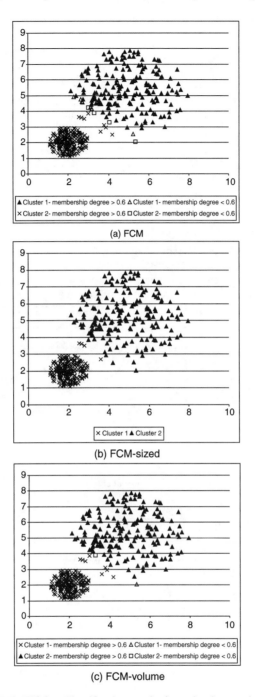

FIGURE IV.8.2 Classification results for a circular test data set.

of the GK and our new size-adaptable approach for this algorithm. They are worse than those of the restricted versions, GK parallel and GK parallel-sized. Nevertheless, if the principal chosen model is well suited for the considered data set, size adaptation has the ability to improve the classification results.

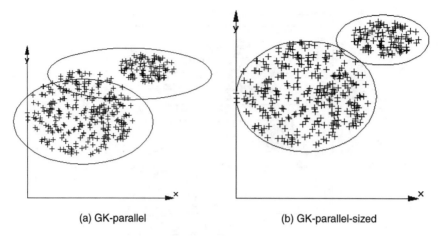

(a) GK-parallel   (b) GK-parallel-sized

FIGURE IV.8.3   Classification results for an ellipsoidal test data set.

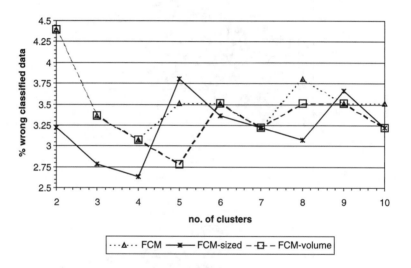

FIGURE IV.8.4   Classification results for Wisconsin breast cancer database — part A.

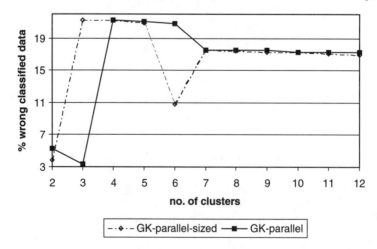

FIGURE IV.8.5   Classification results for Wisconsin breast cancer database — part B.

TABLE IV.8.1   Influence of Parameter l in FCM-Sized — Part A

c	$l = 0.2$	$l = 0.5$	$l = 0.8$	$l = 1.0$	$l = 3.0$	$l = 5.0$
2	25.5	2.8	3.2	3.1	4.0	4.2
3	30.9	2.8	2.8	2.9	3.2	3.2
4	36.6	2.9	2.6	2.6	2.9	3.1
5	47.4	4.4	3.8	3.8	3.7	3.5
6	34.8	2.9	3.4	3.2	3.5	3.5
7	13.6	2.9	3.2	3.4	3.2	3.2
8	17.4	2.8	3.1	3.2	3.8	3.8
9	22.3	4.2	3.7	3.5	3.5	3.5
10	34.6	3.4	3.2	3.2	3.2	3.4
11	12.2	2.9	3.8	3.8	3.8	3.8
12	13.3	3.1	3.5	3.5	3.5	3.5

TABLE IV.8.2   Influence of Parameter l in FCM-Sized — Part B

$l = 0.2$	0.5964
$l = 0.5$	0.9078
$l = 0.8$	0.7045
$l = 1.0$	0.7105
$l = 3.0$	0.7112
$l = 5.0$	0.7119

In Table IV.8.1, clustering results in the case of the size-adaptable fuzzy c-means for different values of parameter l are shown. The values denote the percentage of misclassified data. To calculate this value first for all clusters c, the class which is represented by one particular cluster is determined. Then, the data points corresponding to that cluster but originally belonging to a different class than that cluster's are counted. The sum of misclassified data over all clusters in ratio to the total number of data gives the percentage of misclassified data, also called error rate. It can be seen that the result depends on the choice of the exponent l. For Figure IV.8.4, the value for $l = 0.8$ obtaining the best results has been chosen.

For Table IV.8.2, we have chosen those c-partitions for each l-value from Table IV.8.1, where the least error values occurred, and calculated the maximal membership degree for each datum separately. In Table IV.8.2, the average of these maximal membership degrees is shown for each partition. It is obvious that this value in probabilistic clustering depends on the total number of clusters for the partition, e.g., whether the value for $l = 0.2$ and $c = 11$ clusters is less than the result for $l = 0.5$ and $c = 2$ clusters. The last four entries illustrate the influence of parameter l on the membership degrees. Here, for the l-values 0.8, 1.0, 3.0, and 5.0, the number of clusters was, in all cases, $c = 4$. The calculated values are slightly increasing for increasing l-values.

8.5 Conclusions

Our approach seems to be well suited to adapt to different sizes of clusters. One remaining problem concerning the original versions of the algorithms presented in Sections 8.2 and 8.4 is that all these approaches presuppose uniformly distributed data over all clusters, i.e., the number of data points per cluster are assumed to be equal for all clusters. To cluster data with varying sizes and numerical differences regarding the data points per structure correctly, adaptation to the density has to be taken into account.

Especially in applying fuzzy clustering techniques to the task of rule learning, it is not necessary to implement highly form-adaptable algorithms since more flexible cluster algorithms referring to the cluster shape generally result in a higher loss of information by rule generation. Several approaches to apply fuzzy clustering algorithms to the task of rule learning have been developed in recent years.[11,13,14,20,23] However, a loss of information by the process of rule generation is unavoidable.

As long as we stay with such simple clustering algorithms as FCM or the parallel version of GK, loss of information in the case of rule learning can be basically avoided. This is also valid for the size-adaptable versions of these algorithms since the form describing distance measure is the same. In the case of rule learning, the proposed modified versions of the described algorithms are a good alternative to more complex algorithms like the method introduced by Gath and Geva.

References

1. Bezdek, J., *Pattern Recognition with Fuzzy Objective Function Algorithms*, Plenum Press, New York, 1981.
2. Blake, C., Keogh, E., and Merz, C., UCI of Machine Learning Repository, 1998. On the World Wide Web.
3. Davé, R., Characterization and detection of noise in clustering, *Pattern Recognition Lett.*, 12, 657, 1991.
4. Davé, R. and Krishnapuram, R., Robust clustering methods: a unified view, *IEEE Trans. Fuzzy Sets Syst.*, 5(2), 270, 1997.
5. Gath, I. and Geva, A., Unsupervised optimal fuzzy clustering, *IEEE Trans. Pattern Anal. Machine Intelligence*, 11, 773, 1989.
6. Genther, H. and Glesner, M., Automatic generation of a fuzzy classification system using fuzzy clustering methods, ACM Symposium on Applied Computing (SAC '94), Phoenix, AZ, 180, 1994.
7. Gustafson, D. and Kessel, W., Fuzzy clustering with a fuzzy covariance matrix, in *IEEE CDC*, San Diego, CA, 761, 1979.
8. Höppner, F. et al., *Fuzzy Cluster Analysis*, Wiley, Chichester, 1999.
9. Keller, A. and Klawonn, F., Clustering with volume adaptation for rule learning, *EUFIT '99*, Aachen, 1999.
10. Klawonn, F. and Keller, A., Fuzzy clustering and fuzzy rules, in *7th International Fuzzy Systems Association World Congress (IFSA '97)*, vol. 1, Academia 193, 1997.
11. Klawonn, F. and Kruse, R., Automatic generation of fuzzy controllers by fuzzy clustering, *1995 IEEE International Conference on Systems, Man, and Cybernetics*, Vancouver, Canada, 2040, 1995.
12. Klawonn, F. and Kruse, R., Clustering methods in fuzzy control, in *From Data to Knowledge: Theoretical and Practical Aspects of Classification, Data Analysis and Knowledge Organization*, Gaul, W. and Pfeifer, D., Eds., Springer-Verlag, Berlin, 195, 1995.
13. Klawonn, F. and Kruse, R., Derivation of fuzzy classification rules from multidimensional data, in *Advances in Intelligent Data Analysis*, Lasker, G. and Liu, X., Eds., The International Institute for Advanced Studies in Systems Research and Cybernetics, Windsor, Ontario, 1995.
14. Klawonn, F. and Kruse, R., Constructing a fuzzy controller from data, *Fuzzy Sets Syst.*, 85, 177, 1997.

15. Krishnapuram, R. and Keller, J., A possibilistic approach to clustering, *IEEE Trans. Fuzzy Syst.*, 1, 98, 1993.

16. Ohashi, Y., Fuzzy clustering and robust estimation, 9th Meeting of the SAS Users Group International, Hollywood Beach, FL, 1984.

17. Runkler, T. and Bezdek, J.C., Alternating cluster estimation: a new tool for clustering and function approximation, *IEEE Trans. Fuzzy Syst.*, 7(4), 377, 1999.

18. Selim, S.Z. and Ismail, M.A., Soft clustering of multidimensional data: a semi-fuzzy approach, *Pattern Recognition*, 17(5), 559, 1984.

19. Setnes, M. and Kaymak, U., Extended fuzzy c-means with volume prototypes and cluster merging, EUFIT '98, Aachen, 1360, 1998.

20. Sugeno, M. and Yasukawa, M., A fuzzy-logic-based approach to qualitative modelling, *IEEE Trans. Fuzzy Syst*, 1, 7, 1993.

21. Trauwaert, E., Kaufmann, L., and Rousseeuw, P., Fuzzy clustering algorithms based on the maximum likelihood principle, *Fuzzy Sets Syst.*, 42, 213, 1991.

22. Wolberg, W.H. and Mangasarian, O.L., Multisurface method of pattern separation for medical diagnosis applied to breast cytology, *Natl. Acad. Sci.*, 87, 9193, 1990.

23. Yoshinari, Y., Pedrycz, W., and Hirota, K., Construction of fuzzy models through fuzzy clustering techniques, *Fuzzy Sets Syst.*, 54, 157, 1993.

9

Learning Rules about the Development of Variables over Time

Frank Höppner
University of Applied Sciences
Wolfenbüttel

Frank Klawonn
University of Applied Sciences
Wolfenbüttel

Abstract.   The human approach to learning from time series is guided by the impressive pattern recognition capabilities of the human brain. Rather than looking at quantitative values, a human automatically segments the time series appropriately and characterizes or classifies the segments. Similarity of time series is then decided on the basis of this abstracted representation. In this chapter, we propose an algorithm to support a human in this approach to dealing with large data sets. As an example, we consider the problem of deriving local weather forecasting rules that allow us to conclude from the qualitative behavior of the air-pressure curve to the wind strength. Formally, we learn these rules from a single series of labeled intervals, which has been obtained from (multivariate) time series by extracting various features of interest, for instance, segments of increasing and decreasing local trends. We seek the identification of frequent local patterns in this state series. A temporal pattern is defined as a set of states together with their interval relationships described in terms of Allen's interval logic. In the spirit of association rule mining, we propose an algorithm to discover frequent temporal patterns and to generate temporal rules. We discuss an efficient algorithm for frequent pattern discovery in detail.

9.1   Introduction

The visual pattern recognition capabilities of the human brain are truly remarkable and still outperform most artificial approaches to pattern recognition available up to now. It is, therefore, quite natural for a human to turn temporal information into a graphical representation (e.g., time series plot)

0-8493-1121-7/03/$0.00+$1.50

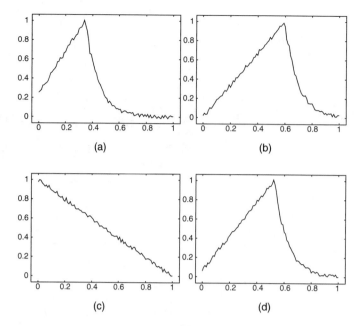

FIGURE IV.9.1   Similarity of time series. In terms of Euclidean distance, (a) is closer to (c) than to (b) or (d).

and to apply this impressive ability to the analysis of time-varying data.* When human experts argue about time series, they frequently use terms like linearly increasing segment or exponentially decreasing segment to describe the data. It is not only true, however, that humans are used to a shape-oriented perception of time series; this approach also seems to be beneficial from a technical or practical point of view. As an example, consider a technical system that is run several times. Usually, such systems are not fully deterministic in their behavior. Even if we can observe, at some points in the system, Gaussian distributed measurements (e.g., to which degree a valve has been opened), this does not necessarily manifest in a Gaussian distribution of the output value at a fixed time. The degree of opening controls the amount of water that flows through a pipe and thereby influences the point in time when the tub is full, which then may cause some other action. Many variables are therefore subject to translation or dilation in time. Those measures used traditionally for estimating similarity (e.g., pointwise Euclidean norm) will fail to provide useful hints about the similarity of the underlying processes. Figure IV.9.1 shows an example motivated by Keogh and Pazzani.[14] Clustering of the profiles by a human would lead to the partition $\{\{a, b, d\}, \{c\}\}$, but using the Euclidean norm would lead to $\{\{a, c\}, \{b, d\}\}$. With respect to the remarks we made before, it seems much more likely that the profiles a, b, and d characterize similar situations of the process than profiles a and c.

However, detecting similar patterns in a large number of plots or local self-similarity in a very long series overcharges a human very quickly, even if the data volume is just moderate. Despite the fact that the detailed inspection of large data sets is often skipped, a human still performs well in developing hypotheses, verifying them, drawing conclusions, designing experiments, etc., that is, working on the basis of a shape-based representation in general. For instance, skippers have developed short-term weather forecasting rules relying (amongst others) on the visual appearance

---

*We exclude highly periodic signals from our considerations, where a frequency spectrum is usually more informative than the time profile itself.

or shape of the air-pressure curve. Other examples can be found in the area of computer-aided monitoring and control.[15]

The aim of this work is to support a human in the weakest link of this chain, namely, the treatment of large data volumes, without being forced to adapt to new representations of the data but instead keeping the representation a human is used to. In this way, we hope to maximize the benefits of human expertise in visual pattern recognition and processing. We want to support the human data analyst by pointing out the most frequent local patterns in multivariate time series and by providing the most informative rules about the qualitative behavior of the series. These are very tedious tasks that humans usually do not like, since ranking of different hypotheses (temporal patterns or rules) requires carefully scanning through all the data at hand.

There are two different types of data we can consider in this setting: we have either a large number of short time series (e.g., histories of different patients) or a small number of very long time series (measurements of a system that runs continuously). We will concentrate on the latter case in this chapter, but the developed method can easily be generalized to the first case. While the focus is on numeric time series, temporal information on symbolic attributes can canonically be incorporated in the analysis (for instance, symptoms or diseases of patients, music notes, syntactical patterns in written language, etc.).

The outline of the chapter is as follows. In order to adopt the human method of time series perception, we develop a suitable notion of temporal patterns in Section 9.2. Next, we have to fix a measure of temporal pattern frequency in order to rank the patterns as well as rules (Section 9.3). With these definitions, the task of frequent pattern and informative rule enumeration is well-defined, however, an efficient implementation is mandatory to overcome the complexity of the problem. In Section 9.4 we propose an efficient algorithm that adopts techniques from association rule mining[3,16] for this task. Section 9.5 shows some results of the algorithm. Section 9.6 discusses possible areas of future research, and conclusions are presented in Section 9.7.

9.2 Temporal Patterns

Let us revisit the example in Figure IV.9.1. Dynamic time warping[6,18] (DTW) is a promising approach to overcome these difficulties when using the Euclidean norm: using dynamic programming techniques, an optimal warping path is searched (local stretching of the time axis) such that the pointwise Euclidean distance of two time series is minimized. Although DTW is computationally quite expensive, by using piecewise linear representations instead of the raw series, a considerable speed-up can be obtained.[14] Since we do not know in advance which parts of the series might be similar, and DTW yields an individual warping path for any two series, we have to perform a pairwise comparison of a huge number of possible segments ($\binom{n}{2}$ calls of DTW if n is the number of segments). As a result, we obtain a gigantic dissimilarity matrix in which we then try to identify similar series by means of relational clustering algorithms. This covers the univariate case only. Therefore, we consider the DTW approach as infeasible for our purposes.

We do not think that a human actually performs some kind of time warping to match different parts of a time series. It seems more reasonable that a human breaks a time series into suitable segments, such that all points in each segment behave similarly or follow the same local trend. Each of these segments is usually simple in shape and easy to grasp. The human labels or classifies the segments into a small number of primitive shapes or patterns. Matching of time series is then performed on the basis of these labeled segments rather than on the raw time series. The primitive patterns can be defined *a priori* (for example, slightly increasing segment),[5,7,17] can be learned from a set of examples (labeled training set),[11] or can be found automatically by means of clustering short subsequences.[8] In all these cases, we finally arrive at a sequence of labeled intervals: time intervals in which a certain condition holds in the original time series. In the example of Figure IV.9.1, we have two descriptions: linear increase followed by exponential decrease, matching *a*, *b*, and *d*, and linear decrease, matching *c*. More complicated patterns are composed by capturing the temporal relationship of the time intervals in which the descriptions hold. Even more, this concept easily generalizes to the case of multivariate temporal patterns, e.g., "water level increases linearly while valve is open and then water level decreases exponentially while tub is tilted."

We will not discuss the various approaches one can consider to transform a time series into a sequence of labeled intervals, but assume that this transformation has been done along with other steps of preprocessing like feature selection. Although we do not discuss these steps here in detail, they are crucial steps in successful data analysis and most influential on the results we may achieve by the subsequent steps.

9.2.1 Labeled Interval Sequence

Let S denote the set of all possible trends, properties, or states that we want to distinguish, for example, "pressure goes down" or "water level is constant." A state $s \in S$ holds during a period of time (b, f), where b and f denote the initial point in time when we enter the state (or start its observation) and the final point in time when the state no longer holds. A labeled interval sequence (or state sequence) on S is a series of triples:

$$(b_1, f_1, s_1), (b_2, f_2, s_2), (b_3, f_3, s_3), (b_4, f_4, s_4), \ldots$$

where $b_i < f_i$ holds. We do not require that one state interval has ended before another state interval starts. This enables us to mix up several state sequences (possibly obtained from different sources) into a single state sequence. Thus, there is no need to distinguish between uni- and multivariate analysis in the case of labeled interval sequences.*

However, we do require that every triple (b_i, f_i, s) is maximal in the sense that there is no (b_j, f_j, s) in the series such that (b_i, f_i) and (b_j, f_j) overlap or meet each other:

$$\forall (b_i, f_i, s_i), (b_j, f_j, s_j), i < j : \qquad f_i \leq b_j \Rightarrow s_i \neq s_j \qquad \text{(IV.9.1)}$$

If Equation IV.9.1 is violated, we can merge both state intervals and replace them by their union $(\min(b_i, b_j), \max(f_i, f_j), s)$. We will refer to Equation IV.9.1 as the *maximality assumption*.

9.2.2 Temporal Patterns

To define a temporal pattern composed of temporal attributes, we use Allen's temporal interval logic[4] to describe the relation between the time intervals. For any pair of intervals, we have 13 possible relationships; they are illustrated in Figure IV.9.2. For example, we say "*A meets B*" if interval *A* terminates at the same point in time at which *B* starts. The inverse relationship is "*B is met by A.*" In the following, we will abbreviate the interval relations as shown in the figure. The set of all interval relationships is denoted by $\mathcal{I}$. For two intervals I_1 and I_2, we denote their temporal relationship by $\text{ir}(I_1, I_2) \in \mathcal{I}$.

Given n state intervals (b_i, f_i, s_i), $1 \leq i \leq n$, we can capture their relative positions to each other by an $n \times n$ matrix R whose elements $R[i, j] \in \mathcal{I}$ describe the relationship between state interval i and j. As an example, let us consider the state sequence in Figure IV.9.3. Obviously state A is always followed by B. The gap between A and B is covered by state C. Below the state interval sequence, both of these patterns are written as a matrix of interval relations.

---

*Note that most of the traditional methods for time series analysis cover the uni- and bivariate cases only.

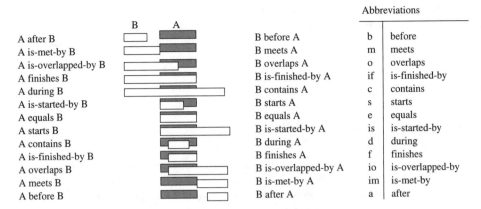

FIGURE IV.9.2 Allen's interval relationships.

state interval sequence:

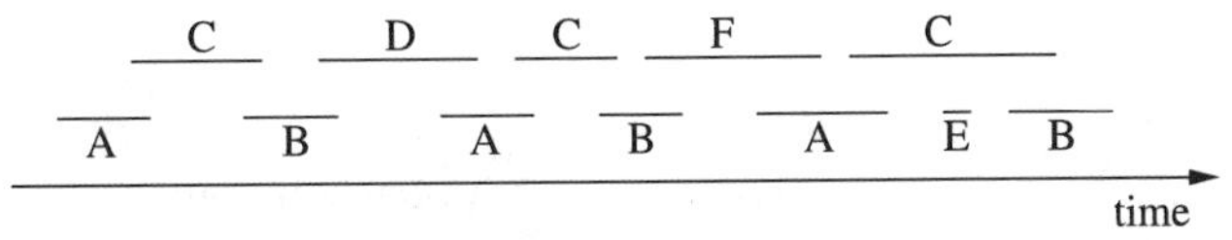

temporal relations:

FIGURE IV.9.3 Example for state interval patterns expressed as temporal relationships.

Definition 1 (temporal pattern): A pair $P := (s, R)$, where $s : \{1, \ldots, n\} \to S$ and $R \in \mathcal{I}^{n \times n}$, $n \in \mathbb{N}$, is called a temporal pattern of size n if there is a state sequence $(b_i, f_i, s_i)_{1 \leq i \leq n}$ such that $s(i) = s_i$ and $R[i, j] = \mathrm{ir}([b_i, f_i], [b_j, f_j])$. By $\dim(P)$ we denote the dimension (number n of intervals) of the pattern P. If $\dim(P) = n$, we say that P is an n-pattern.

Of course, many different state sequences map to the same temporal pattern. We call these sequences *instances* of the temporal pattern. By $TP(S)$ we denote the space of all temporal patterns of arbitrary dimension using symbols S.

9.2.3 Temporal Rules

Next, we define a partial order $\sqsubseteq$ on temporal patterns. Informally, from a temporal pattern Y we obtain a pattern $X \sqsubseteq Y$ by removing any number of states in an instance of Y.

Definition 2 (subpattern relation): We say that the temporal pattern (s_X, R_X) is a subpattern of (s_Y, R_Y) (or $(s_X, R_X) \sqsubseteq (s_Y, R_Y)$), if $\dim(s_X, R_X) \leq \dim(s_Y, R_Y)$ and there is an injective mapping $\pi : \{1, \ldots, \dim(s_X, R_X)\} \to \{1, \ldots, \dim(s_Y, R_Y)\}$ such that

$$\forall i, j \in \{1, \ldots, \dim(s_X, R_X)\}:$$
$$s_X(i) = s_Y(\pi(i)) \wedge R_X[i, j] = R_Y[\pi(i), \pi(j)] \qquad \text{(IV.9.2)}$$

Given two temporal patterns $X \sqsubseteq Y$, we can formulate rules

$$\text{if } X \text{ then } Y \text{ with probability } p \qquad \text{(IV.9.3)}$$

modeling temporal dependencies between states in X and Y. We call the pattern X in the premise the *premise pattern* and the pattern Y in the conclusion the *rule pattern*. The rule pattern also comprises the premise pattern ($X \sqsubseteq Y$). If we remove the premise from the rule pattern, we obtain the *conclusion pattern*. Note that it is not possible to reconstruct the rule pattern from the premise and conclusion patterns only, since the rule pattern additionally determines the temporal relationship between any two intervals where one of them stems from the conclusion and the other from the premise.

9.3   Frequency of Temporal Patterns

Having defined temporal patterns, we are now interested in means to rate them with respect to the frequency of their occurrences: the *support* of a pattern denotes how often a pattern occurs. What is a suitable definition of support in the context of temporal patterns? Perhaps the most intuitive definition is the following: the support of a temporal pattern is the number of instances of the temporal pattern in the state sequence. Let us examine this definition in the context of the following example (see Figure IV.9.2 for abbreviated interval relationships; the patterns are also depicted below the relation matrix):

$$
\text{if } \begin{array}{c|cc} & A & B \\ \hline A & = & b \\ B & a & = \end{array} \text{ then } \begin{array}{c|cccc} & A & B & A & B \\ \hline A & = & b & b & b \\ B & a & = & b & b \\ A & a & a & = & m \\ B & a & a & im & = \end{array} \text{ with probability } p \qquad (IV.9.4)
$$

How often does the pattern in the conclusion occur in the state series in Figure IV.9.4a? We can easily find 3 occurrences, as shown in Figure IV.9.4b. The remaining (unused) states do not form a fourth pattern. How often does the premise pattern occur? By pairing states (1, 4), (2, 6), (3, 7), etc. we obtain a total number of 7. So, we have $p = \frac{3}{7}$. This may correspond to our intuitive understanding of the rule, but we can improve p to $\frac{4}{7}$ when using the rule pattern assignment in Figure IV.9.4c. The latter assignment is perhaps less intuitive than the first one, because the pattern's extension in time has increased. Now, however, we have a state sequence that is assembled completely out of rule patterns; there is no superfluous state. Then, would it not be more natural to have a rule probability near 1 instead of $\frac{4}{7}$?

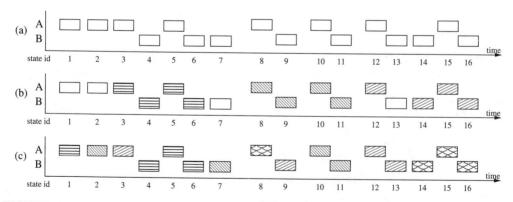

FIGURE IV.9.4   Counting the occurrences of temporal patterns (states with different labels (A and B) are drawn on different levels. Note that the pattern of interest (IV.9.4) requires a *meets* relation in the conclusion).

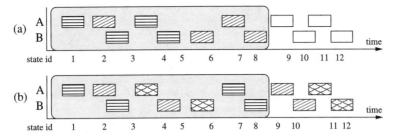

FIGURE IV.9.5   Counting the occurrences of temporal patterns within a sliding window.

The purpose of the example is to alert the reader that rule semantics is not as clear as might be expected. Furthermore, determining the maximum number of pattern occurrences is a complex task and does not necessarily correspond to our intuitive counting. The latter problem can be accomplished by introducing a mask or window such that we count only local patterns that are sufficiently small in their temporal extent. We therefore choose a maximum duration Δt_{win}, which serves as the width of a sliding window which is moved along the state sequence. We consider only those pattern instances that can be observed within this window. In a monitoring and control application, the threshold Δt_{win} could be taken from the maximum history length that can be displayed on the monitor and, thus, be inspected by the operator. This is a reasonable assumption: the whole pattern has to be small enough to be observed by a (forgetful) operator.

However, introducing a constraint on the pattern extension does not simplify the pattern counting; we still have to take the complete series into account in order to find the maximum number of pattern occurrences. Figure IV.9.5a illustrates this with an example. The sliding window is indicated by a shaded rectangle. Maximizing the number of patterns within the window leads us to the assignment in Figure IV.9.5a. The remaining states do not form a third occurrence. However, if we choose a different assignment at the shown position (yielding only one pattern occurrence in the first window), we obtain two additional occurrences when shifting the window along the series (Figure IV.9.5b).

If we denote the number of states in the sequence by L and the (average) number of states in the window by R, the example has shown that the complexity of pattern counting is still $O(c(L))$ and not $O(c(R))$ (where $c(n)$ denotes the complexity of the counting process in a series of n states). Due to the variability of temporal patterns, c increases quickly with n. Therefore, counting the number of pattern occurrences becomes computationally intractable.

This problem can be avoided by choosing a different support definition. The right bound of the sliding window serves as a reference point and will be denoted by t_{act}.

Definition 3 (support of temporal pattern):   We define the total time in which (one or more) instances of P can be observed in the sliding window as the support of the pattern:

$$\text{supp}\,(P) = \text{len}\{t \mid P \text{ is observable in sliding window at position } t_{act} = t\}$$

where len accumulates the lengths of the observation periods. (We will soon see that the time points in which P is observable form a sequence of intervals $S = \{[t_1, t_2], [t_3, t_4], \ldots, [t_{2n-1}, t_{2n}]\}$. Then len$(S)$ is given by $\sum_{i=1}^{n} t_{2i} - t_{2i-1}$.)

Now, for each window, we have a complexity of $O(t(R))$, where $t(n)$ is the complexity of testing whether a pattern is contained in a sequence of n states. Testing for all window positions is then $O(L \cdot t(R))$, which is usually much smaller than $O(c(L))$ because functions c and t grow quickly in their arguments and $R \ll L$.

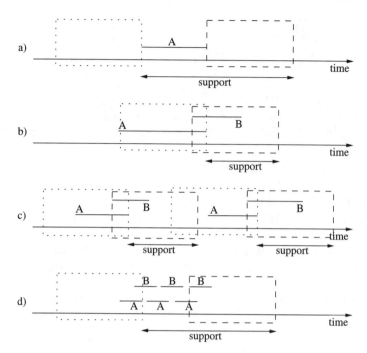

FIGURE IV.9.6   Illustration of support.

However, changing the support definition always changes the semantics of a rule. What is the semantics of the rule probability p? Obviously, we have:

$$\mathrm{supp}(P) \leq \Delta t_{win} + \underbrace{\max\{f_i \mid 1 \leq i \leq L\} - \min\{b_i \mid 1 \leq i \leq L\}}_{\text{sequence length in time}}$$

If we divide the support of a pattern by the sequence length in time plus the window width Δt_{win} we obtain the relative frequency p of the pattern: if we randomly select a window position we can observe the pattern with probability p. This is a nice interpretation, which, in some cases, might be superior to the pattern counting semantics. If we choose a sliding window width that is sufficiently large in Figure IV.9.4, we will always observe at least one occurrence of both the premise and conclusion pattern in our exemplary rule. Thus, we obtain a rule confidence near 1, which is somewhat more desirable than $\frac{4}{7}$.

To illustrate the definition of support with some examples, consider Figure IV.9.6. In Figure IV.9.6a we have a single state A. We see the pattern for the first time, when the right bound of the sliding window touches the initial time of the state interval (dotted position of sliding window). We can observe A until the sliding window reaches the position that is drawn with dashed lines. The total observation time is therefore the length of the sliding window Δt_{win} plus the length of state interval A. The support (observation duration) is depicted at the bottom of the figure. Figure IV.9.6b shows another example, "A overlaps B." We observe an instance of the pattern as soon as we can decide the temporal relationship of A and B and we lose it when we lose the possibility to compare the interval bounds. If the pattern occurs multiple times, two things may happen: If there is a gap between the pattern instances, such that we lose the pattern in the meanwhile, then the support of the individual instances adds up to the support of the pattern, as shown in Figure IV.9.6c. If there is no such gap (Figure IV.9.6d), we see the pattern as soon as a first instance enters the sliding window until the last instance leaves the window. In the meantime, it does not matter how many instances are present, as long as there is at least one.

For a pattern P, we denote the set of reference points t_{act} for which P has been observed in the window by O_P. Figure IV.9.6 also illustrates that the reference points that contribute to the support of a pattern themselves can be characterized as a sequence of intervals. We will make use of this fact later in Section 9.4.3.

9.4 Mining Frequent Temporal Patterns

A pattern is called *frequent* if its support exceeds a threshold $supp_{min}$. The task is to find all frequent temporal patterns, from which we then create the temporal rules. A naive implementation, however, which enumerates all possible patterns and checks each of them individually for being frequent, is computationally too expensive: for an increasing size k, the number of possible temporal patterns grows exponentially, as shown in Figure IV.9.7. For $k = 7$, we already have more than 1.75 million different patterns. For the figure, we have just counted the number of different relationship matrices of size $k \in \{1, \ldots, 8\}$, but we have not yet considered label selection and/or permutation. If there are m different labels ($m > k$) to select from, the number of possible selections $m!/(m - k)!$ (ordering is relevant) has to be multiplied with the number shown in Figure IV.9.7. For a subset of the patterns, we could also select partially identical labels (as in "*A before A*"). This gives us an impression of the tremendous number of possible temporal patterns. The remainder of this section is devoted to an efficient implementation of frequent pattern enumeration.

Apparently, the space of frequent temporal patterns has to be pruned effectively to enumerate them efficiently. We therefore adopt techniques from association rule mining[3] for the enumeration procedure in Figure IV.9.8. To find all frequent patterns, we start in a first database pass with the estimation of the support of every single state (also called candidate 1-patterns). After the kth run, we remove all candidates that have missed the minimum support and create out of the remaining frequent k-patterns a set of candidate $(k + 1)$-patterns whose support will be estimated in the next pass. This procedure is repeated until no more frequent patterns can be found. The fact that the support of a pattern is always greater than or equal to the support of any of its subpatterns guarantees that we do not miss any frequent patterns:

$$\forall \text{patterns } P, Q : \quad Q \sqsubseteq P \ \Rightarrow \ supp(Q) \geq supp(P) \qquad \text{(IV.9.5)}$$

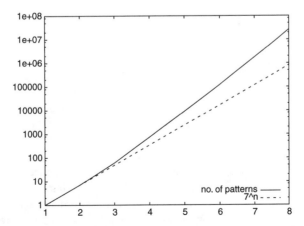

FIGURE IV.9.7 Number of distinct normalized temporal patterns (log scale) vs. the size n of the temporal patterns (using n fixed, distinct labels).

```
 1  proc find_freq_patterns(→ S, → (I_i)_i, ← (F_i)_i)
 2     k := 1;
 3     C_k := {s ∈ S};                                    1-candidates
 4     repeat
 5        support_estimation(C_k, (I_i)_i);               pass over sequence (I_i)_i
 6        F_k := {P ∈ C_k | supp(P) > supp_min};          freq. k-patterns
 7        C_{k+1} := candidate_generation(F_k);           build (k + 1)-candidates
 8        k := k + 1;
 9     until C_k = {};                                    until no new candidates
10   .
```

FIGURE IV.9.8   High-level description of frequent pattern enumeration algorithm. The input parameters are the set of symbols S and the labeled interval sequence $(I_i)_i$; the output is the sequence of frequent pattern sets.

9.4.1   Normalized Patterns

Before we start enumerating patterns, we want to eliminate some combinatorial complexity in the definition of the temporal pattern itself. The relation $\sqsubseteq$ is reflexive and transitive, but not antisymmetric: we can have $(s_A, R_A) \sqsubseteq (s_B, R_B)$ and $(s_B, R_B) \sqsubseteq (s_A, R_A)$ without $s_A = s_B$ and $R_A = R_B$ due to a different state ordering, which complicates matching of temporal patterns. Permutating the states does not, however, change the semantics of the temporal pattern (a pictorial representation would be identical). Therefore, we define $(s_A, R_A) \equiv (s_B, R_B) :\Leftrightarrow (s_A, R_A) \sqsubseteq (s_B, R_B) \wedge (s_B, R_B) \sqsubseteq (s_A, R_A)$ and consider the factorization $({}^{TP(S)}/_{\equiv}, {}^{\sqsubseteq}/_{\equiv})$, where $\sqsubseteq$ has been generalized canonically to equivalence classes. Then, ${}^{\sqsubseteq}/_{\equiv}$ is also antisymmetric and, thus, a partial order on (equivalence classes of) temporal patterns.

To simplify notation, we pick a subset $NTP(S) \subset TP(S)$ of normalized temporal patterns such that $NTP(S)$ contains one temporal pattern for each equivalence class of ${}^{TP(S)}/_{\equiv}$ and $(NTP(S), \sqsubseteq)$ is isomorphic to $({}^{TP(S)}/_{\equiv}, {}^{\sqsubseteq}/_{\equiv})$. In the remainder, we will then use $(NTP(S), \sqsubseteq)$ synonymously to $({}^{TP(S)}/_{\equiv}, {}^{\sqsubseteq}/_{\equiv})$. (In Figure IV.9.7, we have counted normalized patterns only.) The intuitive idea is to order the state intervals in time with increasing index. However, this ordering is slightly more complex with arbitrary intervals than with points.

Definition 4: A temporal pattern (s, R) is said to be in normalized form, if for all i, j the following condition holds:

$$i < j \Leftrightarrow (s(i) = s(j) \wedge R[i, j] = \text{before})$$
$$\vee (s(i) < s(j) \wedge R[i, j] = \text{equals})$$
$$\vee (s(i) \neq s(j) \wedge \left(R[i, j] \in \begin{matrix} \{\text{starts, contains, is-finished-by,} \\ \text{overlaps, meets, before}\} \end{matrix} \right) \quad \text{(IV.9.6)}$$

One can easily show that every temporal pattern has a unique normalized form, which is obtained by sorting the intervals of a pattern instance lexicographically. Having uniqueness, we can check for equivalence of temporal patterns (s_A, R_A) and (s_B, R_B) in their normalized form simply by checking $s_A = s_B \wedge R_A = R_B$. In the remainder, we will use the normalized patterns as representatives for the classes of equivalent temporal patterns. We also assume that the state sequence under consideration is always sorted lexicographically.

9.4.2 Observability of Temporal Patterns

Note that our definition of support does not imply that the whole pattern instance $P :=$ $\{(b_i, f_i, s_i) \mid 1 \le i \le n\}$ must fit into the sliding window. Denoting the temporal extent (or duration) of P by

$$D(P) := \max\{f \mid (b, s, f) \in P\} - \min\{b \mid (b, s, f) \in P\}$$

we do not have $D(P) \le \Delta t_{win}$ in general. Figure IV.9.9 illustrates this fact in case there are intervals with a length that exceeds Δt_{win} (window drawn with solid lines). State A lasts for a time period that is longer than Δt_{win}, nevertheless we can observe the pattern "D after C, A contains C and D" within the window. The pattern "B before C" in the window drawn with dashed lines is another example where we can observe the pattern although $D(P) > \Delta t_{win}$. However, we cannot (yet) observe "A contains C" in the dashed window because the final time of C is not yet visible—within the limited scope of the sliding window, we cannot decide whether there is an an *overlaps* or *finishes* relation to come.

To be able to classify or distinguish a set of labeled intervals as a certain pattern, we have to compare the interval borders with each other. For instance, to distinguish "A starts B" from "A overlaps B," it is sufficient to observe A a little before B, but it does not play a role when A has started exactly. Therefore, the exact length of the very first and last state interval is not important for the classification of the whole set. In general, if we write down the initial and final points of all the participating state intervals in ascending order (see Figure IV.9.10) and then skip the very first (t_{min}) and last ones (t_{max}), we obtain the length of the pattern by subtracting the latest ($t_{nearmax}$) from the latest point shifted by the window width ($t_{nearmin} + \Delta t_{win}$). If some interval borders are identical (as in "A is finished by B"), then we have to write them down multiple times. If this happens for the very first or last point in time, we remove only one occurrence (in the example, we observe the pattern as long as B).

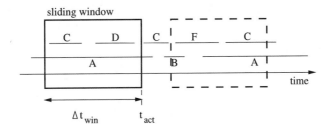

FIGURE IV.9.9 Sliding a window of width Δt_{win} along the state sequence.

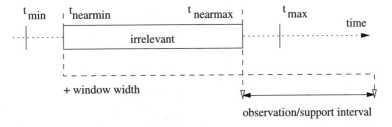

FIGURE IV.9.10 Observation interval.

9.4.3 Candidate Generation

As we have seen, the number of temporal patterns grows exponentially with the size k of the patterns. Efficient pruning techniques are therefore necessary to keep the increase in the number of candidates moderate. We use three different pruning techniques. Probably most important, the technique that is used for the discovery of association rules[3] can still be applied to temporal patterns. Due to Equation IV.9.5, every k-subpattern of a $(k+1)$-candidate must be frequent, otherwise the candidate itself cannot be frequent. We can make use of this fact during candidate generation: to enumerate as few non-candidate $(k+1)$-patterns as possible, we join any two frequent k-patterns P and Q that share a common $(k-1)$-pattern as a prefix (that is, they differ only in the last single state). We will not miss any candidates by considering only the frequent k-patterns that differ in the last component, since every k-subpattern of the new $(k+1)$-candidate X must be frequent—especially those obtained by removing one of the last two states of X.

Let us denote the remaining states in P and Q besides those in the prefix as p and q, respectively. We denote the interval relationship between p and q in the candidate pattern $X = (s_X, R_X)$ as $R_X[k, k+1] = r$. Figure IV.9.11 illustrates how to build the $(k+1)$-pattern matrix R_X out of R_P and R_Q. Since R_P and R_Q are identical with respect to the first $k-1$ states in normalized form, the same is true for the new pattern X (indicated by the same submatrix A). The relationship between p and q and the first $k-1$ states can also be taken from R_P and R_Q. Thus, as we can see in Figure IV.9.11c, the only degree of freedom is r. From the $(k-1)$-pattern prefix and the two states p and q, we can build up a $(k+1)$-pattern which is completely specified up to the relation between p and q.

The freedom in choosing r yields 13 different patterns that might become candidate $(k+1)$-patterns, because there are 13 possible interval relationships. Since we can restrict ourselves without

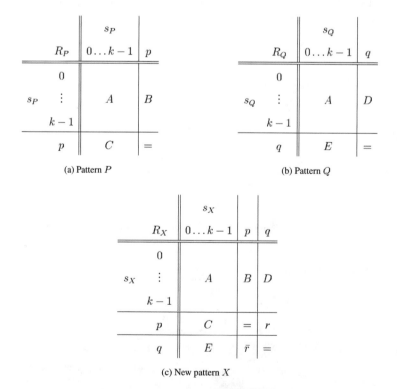

(a) Pattern P (b) Pattern Q

(c) New pattern X

FIGURE IV.9.11 Generating a candidate $(k+1)$-pattern X out of two k-patterns P and Q that are identical when restricted to the first $k-1$ states. The inverse interval relationship of r is denoted by $\bar{r}$.

loss of generality to normalized patterns, the number of possible values for r reduces to a maximal number of 7.

Before we check each of the seven $(k + 1)$-patterns for frequent k-subpatterns, we apply another pruning technique based on the law of transitivity. For example, the two 2-patterns "A meets B" and "A meets C" share the primitive 1-pattern A as a common prefix. We have to fix the missing relationship between B and C to obtain a 3-candidate. The law of transitivity for interval relations[4] tells us that the possible set of interval relations is $\{is\ started\ by,\ equals,\ starts\}$. In normalized form, only 2 out of 7 possible relationships remain. In general, for each state $s(i)$ of the first $k - 1$ states, we apply Allen's transitivity table to the relationship between p and $s(i)$ ($R_P[k, i]$) and $s(i)$ and q ($R_Q[i, k]$). Only those values for r that do not contradict the results of the $k - 1$ applications of the transitivity table yield a candidate pattern.

Finally, for every temporal pattern Q, we maintain an observed and expected support set O_Q and E_Q, respectively. The set O_Q contains all points in time that contribute to the support of the pattern Q, that is, all window positions t_{act} in which the pattern can be observed in the sliding window. Before we consider a $(k + 1)$-pattern P as a candidate pattern, we intersect* all sets O_Q of all k-subpatterns Q of P. The result gives us the expected support of P in E_P. The accumulated length of E_P serves as a tighter upper bound of the support of P than $\min\{E_Q \mid Q \sqsubseteq P,\ \dim(Q) = k\}$ does. If it stays below supp_{min}, the pattern cannot become a frequent pattern; therefore, we do not consider it as a candidate.

9.4.3.1   Candidate Generation Algorithm

It turns out to be useful to employ a sequential order on temporal patterns. Transforming a pattern $P = (s, R)$ into a vector

$$(\underbrace{s(1)}_{\textit{1-pattern}}, s(2), R[2, 1], s(3), R[3, 1], R[3, 2], \dots)$$

1-pattern

2-pattern

3-pattern

and using lexicographical comparison on the tuples employs a total order on temporal patterns. From the tuples, the original patterns P can be reconstructed. During candidate generation, however, we only compare patterns of identical sizes, leading to tuples of identical dimension.

As mentioned before, we create candidates from patterns that share a common $(k - 1)$-prefix. Within a sorted list of frequent k-patterns, blocks with identical $(k - 1)$-prefixes can be identified easily.[16] Two patterns belong to the same block if they differ only in the last state (in label and interval relationships), which corresponds to a difference in the last k dimensions only. By B_P, we denote the index of the first pattern in the list that belongs to the same block as P. If $B_P = B_Q$, both patterns P and Q belong to the same block. We enter a new block if $B_P \neq B_Q$ for consecutive patterns P and Q in the list.

For the first pruning technique, we have to search for all $(k - 1)$-subpatterns of the newly generated $(k + 1)$-pattern. Only $k - 1$ out of $k + 1$ patterns have to be searched; the remaining two were used to build the new candidate. The search for the remaining subpatterns can be done efficiently using binary search in the sorted list of frequent patterns.

Figure IV.9.12 shows the algorithm for candidate generation. Making use of the block start B_P, any two patterns with identical $(k - 1)$-prefix can be identified easily (loops in lines 3 and 5). For primitive 1-patterns P, we have always $B_P \equiv 1$. New candidate patterns C will be generated in

---

*The sets O_Q and E_Q can be organized as lists of intervals. The intersection is also a list of intervals. We only have to add up the interval lengths to obtain the accumulated length.

1 **proc** candidate_generation$(\to F_k, \to (O_P)_{P \in F_k}, \leftarrow C_{k+1}, \leftarrow (E_P)_{P \in C_{k+1}})$

2 $\quad C_{k+1} := \{\}; \; n := 0;$ n is actual value of $|C_{k+1}|$

3 $\quad$ **for** $i := 1$ **to** $|F_k|$ **do**

4 $\quad\quad\quad blockstart := n + 1; \; j := B\{F_k[i]\}$

5 $\quad\quad\quad$ **while** $B_{F_k[i]} = B_{F_k[j]}$ **do** $F_k[i]$ and $F_k[j]$ in same block

6 $\quad\quad\quad\quad\quad$ extend $F_k[i]$ to normalized candidate X using $F_k[j]$ (up to r);

7 $\quad\quad\quad\quad\quad R := \{r \in \mathcal{I} \mid X$ satisﬁes law of transitivity with $R_X[k, k-1]\};$

8 $\quad\quad\quad\quad\quad$ **for** $r \in R$ **do**

9 $\quad\quad\quad\quad\quad\quad\quad$ build k-subpatterns S_l of X (besides $F_k[i]$ and $F_k[j]$);

10 $\quad\quad\quad\quad\quad\quad\quad$ **if** $(\forall 1 \leq l \leq k-1 : S_l \in F_k)$

11 $\quad\quad\quad\quad\quad\quad\quad\quad\quad$ **then** $E_X := O_{F_k[i]} \cap O_{F_k[j]} \cap O_{S_1} \cap \ldots \cap O_{S_{k-1}};$

12 $\quad\quad\quad\quad\quad\quad\quad\quad\quad\quad\quad$ **if** $(\mathrm{card}(E_X) > \mathit{supp}_{min})$

13 $\quad\quad\quad\quad\quad\quad\quad\quad\quad\quad\quad\quad\quad$ **then** $n := n + 1; \; C_{k+1}[n] := X;$

14 $\quad\quad\quad\quad\quad\quad\quad\quad\quad\quad\quad\quad\quad\quad\quad B_{C_{k+1}[n]} := blockstart;$

15 $\quad\quad\quad\quad\quad\quad\quad\quad\quad\quad\quad\quad\quad\quad$ —

16 $\quad\quad\quad\quad\quad\quad\quad\quad$ —

17 $\quad\quad\quad\quad\quad$ **od**

18 $\quad\quad\quad\quad\quad j := j + 1;$ next pattern in block

19 $\quad\quad\quad$ **od**

20 $\quad$ **od**

21 .

FIGURE IV.9.12 High-level description of candidate generation. The procedure takes the set of frequent k-patterns F_k together with the observed support sets O_P and yields a set of candidate $(k + 1)$-patterns C_{k+1} together with the expected support sets E_P.

accordance with their ordering (such that no explicit sorting becomes necessary), and the values B_C are also set during candidate generation (lines 2, 4, and 14). The application of the law of transitivity for the second pruning technique (line 7) is very cheap in terms of complexity ($k - 1$ table look-ups). With the third pruning technique, the intersection of sets of intervals (support sets) is linear in the number of intervals, however, the actual number of intervals is difficult to estimate and is significantly influenced by the overall number L of state intervals in the sequence and the window width Δt_{win}.

Theorem 1: The time complexity of the algorithm in Figure IV.9.12 is

$$O(k \cdot |F_k| \cdot |\mathcal{S}| \cdot (k^2 + \log |F_k| + k \cdot L))$$

Proof of Theorem 1: The iteration over all candidate pairs with identical $(k - 1)$-prefixes is $O(|F_k|M)$, where M is the maximum block size, which is bounded by $7|\mathcal{S}|$ (the outer loop in line 3 is iterated $|F_k|$ times and the inner loop in line 5 is iterated $M \leq 7|\mathcal{S}|$ times). The candidate pattern X is built in $O(k + k^2)$ and the $k - 1$ remaining k-subpatterns of X are built in $O((k - 1)(k + k^2)) = O(k^3)$. The $k - 1$ subpatterns S_l have to be searched in F_k, which can be done in $O((k - 1) \log |F_k|)$ using binary search. Next, for each of the possible 7 interval relationships in a normalized pattern (line 8), the law of transitivity is applied $k - 1$ times (line 7) in $O(7(k - 1))$. The length of any O_P is bounded by the number L of intervals in the sequence (more precisely, by the maximum number of intervals in the sequence with the same label). The intersection of two interval sequences of length L is done in $O(2L)$, therefore line 11 is done in $O((k + 2)L)$. The sum of interval lengths of E_X is also obtained in $O(L)$. This gives us the overall complexity

$$O(\underbrace{|F_k|}_{loop\,1.\,3} \cdot \underbrace{|\mathcal{S}|}_{loop\,1.\,5} \cdot (\underbrace{k^3}_{cand.\,gen.} + \underbrace{k \log |F_k|}_{subp.\,search} + \underbrace{k^2 L}_{intersect}))$$

The complexity of the algorithm in Figure IV.9.12 is dominated by $|F_k|$ and L. In practice, while the sequence length L may become very large (size of the database), the actual complexity of the support interval intersection does not really depend on L but on the fragmentation of the support (number of support intervals in O_P for a pattern P). When starting with primitive 1-patterns, support intervals get merged if their (Hausdorff) distance is below the window width (compare with Figure IV.9.6d). With frequent patterns, O_P quite often consists of a very small number of large support intervals. When the patterns get more complex, the total support decreases but the fragmentation of the support may increase. For high values of k, the more and more complex patterns occur rarely and the number of support intervals becomes identical to the number of pattern instances. The number of pattern instances finally approaches zero when the minimum support is no longer reached by the pattern. Therefore, the average complexity varies depending on k, having local minima at $k = 0$ and $k = k_{max}$, reaching a local maximum somewhere in between (by k_{max} we denote the maximum size k of a frequent pattern found by the algorithm in Figure IV.9.8).

9.4.4 Support Estimation

In order to estimate the support of the candidate patterns, we scan through the state sequence and incrementally update the list of states which are currently visible in the sliding window. We also update the relation matrix for the states in the sliding window incrementally, thereby representing the content of the sliding window itself as a temporal pattern. We denote the content of the sliding window as a normalized pattern W.

In the first database pass, we build up the set of states S and, simultaneously, the sets of observed support $O_{\{s\}}$, $s \in S$. This is done in $O(L)$, where L denotes the number of labeled intervals in the analyzed sequence. While deciding whether a 1-pattern occurs in the sliding window or not is trivial,[*] checking k-candidate patterns with $k > 1$ becomes computationally more expensive (as we will see in Section 9.4.5). Whenever the right or left bound of the sliding window meets an interval bound, the changing content of the sliding window might require some action since temporal relationships get resolved or lost or new intervals are introduced in the window. In principle, whenever the content of the window changes, we have to check every candidate for being a subpattern of the sliding window. Since the size of C_k may become very large, we are again in need of pruning mechanisms that prevent us from unnecessarily calling the subpattern checking routine.

Therefore, the set of candidate patterns is partitioned into three subsets, which we call the sets of passive, active, and potential candidates. Recall that by t_{act} we denote the right bound of the sliding window. The set of passive candidates contains those candidates P that we do not expect in the current sliding window because the expected support does not contain the time of the current window position, that is, $t_{act} \notin E_P$. The set of potential candidates contains those candidates for which we have $t_{act} \in E_P$, that is, there is a chance of observing P in the window and it is worth checking P against the sliding window. Finally, the set of active patterns contains those patterns that are currently observable in the sliding window, that is, $t_{act} \in O_P$.

The rationale behind this subdivision is the following:

- Passive patterns P do not have to be checked against $P \sqsubseteq W$. For any passive pattern, we know that one of its subpatterns is currently not observable, so the pattern itself cannot be visible.
- Active patterns have been observed in the (recent) past, so it is not necessary to check them again as long as the instance found before is still visible, because our notion of support is independent

---

[*]In fact, for a single state s we do not even have to build up the sliding window. It is sufficient to unify all intervals $[b_i, f_i + \Delta t_{win}]$ for every (b_i, f_i, s) in the sequence.

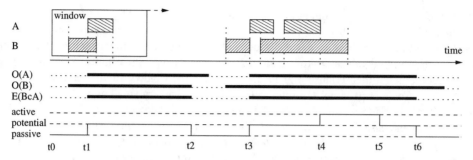

FIGURE IV.9.13 History of belonging-ness to the sets of passive, potential, and active patterns of an example pattern "*B* contains *A*."

of the number of instances simultaneously visible. Since we can compute in advance how long a given pattern instance will be visible, it is possible to postpone the next call of the pattern matching subroutine for P until the instance disappears.

- Potential patterns are the only patterns for which we have to continuously check whether a new instance can be found in the window.

Figure IV.9.13 traces an example pattern P = "*B* contains *A*" through the support estimation. In the figure, by $O(A)$ and $O(B)$, the set of observed support O_A and O_B is depicted. During candidate generation, the set of expected support E_P (shown as E(BcA) in the figure) has been generated. At the beginning (t_0), the pattern is passive. From the observation of A and B, we know that P might be observable at t_1 since both subpatterns are visible from then on. Therefore, P becomes a potential pattern at t_1. Again from the intersection of O_A and O_B, we know that at time t_2 we cannot observe A and B any longer, therefore at t_2 the pattern P is removed from the set of potential patterns and falls back into the set of passive patterns. We schedule the reconsideration of P at time t_3, where the next interval in E_P starts. This time we observe an instance of P at t_4. We calculate the observation duration and let P become an active pattern. From the observation duration of P, we know that this instance will vanish at t_5, and any further checking of P against the sliding window is postponed until t_5. At t_6, the expected support interval in E_P terminates, which causes P to fall back to the set of passive patterns. The intervals for which P became an active pattern then become the set of observed support O_P, which will be used by the candidate generation in the next loop.

Let us discuss in greater detail in which situations a pattern moves from one set of the partition into another. We speak of a LEAVE-PASSIVE event if a pattern is removed from the set of passive patterns to the set of active or potential patterns. In the same way, we define LEAVE-ACTIVE and LEAVE-POTENTIAL events. As already discussed, if interval and window bounds coincide, this will cause such events. There are four possible situations, as depicted in Figure IV.9.14:

- **Case (a)—*Right bound of sliding window meets left bound of interval*:** In this situation, a new interval enters the sliding window. We call this a NEW-INTERVAL event. One or more LEAVE-POTENTIAL events may be caused by the NEW-INTERVAL event since potential patterns may become active. In the depicted example, the pattern "*C* before *A*" becomes active.
- **Case (b)—*Right bound of sliding window meets right bound of interval*:** An interval becomes completely visible, which is called a FULL-INTERVAL event. Again, one or more LEAVE-POTENTIAL events may be caused by the FULL-INTERVAL event since the temporal relationship to other intervals is resolved. In the depicted example, the pattern "*B* contains *A*" becomes active; before this window position, we were not sure about the exact relationship; a "*B* overlaps *A*" pattern was also possible.

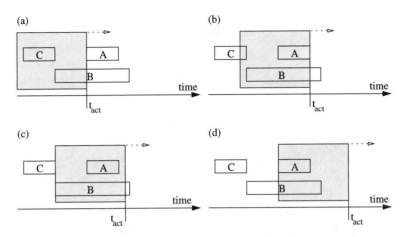

FIGURE IV.9.14 Important events when sliding the window along the sequence.

- **Case (c)—*Left bound of sliding window meets right bound of interval*:** One or more LEAVE-ACTIVE events may be caused. In the example, the pattern "*C* before *A*" disappears from the sliding window.
- **Case (d)—*Left bound of sliding window meets left bound of interval*:** Similar to Case (c), one or more LEAVE-ACTIVE events may be caused in this situation. In the example, the pattern "*B* contains *A*" disappears from the sliding window.

Cases (a) and (b) deal with the appearance, and (c) and (d) deal with the disappearance of pattern instances, since in (a) and (b), new interval bounds enter the window, but no interval bounds vanish, and with (c) and (d), we have the reverse. Once a new interval has encountered the sliding window (NEW-INTERVAL event), its length implicitly defines the associated FULL-INTERVAL event. Cases (c) and (d) are not necessarily relevant for every state interval, rather only if the state interval is currently used by a detected pattern instance. Once we have detected an instance, from its observation interval we also know in advance when we will have to reschedule the next check for this pattern—these points in time correspond to cases (c) and (d). A found instance therefore implicitly defines a LEAVE-ACTIVE event.

Figure IV.9.15 illustrates the pattern state transitions. Initially, all patterns are passive. The intervals in E_P are organized as a stack and the actual top element is denoted by $[a_P, d_P]$, where a_P is the activation time of pattern P and d_P is the deactivation time. When the activation time a_P is reached, depending on whether an instance can be observed in the window or not, P becomes a potential or even active pattern. A potential or active pattern falls back to the set of passive patterns if the deactivation time d_P is reached. These four transitions always occur at times a_P or d_P. A potential pattern becomes active as soon as a pattern instance is observed, which can only happen after a NEW-INTERVAL or FULL-INTERVAL event. From the observation interval of the instance, we know when the instance vanishes. If we cannot find another instance at that point in time, the pattern falls back into the set of potential patterns.

Whenever a pattern instance has been found, the support of the pattern is updated incrementally, that is, we insert the period of pattern observation (the support) into O_P and remove the expected support interval $[a_P, d_P]$ from E_P. Since the sum of interval lengths in E_P is an upper bound of the remaining support, we can perform a fourth online pruning test. If the support achieved so far ($\text{len}(O_P)$) plus the maximally remaining support ($\text{len}(E_P)$) drops below supp_{min}, we do not consider the pattern any longer; it becomes a *pruned pattern*. At the end of each database pass, the set E_P is empty and O_P contains the support of P. All non-frequent patterns will be pruned.

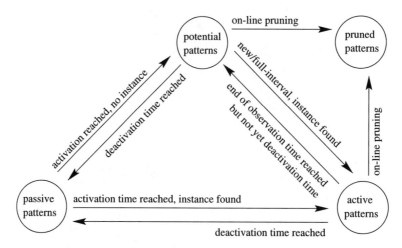

FIGURE IV.9.15   Pattern transition diagram.

Figure IV.9.16 depicts a high-level description of the support estimation algorithm. The algorithm iterates over the labeled intervals of the sequence (line 3) and schedules a NEW-INTERVAL and FULL-INTERVAL event. All events will be processed in the order of their temporal occurrence (lines 6–21). When it is time to process a

$$\left\{ \begin{array}{l} \text{NEW-INTERVAL} \\ \text{FULL-INTERVAL} \\ \text{LEAVE-PASSIVE} \\ \text{LEAVE-POTENTIAL} \\ \text{LEAVE-ACTIVE} \end{array} \right\}$$

event, the affected

$$\left\{ \begin{array}{l} \text{potential} \\ \text{potential} \\ \text{passive} \\ \text{potential} \\ \text{active} \end{array} \right\}$$

patterns have to be processed in order to keep the partition up to date. After each increment of t_{act}, the content of the sliding window has to be updated. Since the state sequence is assumed to be lexicographically ordered, the sliding window can be updated incrementally (adding intervals in line 22, removing intervals in line 20).

Theorem 2: The support estimation algorithm in Figure IV.9.16 can be implemented with time-complexity

$$O(|C_k| \log |C_k| + (n + m) \cdot (T + \log |C_k|)) \qquad (\text{IV.9.7})$$

where $n = \sum_{P \in C_k} \text{card}(E_P)$, $m = \sum_{i=1}^{L} |\text{POTENTIAL}_i|$, card is the number of intervals in a set of intervals, T is the time complexity of the subpattern test for k-patterns, and $|\text{POTENTIAL}_i|$ denotes the number of potential patterns in the ith iteration.

```
 1   proc support_estimation(→ C, → (E_P)_{P∈C}, ← (O_P)_{P∈C})
 2      ACTIVE := {}; POTENTIAL := {}; PASSIVE := C; t_{act} := −∞;
 3      for all labeled intervals (b, s, f) in the sequence do          in lex. order
 4          schedule a NEW-INTERVAL event at time b;
 5          schedule a FULL-INTERVAL event at time f;
 6          while there are unprocessed events scheduled before b do
 7              e := earliest event; t_{act} := time of earliest event;
 8              e.P := pattern associated with e;
 9              e.s := state associated with e;
10              if e = FULL-INTERVAL
11                  then process_all_potential(e.s);
12              elsif e = LEAVE-POTENTIAL
13                  then process_potential(e.P);
14              elsif e = LEAVE-PASSIVE
15                  then process_passive(e.P);
16              elsif e = LEAVE-ACTIVE
17                  then process_active(e.P);
18              fi
19              remove e from schedule;
20              shrink_window();
21          od
22          add_interval(b, s, f);                          process NEW-INTERVAL
23          process_all_potential(e.s);
24      od
25   .
```

FIGURE IV.9.16   The main loop of the support estimation algorithm. The procedure takes the set of candidate patterns C together with the expected support sets E_P and yields the observed support sets O_P.

Proof of Theorem 2: To support efficient operations, the sets E_P and O_P are organized as sorted lists of intervals, and the minimum is simply the left bound of the first interval in the list (and can be obtained in $O(1)$). We keep the set of passive patterns sorted by their activation time a_P. Then, accessing the earliest pattern is $O(1)$ and insertion/deletion is $O(\log(|\text{PASSIVE}|))$. In a similar way, the set of potential patterns is sorted by deactivation times d_P. For any active pattern, there was a positive subrelation test, from which we have obtained an observation interval $[b, f]$, and f is used to sort the set of active patterns. Let T denote the complexity of testing a pattern P of dimension k for $P \sqsubseteq W$. Since k is fixed during support estimation, we consider it as a constant. Let us first examine the subroutines in Figure IV.9.17.

Subroutine *check_pattern*: Testing $P \sqsubseteq W$ is $O(T)$. From an instance π, the observation interval is calculated in $O(k)$. Adding a new observation interval to the sorted list of intervals is $O(1)$, since any new observation has to be inserted at the end of the list. This gives use the overall complexity $O(T)$.

Subroutine *process_passive*: Since E_P is sorted, popping the first element is $O(1)$. Removing inserting a pattern from a sorted set S is $O(\log|S|)$, therefore the overall complexity is $O(\log|\text{PASSIVE}| + T + \max\{\log|\text{ACTIVE}|, \log|\text{POTENTIAL}|\})$. With $|C_k| \geq \max\{|\text{PASSIVE}|, |\text{ACTIVE}|, |\text{POTENTIAL}|\}$, we have the complexity $O(\log|C_k| + T)$.

Subroutine *process_all_potential*: Due to a NEW-INTERVAL or FULL-INTERVAL event (Cases (a) and (b)), potential patterns might become active. There is no way to foresee when this will happen; we therefore have to test all potential patterns against W. Since we check potential

```
1   funct check_pattern(→ P)                              returns true iff P visible
2       subrelation_check(P, W, π);
3       observation_interval((b_{π(1)}, b_{π(2)}, ..), (f_{π(1)}, f_{π(2)}, ..), Δt_{win}, b, f);
4       if t_{act} ∈ [b, f]
5           then append [b, f] at end of O_P; return true
6           else return false _
7   .
9   proc process_passive(→ P)                       pre: t_{act} has reached a_P
10      pop rst interval [a_P, d_P] from E_P;                         O(1)
11      remove P from PASSIVE
12      if check_pattern(P) then insert P into ACTIVE
13                          else insert P into POTENTIAL _
14  .
16  proc process_all_potential(→ s)        pre: new-order full-interval event
17      for P ∈ POTENTIAL do
18          if s occurs in P
19              then if check_pattern(P)
20                  then move P from POTENTIAL to ACTIVE; _
21              _
22      od
23  .
25  proc process_potential(→ P)                     pre: t_{act} has reached d_P
26      remove P from POTENTIAL;
27      if supp_{min} reachable then insert P into PASSIVE _
28  .
30  proc process_active(→ P)          pre: t_{act} has reached end of observation
31      if (t_{act} = d_P) ∨ (¬check_pattern(P))
32          then remove P from ACTIVE;
33              if supp_{min} reachable then insert P into POTENTIAL _
34      _
35  .
```

FIGURE IV.9.17   Subroutines of support estimation. Input parameter P denotes a temporal pattern, and s denotes the label of the interval associated with the new/full-interval event.

patterns continuously, the newly introduced or fully visible state interval (b, f, s) must be the reason for P becoming active. This allows for a simple pruning test: the algorithm has to be called only in case the state s occurs in P. The overall complexity is then $O(|\text{POTENTIAL}| \cdot (T + \log |C_k|))$.

Subroutine *process_potential*: In Cases (c) and (d), potential patterns might become passive at their deactivation time. The deactivation time is known in advance, and turning potential patterns into passive patterns requires no subpattern test. The online pruning is $O(1)$ since the accumulated length of the continuously updated E_P and O_P can also be updated incrementally. The algorithm complexity is therefore $O(\log |C_k|)$.

Subroutine *process_active*: The complexity is $O(T)$ for the test, $O(\log |\text{PASSIVE}|)$ for removing P from PASSIVE, $O(1)$ for the online pruning, and $O(\log |\text{POTENTIAL}|)$ for the insertion into POTENTIAL: $O(T + \log |C_k|)$.

Main routine: The initialization is $O(|C_k| \log |C_k|)$ since all candidates have to be sorted by their activation time in order to build up PASSIVE. Let us now consider the main loop.

Consider again the state transitions in Figure IV.9.15 How often do the four transitions coming from or leading to node PASSIVE occur? As they are implicitly scheduled by the interval bounds $[a_P, d_P]$ in E_P, the total number amounts to $n = \sum_{P \in C_k} \text{card}(E_P)$. Thus, we have n LEAVE-PASSIVE events and n calls of either *process_active* or *process_potential* with $t_{act} = d_P$ (corresponding to LEAVE-ACTIVE and LEAVE-POTENTIAL events). The total complexity of all these calls is then $O(n \cdot (T + \log |C_k|))$.

The complexity of the actions invoked by the NEW-INTERVAL and FULL-INTERVAL events remains to be measured, corresponding to the transitions between the active and potential node in Figure IV.9.15. For each of the L intervals, a NEW-INTERVAL and FULL-INTERVAL event is scheduled, which leads to $m = \sum_{i=1}^{L} |\text{POTENTIAL}_i|$ *process_all_potential* calls: $O(m \cdot (T + \log |C_k|))$. In the worst case, all potential patterns become active, and the observation interval terminates before the deactivation time is reached. Then, $|\text{ACTIVE}_i| = |\text{POTENTIAL}_i|$ and *process_active* is called for every potential pattern: $O(m \cdot (T + \log |C_k|))$.

This gives the complexity

$$O(|C_k| \log |C_k| + n(T + \log |C_k|) + m \cdot (T + \log |C_k|))$$
$$= O(|C_k| \log |C_k| + (n + m) \cdot (T + \log |C_k|))$$

It is difficult to estimate the practical run-time complexity of the support estimation algorithm, because it depends heavily on the (average) sizes of the three subsets PASSIVE, POTENTIAL, and ACTIVE of C_k. The most expensive operation, however, is checking $P \sqsubseteq W$. In a naive implementation—without partitioning C_k into the three subsets—we would call the subrelation check $4 \cdot L \cdot |C_k|$ times. For each out of L intervals, we have four relevant situations (Cases (a–d)), and in each case we would test every candidate against the sliding window content. Using the partitioning of C_k, we have $n + m$ of these tests. If we understand the transitions from passive to active patterns and vice versa as shortcuts of two consecutive transitions from passive to potential and potential to active, then the number of potential patterns in each iteration increases slightly: The m tests would be performed within the normal processing of potential patterns. Therefore, $\frac{n+m}{L}$ can be considered as the average number of potential patterns during the run. Since we expect from our earlier considerations that the average number of potential patterns will be smaller than the total number of candidates $|C_k|$, we have $n + m < L \cdot |C_k|$ and, therefore, a reduced complexity compared to a naive implementation.

We have claimed that a new subrelation test can be postponed until a found instance vanishes from the sliding window. We implicitly assumed that the subrelation test always returns the earliest instance of a pattern in the window in case there is more than one instance. If that would not be the case, by postponing any further tests we would miss the support of earlier patterns, which will no longer be part of the sliding window when we restart testing.

9.4.5 Testing for Subpatterns

A small but important piece of the algorithm is still missing: the subpattern check. Obviously, a pattern P may occur several times within a long state sequence. We have seen in the previous section that it is necessary for the correctness of the pruning mechanisms that the pattern matching subroutine always yield the first occurrence of the pattern. Do we have to enumerate all possible instances to decide which one occurs first? Fortunately not, since we can show the following.

Theorem 3: Given two normalized patterns P and Q with $P \sqsubseteq Q$ and two instances ϕ and ψ of P, then $\pi = \min(\phi, \psi)$ $(i \mapsto \min(\phi(i), \psi(i)))$ is also an instance of P and $b_\pi \leq \min(b_\phi, b_\psi)$ holds (where b_π denotes the first point in time when the instance ϕ is observed). By an instance ϕ we denote a mapping of pattern interval number i to interval number $\phi(i)$ in the lexicographically sorted state sequence.

```
 1  funct subrelation_check(→ P, → Q, ← π)                    P and Q normalized
 2    if dim(P) < dim(Q) then return false fi;
 3    i := 1; j := 1; π(·) := 0;
 4    return perm(i, j, P, Q, π)
 5  •
 7  funct perm(i, j, P, Q, π)                                 i/j current index in P/Q
 8    found := false
 9    repeat
10      if sₚ(i) = s_Q(j)
11        then π(i) := j; ok := true;
12             for k = 1..i do ok := ok ∧ Rₚ[i, k] = R_Q[j, π[k]] od
13             if ok then
14                   if i < dim(P)
15                     then found := perm(i + 1, j + 1, P, Q, π)
16                     else found := true
17                   fi
18             fi
19      fi
20      j := j + 1;
21    until (j > dim(Q)) ∨ (found)
22    return found;
23  •
```

FIGURE IV.9.18 The subrelation check. The function *subrelation_check* takes two patterns P and Q and returns **true** iff $P \sqsubseteq Q$. It also yields the state mapping π (cf. Definition 2).

To find the earliest instance of a given pattern, it is therefore necessary to find the instance that is the pointwise minimum of all possible instances. A naive implementation of the subpattern test is given in Figure IV.9.18. If the sliding window is very broad and contains more and more intervals, we use a more sophisticated implementation, which will not be discussed here. Obviously, algorithm 18 fulfills the stated requirement, since intervals with lower indices are tested first.

To estimate the support of a pattern P, we have to check continuously for $P \sqsubseteq W$ while sliding W along the state sequence. Can we make use of the fact that the content of the sliding window does not change abruptly but develops slowly over time? Since the subrelation test is expensive, we would like to take advantage of previous tests in order to prune as many unnecessary tests as possible. Consider a certain window position t_{act} and suppose that pattern P is not contained in W, that is, when inspecting the part of the state sequence covered by the sliding window we do not observe P. We distinguish two cases:

- Consider the case that an instance π of P will be visible shortly and $\pi(\mathbb{N})$ is a subset of the intervals contained in W. This means that all intervals participating in the instance are already visible, but only partially. As the window slides further, some temporal relationships will be resolved, but we do not make use of new state intervals. An example for this situation has been given in Figure IV.9.9, with the dashed window position and the pattern "A contains C." Although the occurrence of P is not yet determinable from the observer's point of view, since we perform analysis on historical data, it is possible to forsee the instance from the algorithm's point of view. That is, after more time, an instance π of P will appear in W, making use only of intervals we already know. Since we know that an instance is about to come into view, can we already stop checking for P and make the pattern active? We would not be allowed to do so if there is a chance of observing another instance φ of P that starts earlier than π, as discussed in the previous section. However, Theorem 3 tells us that this cannot be the case: an

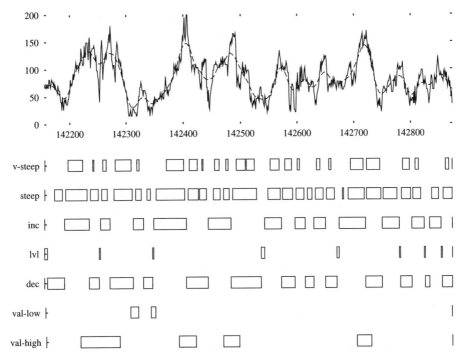

FIGURE IV.9.19   Extracted wind strength features.

instance φ making use of intervals (b_n, f_n, s_n), $n > \max \pi(\mathbb{N})$, cannot start earlier than π since the pointwise minimum of π and φ will exclude n. Therefore, it is impossible that $b_\varphi > b_\pi$. Once we have found an instance—although it may not be visible right now—we can make the potential pattern active. We only have to consider the observation interval properly.

- Suppose we will never be able to observe an instance π of P using intervals in W only. Whenever an instance φ will be visible in W, we can find the smallest index j such that interval j is not yet in W. Since P is normalized, the instance φ is monotone ($i > j \Rightarrow \pi(i) > \pi(j)$). This means that for all $i > j$, we also know that $\varphi(i)$ is not contained in the current window W. In particular, this is always true for the last state in P, $i = \dim(P)$, regardless of the actual value of j. Therefore, if we cannot observe an instance of P right now, we may wait until a new state enters the sliding window that matches the symbol of the last state in P. Unless there is no such state, we cannot find an instance of P. This can be used to postpone the next subrelation test once we had a negative test result.

In summary, when continuously checking P for being a subpattern of the sliding window, (1) we can make use of our knowledge about the interval lengths to find the earliest pattern instances (we do not have to simulate the operators view), and (2) we do not have to check continuously for new instances, but only if a new interval enters the sliding window which matches the symbol of the last state.

9.5   Evaluation and Discussion

We have examined air pressure and wind strength/wind direction data from a small island in the northern sea.* It is well known that local differences in air pressure are the cause for wind, therefore

---

*Helgoland, 54:11N 07:54O.

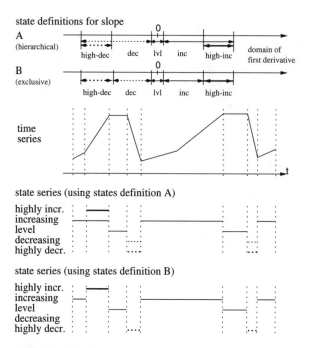

FIGURE IV.9.20   Two ways of partitioning a time series.

we should find some relationships between these variables. Although global weather forecast is (more or less) made perfectly by large-scale weather simulations, it is still not possible to precisely localize where a certain weather phenomenon will occur, to which extent, at what time. Rules about the qualitative behavior of the air-pressure curve help sailors in short-term local weather forecasting.[13]

We have applied kernel smoothing in order to compensate for noise and to get more robust estimates of the first and second derivatives. Then, the smoothed series have been partitioned into primitive patterns. To avoid missing meaningful temporal patterns, we have tried to simulate the way a human would partition the time series. In the first stage, the air-pressure curve has been segmented into increasing, level, and decreasing segments. Among the increasing segments, if the derivative is larger than 50% of all values measured for increasing derivatives, we refine them by an additional state, highly increasing. If it is larger than 80%, we call it very highly increasing. As an alternative to this overlapping state definition, we could have defined an exclusive partition such as very highly increasing, highly increasing, increasing, etc. In both cases, having chosen the threshold values heuristically, we cannot be sure that we have chosen them meaningfully (with respect to some patterns we want to discover). As we will see in the following example, if we are not sure about the threshold values used to define the states, the hierarchical definition is preferable over the exclusive. There is a pattern, "highly increasing segment meets level segment meets highly decreasing segment" in the time series depicted in Figure IV.9.20. If the threshold values for the derivatives have not been chosen appropriately, the increasing flank of the second wave will not be classified as highly increasing. But if we use the state series A in Figure IV.9.20, we will at least discover the pattern, "increasing segment meets level segment meets highly decreasing segment." If we choose the exclusive state definition B in Figure IV.9.20, the depicted state series contributes only partially to the support of both discussed patterns. If we have badly chosen some threshold values, with a hierarchical state definition we can at least be sure that we will find a pattern that is similar (in terms of the employed state hierarchy) to the true relationship.

In addition to states that characterize the slope, we used some states that address the second derivative of the air-pressure curve. High values in the second derivative can be used to distinguish

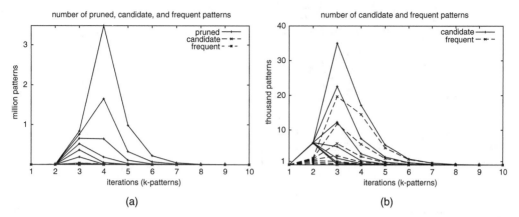

FIGURE IV.9.21 Number of pruned, candidate, and frequent patterns generated during several runs with different numbers of states.

sharp peaks from flat hills, for example. To simplify the notation, we will use the following state abbreviations: dec, lvl, inc for decreasing, constant, and increasing trends, respectively. ccv and cvx denote concave and convex curvature. High gradients and curvature are denoted by grd-high and crv-high. A suffix -w, -p, -d indicates that the segments refer to the wind strength, air pressure, or wind direction curve, respectively.

The window width t_{max} was chosen to be 48 hours, $supp_{min} = 2\%$ of the total time. Three years of hourly measured data have been preprocessed as described above, and the analysis of the resulting state sequence took 18 minutes on a 64 MB laptop computer with a Pentium II mobile processor running Linux.

In general, for a fixed support percentage, we have observed the algorithm to scale linearly in the length of the state sequence as expected. However, the computational costs increase superlinearly with the average state density or average number of states in the sliding window (caused by considering additional variables for instance). If the newly introduced variables are correlated with those used before, the number of frequent patterns increases significantly. From the way we have generated the attributes, we can expect strong patterns such as increasing after decreasing, concave after convex, highly increasing during increasing, etc. Figure IV.9.21 shows the number of pruned, candidate, and frequent patterns for a varying number of states and window widths, where this superlinear increase in the number of patterns can be observed. From Figure IV.9.21b (note the change in scale), we can see that the pruning techniques were quite efficient; more than 50% of all candidate patterns were frequent.

For any two frequent patterns $X \sqsubseteq Y$, we can build a rule (see Agrawal et al.[3] for an efficient algorithm). The confidence value of the rules, defined as the ratio of premise and rule pattern support, was not a very good indicator for rule quality. We have used slightly modified rule semantics[12] to compensate for the fact that the support decreases automatically with the increasing complexity of the patterns. Instead of the confidence value, we have used the J-measure to rank the rules.[20] It measures the difference between the *a priori* and *a posteriori* probability of the conclusion.

Here are some examples of rules we have found. Light boxes indicate premise states, and bold boxes represent conclusion states. A change in the wind direction is an indication for upcoming winds (since the front has passed). (Support of premise 15%, confidence 71%.)

north-east-d	north-west-d	**north-east-d**	grd-high-w

High gradients together with high curvature in the air-pressure curve are good indicators for significant changes in wind strength. (Support of premise 29%, confidence 95%.)

grd-high-p		grd-high-w
crv-high-p		

This rule can be found with a number of different temporal relationships between the gradient/curvature states of air pressure. The quality of the rule can be improved by relaxing the temporal relationship to a disjunctive combination of *contains*, *finishes*, *overlaps*, *meets*, and *after*. The composition of rules can be done easily by merging the observed support sets O_P for premise and rule patterns.

While the previous rule did not distinguish between increasing or decreasing wind strength, the following does. (Support of premise 12%, confidence 73%):

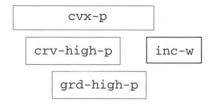

9.6   Related Work

In general, finding temporal patterns in time series is strongly related to subsequence matching in time series. There has recently been a lot of work on subsequence matching.[1,2,6,9,14] Most of these approaches assume that an initial pattern is provided by the expert and so they restrict themselves to find all subsequences that deviate from this initial pattern no more than a certain threshold. These patterns can be used to verify a temporal rule an expert had in mind. However, verification has the limitation that we have to guess the rule before we can verify it, thus we can miss interesting rules. All the cited papers deal with continuous features (the time series itself or its Fourier transform, for example), whereas we analyze a labeled interval sequence.

The discovery of rules in time series has also been considered before.[8,11,19] The main differences from earlier methods are: (1) the use of Allen's interval logic to describe the relation between states (only parallel or adjacent states have been considered before), and (2) the adaption of association rule-mining techniques to overcome the combinatorial explosion of possible rules. Especially in the clinical domain, prior knowledge is often used to limit the space of rules. Although additional knowledge can easily be incorporated to restrict the considered rules, prior knowledge is not a requirement in this approach.

We have adapted techniques from the discovery of association rules[3] to the problem of finding rules on temporal patterns in state sequences. Association rules have been generalized to work with time-stamped data, that is, data ordered in time (but without duration).[16,21] The work of Mannila et al.[16] is probably most closely related to our work as described in this chapter.

9.7   Conclusion

We have proposed a technique for the discovery of temporal rules in state sequences and have applied this technique to the discovery of qualitative dependencies in multivariate time series. The examples in Section 9.5 have shown that the proposed method is capable of finding meaningful rules that can be used not only as rules-of-thumb by a human, but also in a knowledge-based expert system. We see the main application area in knowledge discovery applications, where the system may help experts gain a better understanding of dependencies and relationships between the variables as they develop over time. The rules can be easily interpreted by a domain expert, who can verify the rules by means of his background knowledge or use them as an inspiration for further investigation. Even if there is already considerable background knowledge, the application of this method might be valuable, for example, if the known rules incorporate more variables than available in a specific technical system. For example, weather forecasting rules as discussed by Karnetzki[13] also use information about the general weather outlook (cloudiness) or information from the local weather forecasting station. Such information might be difficult to incorporate or expensive to measure, and in such a case one is interested in how much one can achieve by just using the available variables.

Acknowledgments

This work was supported by the Deutsche Forschungsgemeinschaft (DFG) under grant no. Kl 648/1. Thanks also to the Deutsche Wetterdienst (DWD) for providing the data.

References

1. Agrawal, R., Faloutsos, C., and Swami, A., Efficient similarity search in sequence databases, *Proceedings of the 4th International Conference on Foundations of Data Organizations and Algorithms*, Chicago, 1993, 69.
2. Agrawal, R. et al., Fast similarity search in the presence of noise, scaling, and translation in time-series databases. Proceedings of the 21st International Conference on Very Large Databases, 1995.
3. Agrawal, R. et al., Fast discovery of association rules, in *Advances in Knowledge Discovery and Data Mining*, Fayyad et al., Eds., MIT Press, Cambridge, MA, 1996, 307.
4. Allen, J.F., Maintaining knowledge about temporal intervals, *Commun. ACM*, 26(11), 832, 1983.
5. Bakshi, B.R. and Stephanopoulos, G., Reasoning in time: modelling, analysis, and pattern recognition of temporal process trends, in *Advances in Chemical Engineering*, vol. 22, Stephanopoulos, G. and Han, C., Eds., *Academic Press*, New York, 1995, 485.
6. Berndt, D.J. and Clifford, J., Finding patterns in time series: a dynamic programming approach, in *Advances in Knowledge Discovery and Data Mining*, Fayyad, U.M. et al., Eds., MIT Press, Cambridge, MA, 1996, 229.
7. Capelo, A.C., Ironi, L., and Tentoni, S., Automated mathematical modelling from experimental data: an application to material science, *IEEE Trans. Syst. Man Cybernetics, Part C*, 28(3), 356, 1998.
8. Das, G. et al., Rule discovery from time series, in *Proceedings of the 4th International Conference on Knowledge Discovery and Data Mining*, AAAI Press, Menlo Park, CA, 1998, 16.
9. Faloutsos, C., Ranganathan, M., and Manolopoulos, Y., Fast subsequence matching in time-series databases, Proceedings of the ACM SIGMOD International Conference on Data Management, May 1994.
10. Fayyad, U.M. et al., Eds., *Advances in Knowledge Discovery and Data Mining*, MIT Press, Cambridge, MA, 1996.
11. Guimarães, G. and Ultsch, A., A method for temporal knowledge conversion, in *Proceedings of the 3rd International Symposium on Intelligent Data Analysis*, Hand, D.J., Kok, J.N., and Berthold, M.R., Eds., Springer, Berlin, 1999, 369.
12. Höppner, F. and Klawonn, F., Finding informative rules in interval sequences, in *Proceedings of the 4th International Symposium on Intelligent Data Analysis*, vol. 2189 of *Lecture Notes in Computer Science*, Springer, Berlin, 2001, 123.

13. Karnetzki, D., *Luftdruck und Wetter*, 3rd ed., Delius Klasing, 1999.

14. Keogh, E.J. and Pazzani, M.J., Scaling up dynamic time warping to massive datasets, in *Proceedings of the 3rd European Conference on Principles and Practices of Knowledge Discovery in Databases*, vol. 1704 in *Lecture Notes in Artificial Intelligence*, Springer, Berlin, 1999, 1.

15. Konstantinov, K.B. and Yoshida, T., Real-time qualitative analysis of the temporal shapes of (bio)process variables, *Artif. Intelligence Chem.*, 38(11), 1703, 1992.

16. Mannila, H., Toivonen, H., and Verkamo, A.I., Discovery of frequent episodes in event sequences, Technical Report 15, University of Helsinki, Finland, Feb. 1997.

17. McIlraith, S.A., Qualitative data modeling: application of a mechanism for interpreting graphical data, *Comput. Intelligence*, 5, 111, 1989.

18. Sankoff, D. and Kruskal, J.B. Eds., *Time Warps, String Edits, and Macromolecules: The Theory and Practice of Sequence Comparison*, Addison Wesley, Reading, MA, 1983.

19. Shahar, Y. and Musen, M.A., Knowledge-based temporal abstraction in clinical domains, *Artif. Intelligence Med.*, 8, 267, 1996.

20. Smyth, P. and Goodman, R.M., Rule induction using information theory, in *Knowledge Discovery in Databases*, Piatetsky-Shapiro, G. and Frawley, W.J., Eds., MIT Press, Cambridge, MA, 1991, 159.

21. Srikant, R. and Agrawal, R., Mining sequential patterns: generalizations and performance improvements, *Proceedings of the 5th International Conference on Extending Database Technology*, Avignon, France, March. 1996.

10

Integrating Adaptive and Intelligent Techniques into a Web-Based Environment for Active Learning

Hongchi Shi
University of Missouri, Columbia

Othoniel Rodriguez
University of Missouri, Columbia

Yi Shang
University of Missouri, Columbia

Su-Shing Chen
University of Missouri, Columbia

Abstract. Web-based learning offers many benefits over traditional learning environments: distance independence, time independence, computing platform independence, and classroom size independence. In the last five years, thousands of Web-based courses and other educational applications have been developed. However, the problem is that most of these are nothing more than a network of static hypertext pages. To better support the learning activities of a large number of learners with different learning styles and diverse backgrounds, and learning goals, a challenging research issue is to develop Web-based learning environments that are adaptive and intelligent.

In this chapter, we present an approach to integrating adaptive and intelligent techniques into a Web-based environment to support active learning, the intelligent distributed environment for active learning (IDEAL). IDEAL is integrated with adaptive and intelligent techniques and built upon five elements: learning object technology, instructional design theories, active learning approach, multi-agent technology, and neural, reinforcement, and symbolic machine learning methods. IDEAL uses the learner's learning characteristics, background, and learning goals to select, organize, and present learning material in the most appropriate way to support learner-centered, self-paced, and highly interactive learning.

10.1 Introduction

The WWW, the forthcoming markup-fueled semantic Web in particular,[1] represents a rich environment with a great potential for the deployment of learning resources. The term being adopted for Web-based learning is e-learning. In its current manifestations, e-learning has not reached the stage where it is able to exploit the possibilities inherent in the semantic Web. The introduction of the concept of learning objects (LO) and recent standardization efforts by IEEE-LTSC and the IMS Consortium in areas related to course management systems, such as learner information packaging (LIP),[2] quiz and test interoperability (QTI),[3] and learning object metadata (LOM),[4] however, establish a minimum open infrastructure that can be used for the exploitation of the features inherent to the semantic Web for making possible the so-called e-learning.

E-learning has become one of the fastest growing and most promising markets in the education industry. Web-based learning offers many benefits over traditional learning environments: distance independence, time independence, computing platform independence, and classroom size independence. Thousands of Web-based courses and other educational applications have been developed within the last five years. However, today's e-learning can be characterized by content that is mostly non-interactive, not adaptable or tuned to individual learner idiosyncrasy, oriented toward technical training within corporate environments, and nothing more than a network of static hypertext pages. Web delivery alone will not revolutionize the learning experience, and the very nature of education should be changed from a learn-by-telling model to a learn-by-doing model.[5] To better support the learning activities of a large number of learners with different learning styles and diverse backgrounds and learning goals, a challenging research issue is to develop Web-based learning environments that are adaptive and intelligent.

Many adaptive and intelligent techniques have been developed for Web-based learning.[6] Although these techniques can enhance different aspects of Web-based learning, they have not yet found their place in real courseware used by thousands of distance learners. Most of the existing Web-based intelligent educational systems are typical lab systems that have never been used for teaching real online classes or are only used in a few small classes. A major limitation of these systems is that the knowledge to support system adaptation and intelligence is represented in symbolic form and requires a time-consuming knowledge acquisition and engineering process, which is prohibitive for large-scale Web-based applications.

In this chapter, we present our approach to integrating adaptive and intelligent techniques into a Web-based environment to support active learning, the intelligent distributed environment for active learning (IDEAL).[7-12] IDEAL is integrated with adaptive and intelligent techniques and is built upon five elements: learning object technology, instructional design theories, active learning approach, multi-agent technology, and neural, reinforcement, and symbolic machine learning methods. It uses the learner's learning characteristics, background, and learning goals to select, organize, and present learning material in the most appropriate way to support learner-centered, self-paced, and highly interactive learning. IDEAL is an open scalable e-learning environment, providing comprehensive and integrated e-learning solutions tailored to the learner's needs, allowing each learner to create a personalized learning path.

10.2 Adaptive and Intelligent Techniques and Active Learning Approach

Researchers in different disciplines have studied the active learning approach, characteristics of learners on their learning performance, and adaptive and intelligent techniques that may be used to support Web-based active learning.

10.2.1 Adaptive and Intelligent Techniques for Web-Based Learning

Most adaptive and intelligent techniques for Web-based learning come from two areas: adaptive hypermedia systems and intelligent tutoring systems that use the knowledge about the domain, learner, and teaching strategies to support flexible individualized tutoring.[6] The existing techniques can be classified into four categories: curriculum sequencing, problem solving support, adaptive hypermedia techniques, and learner model matching, as shown in Figure IV.10.1.

Curriculum sequencing (also known as instructional planning) is the oldest intelligent tutoring technology.[13,14] It attempts to provide the learner with the most suitable individually planned sequence of knowledge units to learn and the most suitable sequence of learning tasks to work with. In other words, it helps the learner find an optimal path through the learning material. A classic example is the BIP system.[15] There are two different kinds of sequencing: active and passive. Active sequencing finds the best individual path to achieve a learning goal, i.e., a subset of domain concepts or topics to be mastered. Example systems include ELM-ART-II,[16] AST,[17] ACE,[18] and KBS-Hyperbook.[19] Passive sequencing (also known as remediation) is a reactive technology and does not require an active learning goal. It offers the learner a subset of available learning material, which can fill the gap in a learner's knowledge to resolve a misconception. Example systems include InterBook,[20] PAT-InterBook,[21] CALAT,[22] and Remedial Multimedia System.[23] Sequencing can happen at the knowledge level or task level. Knowledge sequencing determines the next learning subgoal such as the next concept, set of concepts, topics, or lessons to learn. Task sequencing determines the next learning task, such as problem, example, or test, within the current subgoal.

For many years, problem-solving support has been considered the main duty of an intelligent tutoring system. There are three major problem-solving support technologies: intelligent analysis of learner solutions, interactive problem-solving support, and example-based problem-solving support. Intelligent analysis of learner solutions deals with a learner's final answers to learning problems. Systems implementing this technology include ELM-ART,[24] ELM-ART-II,[16] and PAT-InterBook.[21] Interactive problem-solving support is a more recent technology. Instead of waiting for the final solution, this technology can provide a learner with intelligent help at each step of problem solving. The classic example is the LISP-TUTOR.[25] The example-based problem-solving technology is the most recent. It helps learners solve new problems not by articulating their errors, but by suggesting relevant successful problem-solving cases from previous experience. Example systems include ELM-PE,[14] ELM-ART,[24] and ELM-ART-II.[16]

Adaptive hypermedia techniques apply different forms of learner models to adapt the content and the links of hypermedia pages to the learner.[26–28] There are two major technologies in adaptive hypermedia: adaptive presentation and adaptive navigation support. The adaptive presentation technology aims at adapting the content of a hypermedia page to the learner's goals, knowledge, and other information stored in the learner model. For example, advanced learners receive more detailed

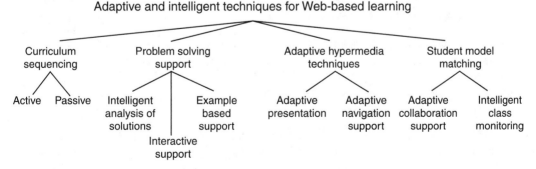

FIGURE IV.10.1   A classification of adaptive and intelligent techniques for Web-based learning.

and deep information, while novices receive additional explanation. Adaptive presentation is very important in the WWW context, where the same page has to suit to very different learners. Systems using adaptive presentation include PT[29] and AHA.[30] The adaptive navigation support technology aims at supporting the learner in hyperspace orientation and navigation by changing the appearance of visible links. It can be viewed as a generalization of the curriculum sequencing technology in the hypermedia context. An example system is Hypadapter.[31]

Learner model matching represents a new group of Web-inspired technologies for Web-based learning. It analyzes and matches learner models of many learners at the same time, whereas traditional intelligent tutoring systems usually work with one learner and one learner model at a time. The two types of learner model matching being developed in recent years are adaptive collaboration support and intelligent class monitoring. Adaptive collaboration support uses a system's knowledge about different learners to form a matching group for different kinds of collaboration.[32] Intelligent class monitoring identifies the learners who have learning records that are essentially different from those of their peers. This technology is effective in finding learners who need special attention in a Web-based classroom.[33] The technology has also been used for just-in-time teaching of physics,[34,35] where Web-based assignments and assessment material are used to provide the regular classroom teacher with feedback about a learner group's performance just before the class meets, allowing the teacher to focus his/her interventions into the identified weak areas.

Although existing adaptive and intelligent technologies can enhance different aspects of Web-based learning, they have not yet found their place in real courseware used by thousands of distance learners.[6] The dozens of commercial and university-grown Web courseware systems used in hundreds of real distance courses apply very little adaptive and intelligent technologies. Most of the existing Web-based intelligent educational systems are typically lab systems that have never been used for teaching real online classes or are only used in a few small classes. A major limitation of these systems is that the knowledge is represented in symbolic forms and requires a time-consuming knowledge acquisition and engineering process, which becomes prohibitive when scaled-up to meet the wide variety of adaptations required for large-scale Web-based applications. In Web-based learning, due to their weakness in content organization and their abilities of automated learning of curriculum sequencing and adaptive navigation knowledge, these systems have difficulty in effectively utilizing the large amount of learning material on the Web and cannot handle a large number of learners with different learning styles and diverse backgrounds and career goals. A new approach is developed in this chapter to overcome some of these problems.

10.2.2 Active Learning Approach

The recent shifts in emphasis in educational research include more emphasis on multi-sensory and multimedia data sources, simulations, performance assessments, sophistication of content models, and viewing the learner as both an individual and social learner within a classroom conceived of as a complex, self-organizing, self-regulating system.

The literature on educational research suggests that learners who are actively engaged in the learning process will be more likely to achieve success.[36] Active learning emphasizes engaging learners in the learning process,[37–39] where learning activities involve some kind of experience or dialog. The two main kinds of dialog are dialog with self (think reflectively) and dialog with others, while the two main kinds of experience are observing and doing.

Active learning can be incorporated into learning in the following ways:

- *Expanding learning experience.* The traditional teaching consists of little more than having learners read a text and listen to a lecture, providing a very limited form of dialog with others. Examples of more dynamic forms are creating small groups of learners and having them make

TABLE IV.10.1   Learning Modes Related to Experience and Dialog

Kind of Dialog	Kind of Experience	
	Observing	Doing
Self	Behavioral	Constructive/exploratory
Others	Receptive	Peer learning

a decision or answer a focused question periodically, or finding ways for learners to engage in authentic dialog with people, other than fellow classmates, who know the subject.

- *Taking advantage of the power of interaction.* There are four modes of learning when the two kinds of dialog are coupled with the two kinds of experience, as shown in Table IV.10.1. Each of the four modes has its own value, and just using more of them should add variety and thereby be more interesting for the learners. However, when properly connected, the various learning activities can have an impact that is more than additive or cumulative. They can interact synergistically and multiply the educational impact.
- *Creating dialectic between experience and dialog.* New experience has the potential to give learners a new perspective on what is true (beliefs) and/or what is good (values) in the world. Dialog has the potential to help learners construct many possible meanings of experience and insights that come from dialogs. People learn faster when new concepts are useful in their present as well as future lives.

The growing emphasis on learner-centered active learning has yielded a veritable revolution in educational theory and practice. A number of current theories of learning and pedagogy revolve around constructivism, which emphasizes the learner's knowledge construction process. The constructivist learning theory holds that learners actively construct and reconstruct knowledge out of their experience in the world.[40] As a set of instructional practice, constructivism favors processes over end products; guided discovery over expository learning; authentic, embedded learning situations over abstracted, artificial ones; portfolio assessments over multiple-choice exams; many educators and cognitive psychologists have applied constructivism to development of learning environments. From these applications, Jonassen[41] has summarized the following design principles:

- Create real-world environments that employ the context in which learning is relevant.
- Focus on realistic approaches to solving real-world problems.
- The instructor is a coach and analyzer of the strategies used to solve these problems.
- Stress conceptual interrelatedness, providing multiple representations or perspectives on the content.
- Instructional goals and objectives should be negotiated and not imposed.
- Evaluation should serve as a self-analysis tool.
- Provide tools and environments that help learners interpret the multiple perspectives of the world.
- Learning should be internally controlled and mediated by the learner.

The constructivist learning theory can be applied to Web-based education by providing opportunities for learners to engage in interactive, creative, and collaborative activities that encourage knowledge construction. It can provide a good learning environment by supporting three elements: choice, diversity, and congeniality. If learners are empowered to choose what to construct or create, they are more likely to be personally engaged and invested in the activity. By recognizing that learners have different skill levels and learning styles, a rich learning environment encourages multiple representations of knowledge.[40] Having a congenial environment free from constraints of time and space is central for creating and sharing knowledge. In the active learning approach, learners should be engaged in active exploration, be intrinsically motivated, and develop an understanding of a domain through challenging and enjoyable problem-solving activities. Several systems and procedures that support active learning have been developed.[42–46]

10.2.3   Learning Characteristics

Learning performance depends on many factors, such as motivational factors, cognitive style, right–left brain hemisphere dominance, and socio-economic factors. Understanding different learning modalities is essential to improving learning performance. Learning modalities typically include physical, environmental, cognitive, affective (emotional), and socio-economic factors. Methods for assessing these factors include:

- Myers–Briggs type indicator, which attempts to measure and identify a person's psychological or personality profile based on Jung's typology of conscious functioning (archetypes)[47–49]
- Personal style inventory by Hogan and Champagne based on Myers–Briggs inventory
- Grasha–Riechmann student learning style scales (GRSLSS)[50] designed for university populations, which focuses on how learners interact with teachers and colleagues in their learning processes
- Auditory–visual–tactile/kinesthetic learning style.[51] In the multiple sensory approach of learning theory, a learner uses his favored style as a point of entry into engaging a specific content area for learning and other styles to reinforce learning.
- Environmental effects and chronopsychology.[52] Learning is more productive when the learner is engaged in an overall healthy and personally appropriate environment.

The learning styles of men and women are quite different.[53–56] Belenky and MacKeracher's models[57] suggest male separate and female connected learning styles. This gender difference is true not only in traditional face-to-face learning environments, but also in online distance learning environments.[58–60] In traditional learning environments, the separate male learning style has long dominated female learners by simply speaking more in the classroom. In online environments, more male learners respond to posted questions than female learners, and male learners tend to post more messages than female learners. The content of male messages often has a tone of certainty, is slightly arrogant, brief, and not tempered by polite words to reduce negative reactions of the reader. In contrast, female learners place emphasis on relationships, are empathetic in nature, and prefer to learn in an environment where cooperation is stressed rather than competition. Males communicate with an underlying purpose of seeking power or status, while females more often communicate striving to help others. Males usually dominate the conversation and effectively silence females.[56,61–63]

The design of an online learning environment should take the differences in learning characteristics into consideration and strive to provide a more equitable learning environment that accommodates various learning styles, supports collaborative learning, and avoids male dominance.[64,65]

10.3   Integrating Adaptive and Intelligent Techniques into Web-Based Active Learning

Applying research results in the areas of information technology, machine learning, cognitive psychology, and educational technology to significantly advance the field of Web-based education, we are developing an intelligent learning environment, IDEAL, for large-scale Web-based active learning. Integrated with adaptive and intelligent techniques, IDEAL dynamically models and updates the learning characteristics and learning performance of learners during the learning process, and intelligently selects and presents the learning material organized into open learning objects (OLOs). Following the active learning approach, OLOs are designed based on proven instructional design theories to accommodate diverse learners. Coarse-grain sequencing of OLOs is determined based on instructional design theories, whereas the knowledge of fine-grain sequencing is learned based on a bottom-up cognitive model using efficient neural network, reinforcement, and symbolic machine learning methods. Multiple sequencing models are learned with respect to different types of learners, e.g., female learners or learners with little background knowledge. After they are learned, these models are used to provide personalized intelligent guidance to individual learners to help them achieve the best learning result in the shortest time. The sequencing models are updated and improved continuously as more learners use the system and more training data become available. Symbolic high-level knowledge extracted from these models provides valuable insights to cognitive psychology about how learners with different learning characteristics respond to different types of learning material and teaching methods, and under what circumstances they learn most effectively.

10.3.1   Open Learning Objects for Content Organization

Information technology is transforming teaching and learning at all levels of education. An instructional technology developed for organizing educational material, learning object (LO), poises to become the technology of choice in the next generation of instructional design, development, and delivery. Technological standards are being actively developed in several international organizations, including IEEE,[66] IMS Global Learning Consortium,[67] and Advanced Distributed Learning Network[68] to facilitate the widespread adoption of the learning object approach.

The idea of learning objects comes from the object-oriented paradigm of computer science. The object-oriented paradigm values the creation of components called objects that can be reused. Informally speaking, learning objects are the small indivisible instructional components that address a single instructional need and can be reused. They are digital entities and can be delivered over the Internet, which enables any number of people to access and use them simultaneously. They provide a means for efficient development of a large amount of computer-based, interactive, multimedia instruction. Organizing learning material into LOs represents an effort to decimate knowledge concepts and topics into self-contained modules that can be easily reused in different learning situations/contexts.

In the IEEE specification, a learning object is defined as any entity that can be used, reused, or referenced during technology-supported learning. Examples of learning objects include multimedia content, instructional content, instructional software, and software tools that are referenced during technology supported learning. In a wider sense, learning objects could include learning objectives, persons, organizations, or events. Relevant properties of learning objects include type of object, author, owner, terms of distribution, and format. Where applicable, learning object metadata may include pedagogical properties such as teaching or interaction style, grade level, mastery level, and prerequisites.

Although these current efforts in learning objects substantially advance the goals of interoperability, they do not address the need for accessibility to the internal description of the LOs themselves. Under the present standards, an LO remains an opaque entity from which valuable information cannot be easily gleaned or harvested nor introduced or adapted. For example, the need for learner modeling both in terms of general interaction preferences and in terms of learning is not very well supported within these opaque objects. Most of the technologies available for authoring LOs[69,70] and many of the current proposals for LOs[71] are based on proprietary technologies.

In this section, we present an approach to open learning objects (OLOs) which is based on open standards, mostly the XML markup language. OLOs designed using this approach still retain the encapsulation and modularity required to abstract those implementation aspects that an author would like to reserve for future enhancement and growth, yet they support a systematic and consistent way for extracting the results of the interaction with the learner. This interaction history or trace information, stored in a centrally located learner information record repository, can be used to better model the learner and also to better test and validate the OLO itself across experiences from many learners.

10.3.1.1 Instructional Design of Open Learning Objects

Two important questions need to be answered in designing OLOs. One is the scope of each OLO, and the other is the order of presenting multiple OLOs. General guidance in answering these questions is available in various instructional design theories[72] including elaboration theory,[73] work model synthesis,[74] and the four-component instructional design (4C/ID) model.[75]

The elaboration theory selects and sequence content in a way that optimizes the attainment of learning goals.[72,76] Its simplifying conditions method (SCM) provides relevant information regarding scope and sequence of instructional content. SCM is composed primarily of two parts: epitomizing and elaborating. Epitomizing means finding the simplest version of the task that is representative of the entire task. Elaborating means teaching learners increasingly complex versions of the task.

Work model synthesis provides a framework in which individual objectives can be combined into meaningful, real-world performance, or work models.[74,77] It also builds on the notion of increasingly complex micro-worlds.[78] Instructional objectives can be mapped to work models in various ways, including many-to-one, one-to-many, and one-to-one mappings. A work model specifies how instructional events can be created. Work models can be used to create a number of instructionally equivalent events that could be traversed (or bypassed) in different orders by different learners. This fits perfectly with the needs of adaptive learning environments. For example, instructionally equivalent but presentation-style diverse learning objects (visually oriented, simulation-oriented, etc.) could be delivered based on learner profiles.

The 4C/ID model outlines an approach that supports the learning of complex cognitive skills.[75] It has four major steps: (1) principled skill decomposition, in which the complex cognitive skill to be learned is broken into a set of recurrent (algorithmic) constituent skills and a set of non-recurrent (heuristic) constituent skills; (2) further analysis of these two sets of constituent skills, which reveals the knowledge that supports these skills; (3) selection of instructional methods for practicing constituent skills and presenting supporting information; and (4) composition of a training strategy. Recurrent constituent skills are skills demonstrated in the same manner each time. They are usually taught using part-task practice, and the prerequisite knowledge is provided just in time. Non-recurrent skills are skills performed differently in different situations. They are usually taught using whole-task practice, and the supporting knowledge is presented to promote elaboration and understanding. Scopes of instructional content are identified at three levels: skill clusters, case types, and specific problems. Correspondingly, content sequencing is done at three levels: macro-level, meso-level, and micro-level. At the macro-level, skill clusters are ordered such that the skills in the first cluster are prerequisites for success in the second cluster, and so on. At the meso-level, case types are ordered according to a whole-task sequence. Finally, at the micro-level, specific problems are ordered based on the cognitive load theory, in which the problem formats (or types) and the whole-task sequence interact.

When applying these instructional design theories to OLO design, the principles can be summarized as follows:

- *Scope.* OLOs do not all need to have the same size and should be large enough to teach either the epitome or elaboration (elaboration theory). OLOs are the instructional events resulting from the instantiation of a work model and should be large enough to teach meaningful, real-world performance (work model synthesis). OLOs can be of three sizes: skill clusters, case types, and specific problems. The first cluster should be small enough for learners to begin practicing a simplified but authentic version of the whole task within a few days. The final cluster must be large enough to rely on all of the constituent skills identified in the preliminary analysis. Specific problems should only be large enough to provide examples or practice of a specific skill (4C/ID).

- *Sequence.* OLOs should be presented in order of increasing complexity, beginning with the epitome or simplest case (elaboration theory). OLOs should be sequenced in an order that simulates the real-world performance with increasing fidelity. Instructionally equivalent learning objects can be used to substitute each other in the sequence (work model synthesis). OLOs should be sequenced according to their levels and types. Macro-level skill clusters should be sequenced in a part-task manner, meaning that skills are taught one at a time and gradually combined. Meso-level case types should be sequenced according to a whole-task order, in which all skills are taught simultaneously. Micro-level specific problems can be sequenced in the common simple-to-complex order or, when feasible, in a random sequence in order to promote skill transfer (4C/ID).

10.3.1.2 Adaptability of Open Learning Objects

Adaptability is a critical requirement for OLOs in order to accommodate the varied classes of learners that access the OLOs through IDEAL. OLOs make learning material more flexible and enhance the ability of adaptive delivery and presentation. The OLO adaptability enables OLOs to accommodate the evolving model of the learner in terms of skill levels and cognitive style preferences.

The first adaptability of OLOs is in language. The OLO XML markup allows the tagging of each text string to be used in the OLO with a language attribute. Although there is the possibility of automated translation, experience has shown that this mechanical translation many times fails to convey the intended meaning of whole sentences or paragraphs. Although a first-level rough translation can be attempted to help identify the equivalent translated vocabulary, keywords, phrases, and other sentence structural elements, full translation should always be completed and validated by the appropriate subject domain expert that knows both the origin and target natural languages. Of course, the learner long-term modeling (LTM) must include the preferred learning language, which, if supported by the OLO, is used for all verbal/textual content.

Learning style adaptability for OLOs focuses on those learning styles from the many styles identified[79,80] that are more amenable to support in the Web-based e-learning environment. Of the many developed over time, the Felder–Silverman[81] learning style classification centers on the answer to five fundamental questions,[82] as shown in Table IV.10.2. All the styles identified by this index of learning styles (ILS) can be significantly impacted through Web-based e-learning. The ILS questionnaire of 44 questions is available on the Web.[83] Although it does not attempt to measure the inductive/deductive style, it can be usable as the basis for guidance in OLO learning style adaptation.

Taking, for example, the sensory modality learning styles described as visual learners and verbal/textual learners, the OLO authoring process should incorporate multimedia supportive of adaptation to these styles. Visual learner adaptation should exploit colors, images, graphics, and animations to convey as much of the content as possible while minimizing text output. The use of iconic languages for interaction with the learner is also advisable. Verbal/textual learners should be provided with audio clips or synthesized audio describing any visual animation or, at least, reading out the text content in a synchronized fashion with the text display. These learners should be presented with well-structured text content and with appropriate letter sizing, coloring, and sequencing to optimize retention. Administration of learning style inventory often identifies no particularly dominant learning style preference for a learner. To accommodate this case, OLO learning style adaptation mechanism should be able to aggregate OLO content geared to a mixed learning style. Similarly the subject matter content of an OLO may by itself limit the modalities in which it may be delivered.

Accessibility adaptation addresses OLO issues related to any special learning barriers or disability limitations associated with the learner as part of the learner model. The OLO approach adopts the

TABLE IV.10.2   Learning Style Classification Based on Answers to Five
Fundamental Questions

Learning Style Classification	Description
What type of information does the student preferentially perceive?	
sensing learners	concrete, practical, oriented toward facts and procedures
intuitive learners	conceptual, innovative, oriented toward theories and meanings
Through what sensory modality is sensory information most effectively perceived?	
visual learners	prefer visual representations of presented material— pictures, diagrams, flow charts
verbal learners	prefer written and spoken explanations
With which organization of information is the student most comfortable?	
inductive learners	prefer presentations that proceed from the specific to the general
deductive learners	prefer presentations that go from the general to the specific
How does the student prefer to process information?	
active learners	learn by trying things out, working with others
reflective learners	learn by thinking things through, working alone
How does the student progress toward understanding?	
sequential learners	linear, orderly, learn in small incremental steps
global learners	holistic, system thinkers, learn in large leaps

WWW Consortium standard on accessibility guidelines.[84] The rules for accessibility adaptation are taken directly from that standard and include the adaptation of font sizes and colors, automatic text-to-speech conversion for visually impaired learners or speech-to-text conversion for hearing-impaired, descriptive text for all images and graphics, and provision for alternative learner input mechanisms.

The skill level (SL) adaptation requirement introduces another view of OLOs where the OLO content is tagged with skill-level attributes for learners of different skill levels. However, the higher skill levels are not each a superset of the lower skill levels necessarily. Instead, each piece of content and synchronized-interaction component of OLOs is marked up with membership into a particular skill level. For some OLOs, these pieces overlap and are marked up as belonging to more than one skill level, while other pieces are used soley for a particular skill level and are marked up as such.

The architecture of instruction (AOI) adaptability requires the content of OLOs to be organized along four architectural lines, which rely on the instructional design strategies already discussed. It has been proposed[85,86] that a particular architecture of instruction is more appropriate to a particular learner, and as the learner skill level evolves, the OLOs adapt to being presented from the corresponding compatible AOI. Otherwise, the AOI view of an OLO is exploited primarily during the OLO design phase. This allows the authoring of OLOs with clear objectives in mind and with a clear delimitation of concerns that the OLO authoring tool can support.

The skill level adaptation correlates with the AOI adaptation that, in turn, is intimately correlated with particular learning theories, as shown in Table IV.10.3.

10.3.2   Learner Profiles and Learner Modeling

The knowledge about personal traits, skill levels, and learning material access patterns of learners is the most important aspect of a learner-centered intelligent learning environment.[26,87] A key requisite for intelligence and adaptation in a learning environment is learner modeling.

TABLE IV.10.3   Architectures of Instruction Related to Skill Levels

Skill Level	Architecture of Instruction	Learning Theory
Novice	Behavioral	Behaviorism
Beginner	Receptive	Rote learning
Intermediate	Behavioral	Behaviorism
Advanced	Guided discovery, exploratory	Constructivism
Expert	Guided discovery, exploratory	Constructivism

IMS has defined a standard model for learners called the IMS learner information packaging (LIP) model.[2] IMS LIP is based on a data model that describes those characteristics of a learner needed for the general purposes of:

- Recording and managing learning-related history, goals, and accomplishments
- Engaging a learner in a learning experience
- Discovering learning opportunities for learners

A learner's profile contains the following IMS LIP elements:

- *Identification:* Biographical and demographic data relevant to learning
- *Goal:* Learning, career, and other objectives and aspirations
- *QCL:* Qualifications, certifications, and licenses granted by recognized authorities
- *Activity:* Any related activity in any state of completion
- *Interest:* Information describing hobbies and recreational activities
- *Competency:* Cognitive, affective, psychomotor skills, knowledge, and abilities
- *Accessibility:* General accessibility to the information as defined through language capabilities, disabilities, eligibilities, and learning preferences including cognitive preferences (e.g., issues of learning style), physical preference (e.g., preference for large print), and technological preference (e.g., preference for a particular computer platform)
- *Transcript:* Record providing summary of academic achievement
- *Affiliation:* Membership in professional organizations, etc.
- *Security key:* A set of passwords and security keys
- *Relationship:* A set of relationships between the core elements or sub-components

IDEAL views learner modeling being performed on two different time scales: long-term and short-term modeling. Long-term modeling (LTM) attempts to model those aspects of a learner that are not expected to change too dynamically during the course of interaction with a group of OLOs. Long-term learner modeling includes the administration of learning style inventory[81] when a learner begins using IDEAL and summative assessments after completing a group of OLOs that represent a course, a unit, or higher-level OLO grouping. These are independently represented in IMS LIP as subactivities within the activity field representing the course or unit.

Short-term modeling (STM) refers to the modeling of the learning achieved by the learner when interacting with a particular OLO. This modeling can be performed indirectly or directly. Indirect short-term modeling occurs when the OLO is able to infer from the interaction with the learner that some learning specific to the OLO has occurred. Indirect short-term modeling includes measuring the number of times a learner reviews the OLO, measuring the total time taken to complete the topic or concept embodied in the OLO, and the results of other local interactive or constructive exercises about the OLO presented to the learner. This information can be used immediately and automatically by the OLO. Direct short-term modeling is carried out by explicit administration of an assessment that evaluates the learner performance as a skill level on the current OLO. The current set of skill levels as defined for IDEAL are:

- Novice
- Beginner
- Intermediate
- Advanced
- Expert

The learning environment can update the skill level of a learner using a simplified Bayesian network from the evidence of a learner's performance on the OLO. Direct and indirect short-term modeling extends the classical overlay model[88] that associates a performance metric with each learned topic.

There are many techniques for generating learner models. Most of them are computationally complex and expensive, for example, the Bayesian networks,[87–90] the Dempster–Shafer theory of evidence,[91] and the fuzzy logic approach.[92] Other techniques such as the model tracing approach,[93] although computationally cheap, can only record what a learner knows but not the learner's behavior and characteristics.

The difficulties in applying Bayesian modeling are the high cost in knowledge acquisition and in the time to update the learner model. The inference in the Bayesian belief networks is NP-hard, and the model requires prior probabilities. In developing practical and efficient Bayesian methods, we trade complexity of knowledge representation and depth of modeling for linear-time belief updating and a small number of model parameters.

In IDEAL, a portion of the short-term learner model is inferred from the performance data using a Bayesian belief network. The measure of how well a skill is learned is represented as a probability distribution over skill levels, such as novice, beginning, intermediate, advanced, and expert. Under the assumption that the performance on questions is independently distributed, to model one skill with n skill levels and q questions for each in a Bayesian network, we need nq probabilities plus the n prior probabilities of the skill levels to calculate the probability distribution of skill levels given all the question scores. To model k skills with the same skill levels for each, we need knq probabilities, which is too large for non-trivial real-world applications.

To reduce the number of probabilities required and improve the efficiency of the algorithm in IDEAL, questions are grouped according to difficulty levels into categories associated with the conditional probabilities of answering each set of questions correctly for the possible skill levels. Now, only knc probabilities are required, where c is the number of categories.

To further reduce the number of probabilities, we match the question categories to the skill levels. For $n + 1$ skill levels, only n question categories are required. If a learner has reached a certain skill level, then he should be able to answer all questions at that skill level as well as all easier questions. Considering that learners sometimes miss questions that they should know or may guess the right answer, the probability of a slip s (e.g., 0.1) and the probability of a lucky guess g (e.g., 0.2) are used in the conditional probabilities for correct answers to questions of increasing difficulty. By using these two probabilities, a simple way to set the conditional probabilities for 5 skill levels is as shown in Table IV.10.4.

Now, the total number of probabilities required is reduced to the prior probabilities for the skill levels plus the probabilities s and g.

TABLE IV.10.4 Conditional Probabilities with Respect to Skill Levels and Question Categories

Question Category	Skill Level				
	Novice	Beginning	Intermediate	Advanced	Expert
Beginning	g	$1-s$	$1-s$	$1-s$	$1-s$
Intermediate	g	g	$1-s$	$1-s$	$1-s$
Advanced	g	g	g	$1-s$	$1-s$
Expert	g	g	g	g	$1-s$

Based on this model, the probability distribution of the skill levels given the performance data can be determined in linear time. Based on the Bayesian theory and the assumption that the performance data are independent, the conditional probability of skill levels is as follows:

$$p(X = x_j|\tilde{e})$$

$$= \frac{1}{p(\tilde{e})} p(X = x_j) p(\tilde{e}|X = x_j)$$

$$= \frac{1}{p(\tilde{e})} p(X = x_j) \prod_{i=1}^{n} p(\tilde{e}|X = x_j)$$

$$= \frac{1}{p(\tilde{e})} p(X = x_j)(1 - s)^{\sum_{i=1}^{j} e_{i+}} s^{\sum_{i=1}^{j} e_{i-}} g^{\sum_{i=j+1}^{n} e_{i+}} (1 - g)^{\sum_{i=j+1}^{n} e_{i-}}$$

where X represents the skill levels; $\tilde{e}$ is the evidence vector of n elements, in which each element e_i contains two numbers, e_{i+} and e_{i-}, corresponding to the number of correct and incorrect answers to questions at difficulty level i, respectively.

The advantages of this model are: (1) questions can be added, dropped, or moved between categories with minimal overhead; (2) the model incorporates uncertainty and allows for both slips and guesses in learner performance; (3) the time complexity is linear in the number of data items, whereas updating the Bayesian belief networks is in general NP-hard; and (4) only a small number of parameters are required. The restrictions of the model are that only binary-valued evidence is used, and only one skill can be modeled at a time.

10.3.3 Open Learning Object Sequencing and Machine Learning for Sequencing

Optimal sequencing of open learning objects is critical in providing effective adaptive learning to learners with diverse previous background knowledge, learning styles, and learning goals. OLO sequencing can be seen as a two-step process: finding relevant OLOs and selecting the best one. Sequencing is based on inter-OLO dependencies and the learner model. Curriculum sequencing, a key component in intelligent tutoring systems,[13,14,94] can be adopted for OLO sequencing.

Coarse-grain sequencing of OLOs is determined based on instructional design theories. The dependency relations between OLOs, as documented within the LOM metadata such as prerequisite, corequisite, related, remedial, and equivalent, are usually approximate, incomplete, and partial. More detailed, specific knowledge for fine-grain sequencing is obtained based on a bottom-up cognitive model using efficient neural networks, reinforcement, and symbolic machine learning methods.

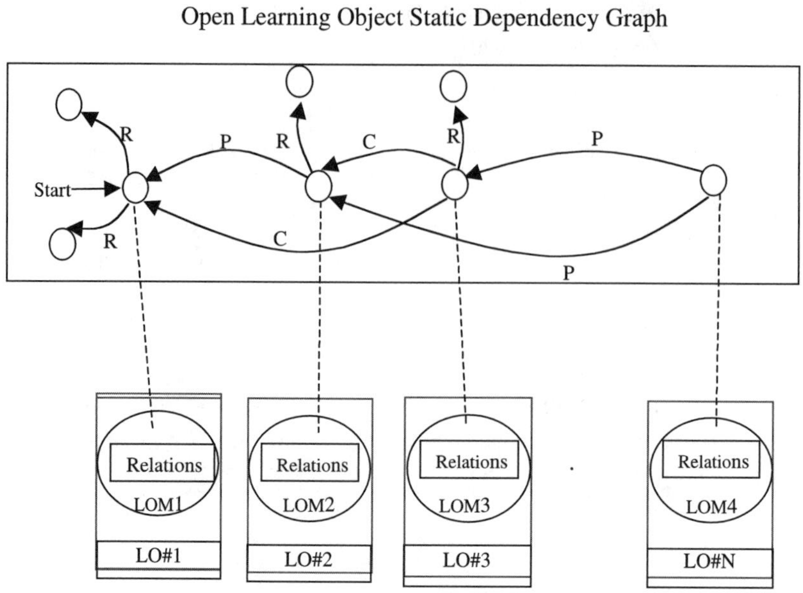

Open Learning Object Static Dependency Graph

FIGURE IV.10.2 Using learning object metadata for defining a static dependency graph. P = prerequisite, C = Corequisite, R = remedial, LO = learning object, LOM = LoMetadata.

10.3.3.1 OLO Dependency Graph

In IDEAL, each OLO has an associated LOM[4] record that contains a relations element identifying several types of relationships with other OLOs. Using this as a bootstrap set of initial constraints, IDEAL allows for adaptive navigation of OLOs.

The OLO dependency graph (OLO-DG) is a directed acyclic (DAG) graph that represents all the relationships made explicit between OLOs by their authors. This information is captured into the LOM records within the relations element for each OLO. The relationships between OLOs include prerequisite, corequisite, related, and remedial. Remedial OLOs such as special topics are not required to be learned by all the learners. An OLO-DG is extracted from the set of OLOs that are designated members of a course from the pool of available OLOs. Additional relations with OLOs outside the course are considered as remedial relations and are only captured for a distance of one link outside the core set of OLOs in the course. Once a set of interrelated OLOs are selected to belong within a course, this set defines the set of vertices, while all the relations records on these OLOs define the set of directed edges of the OLO-DG, as graphically depicted in Figure IV.10.2.

10.3.3.2 OLO Sequencing

Intelligent OLO sequencing represents a high-level (course-level) adaptation that seeks to present the OLOs associated with a course in an optimal order, where the optimization criteria take into consideration not only the content dependencies among the OLOs but also the learner background and performance on related OLOs. Previous work on adaptive navigation has modeled the learner's knowledge about an OLO as falling within four categories: not-ready-to-be-learned, ready-to-be-learned, in-work, and learned.[24] IDEAL extends this discrete metric by computing a learned score for each prerequisite OLO as a function of several other basic metrics, such as OLO review time and performance scores, combined into a weighted sum.

A learner is ready to learn an OLO only if he has performed sufficiently well on its prerequisites. How well a learner has learned an OLO is judged by the learner's performance of and his access patterns to the course material. The access patterns include how much time he has spent studying an OLO, whether he used corresponding multimedia material such as audio and video, and if he has reviewed the OLO multiple times. Specifically, the performance on an OLO is determined based on the following three factors:

- *Quiz performance.* Quizzes give the learning environment the most direct information about the learner's knowledge. Quizzes can be dynamically constructed based on the learner model. Questions are provided to cover the OLO most recently completed as well as OLOs that should be reviewed. Each question has a level of difficulty matching the skill level category. Correctly answering a harder or higher skill level question demonstrates a higher ability than correctly answering an easier or lower skill level one. The quiz scores are calculated using the following formula:

$$Quiz\_Score^{k+1} = \begin{cases} \alpha^* \, Quiz\_Score^k + (1-\alpha)\frac{d}{L} & \text{if answered correctly} \\ \max(\alpha^* Quiz\_Score^k - (1-\alpha)\frac{L+1-d}{L}, 0) & \text{if answered incorrectly} \end{cases}$$

where $0 < \alpha < 1$ is a constant corresponding to the updating rate, k is the index of the updating iteration, $1 \leq d \leq L$ is the level of difficulty, and L is the total number of skill levels. The score is bounded between 0 and 1. Each topic has an initial score of 0 or some heuristic value derived from prior knowledge.

- *Study performance.* The main interaction that learners have with the learning environment is through viewing, listening, or interacting with the course material in multimedia forms. The study score is used to judge how much comprehension the learner has gained through these activities. An OLO is usually presented in multiple pages, and each page is assigned a weight corresponding to its importance. Then, the study score in the range between 0 and 1 is calculated based on the pages visited and the amount of time spent on each page. An optimal time for each page is used as the baseline. If the learner spends this optimal amount of time, then the study score for the page is 1. As he moves away from this point, his score decreases. The study score of a learner on an OLO is the weighted sum of the scores on the pages, calculated as follows:

$$Study\_Score = \sum_{i=1}^{N} w_i S_i$$

where N is the number of pages, w_i is the weight on the ith page with $\sum_{i=1}^{N} w_i = 1$, and S_i is the score on the ith page. If page i is studied more than once, the total time on the page is used for S_i.

- *Reviewed topics.* The review score on an OLO records how often the learner has returned to review the OLO again. It is based on how many times the OLO is reviewed and how much of the material is viewed each time. If a learner is reviewing frequently, then he has not learned the material. The review score is in the range between 0 and 1 and starts at 1 for each OLO. Each time the learner reviews the OLO, the review score is updated by multiplying the value calculated using the following equation.

$$Review\_Score^n = \begin{cases} 1 & \text{if } n = 1 \\ Review\_Score^{n-1} - 0.5^n & \text{if } n > 1 \end{cases}$$

These three scores on quiz performance, study performance, and reviewed topics are part of the learner model. For OLO sequencing, the scores are combined into a single value, the *learned score*, indicating how well the learner has learned the OLO. The quiz score is the most important among the three. When a learner has a reasonably high quiz score, such as over 0.8, then the other scores do not matter much, and the final value is the quiz score. However, if the learner's quiz score is less than 0.8, the other factors become important, and the final value is the weighted sum of the three scores with weights such as 0.7, 0.2, and 0.1, respectively.

The *ready score* of an OLO indicates whether the learner is ready to learn the OLO or not. It is calculated based on the OLO's learned score and its prerequisite OLO's learned scores. If a given OLO's learned score is too low, it should be presented again with a pedagogic strategy appropriate to a lower skill level. In order to start a new OLO, a learner should show sufficient scores in its prerequisite OLOs. One formula of the ready score is the weighted sum of the OLO's learned score and its prerequisite OLOs' learned scores with predetermined weights.

In IDEAL, the learner has the option of letting the learning environment choose the next OLO or choosing it himself. For learners with novice and beginner skill levels, this alternative is normally hidden and must be explicitly requested. For advanced and expert skill levels, this alternative is the default as the system grants these learners larger navigation autonomy. In both cases, the learner must achieve a sufficiently high ready score on the OLO to proceed. If IDEAL is asked to choose the next OLO, it will choose the one with the highest ready score. If the learner decides to choose the next OLO, he is presented with the OLO dependency graph neighborhood for the last completed OLO annotated with suggestions on which OLOs to repeat and/or which new OLOs to study.

10.3.3.3 Machine Learning for Sequencing of Open Learning Objects

The principles derived from instructional design theories are useful in OLO coarse-grain sequencing. They may be sufficient for sequencing a small number of OLOs to support rudimentary adaptive learning. However, as online e-learning becomes a huge market, and more and more international organizations and commercial companies develop learning objects, the number of learning objects available on the Web will grow tremendously. In addition, many instructionally equivalent or similar learning objects will be developed for different target audiences. The ability to take advantage of the rich set of learning objects and, at the same time, finding the optimal path through the learning material for individual learners becomes an important feature of advanced Web-based intelligent learning environments. Due to the fact that fine-grain learning object sequencing is dynamic in nature and usually subject- and domain-dependent, any manual approach is impractical, and automatic learning of sequencing knowledge using machine learning techniques becomes a necessity of an effective adaptive learning environment.

10.3.3.3.1 *Cognitive Skill Learning Models*

Learning sequencing knowledge is very similar to learning skills. In IDEAL, previous work on skill learning models in cognitive psychology is applied to the development of computational models for learning sequencing knowledge. In general, skill knowledge can be classified into two categories: procedural and declarative. The distinction between procedural knowledge and declarative knowledge has been made in many theories of learning and cognition.[95–98] It is believed that both procedural knowledge and declarative knowledge are essential to cognitive agents in complex environments. Anderson[96] proposed the distinction based on data from a variety of skill learning studies ranging from arithmetic to geometric theorem proving, to account for changes resulting from extensive practice. According to Anderson, the initial stage of skill development is characterized by the acquisition of declarative knowledge. Through practice, a set of specific procedures is developed, which allows aspects of the skill to be performed without using declarative knowledge.

Several other distinctions made by researchers capture a similar difference between different types of processing. For example, Smolensky[99] pointed out a distinction between conceptual (publicly accessible) and subconceptual (inaccessible) processing. In his framework, the use of declarative knowledge is based on conceptual processing, while the use of procedural knowledge is based on subconceptual processing and, thus, is inaccessible. The inaccessibility of procedural knowledge is accepted by most researchers and embodied in most computational models that capture procedural skills.

Machine and human learning can proceed from procedural to declarative knowledge (bottom-up learning), as well as the reverse (top-down learning). Most of the work in skill learning that makes declarative/procedural distinction assumes a top-down approach, that is, learners first acquire a great deal of explicit declarative knowledge in a domain and then, through practice, turn this knowledge into a procedural form (called proceduralization), which leads to skilled performance.[97,100–102] Other work demonstrates that individuals may learn to perform complex skills without first obtaining a large amount of explicit declarative knowledge.[103–106] When there is no sufficient *a priori* explicit knowledge available, learning is usually bottom-up. In IDEAL, learning of fine-grain sequencing knowledge falls into this category.

Many computational models have been developed based on cognitive learning models. Representative models of the top-down approach include ACT and ACT-R.[96,97,100] ACT is made up of a semantic network for declarative knowledge and a production system for procedural knowledge. Productions are formed through proceduralization of declarative knowledge, modified through generalization and discrimination (i.e., specialization), and have strengths associated with them that are used for firing. ACT-R is a descendant of ACT in which procedural learning is limited to production formation through mimicking, and production firing is based on odds of success. The ACT models form the foundation of many traditional intelligent tutoring systems.

A representative model of the bottom-up learning approach is CLARION.[106–108] This model integrates neural network, reinforcement, and symbolic learning methods. Procedural knowledge is captured by a sub-symbolic distributed representation such as a neural network and is learned using reinforcement learning when there is no direct input/output mapping provided externally. Declarative knowledge is captured in a symbolic representation and is mainly learned through utilizing procedural knowledge acquired at the bottom level. The model taps into the synergy of both procedural and declarative knowledge and has demonstrated its effectiveness in learning sequential decision tasks such as minefield navigation.

CLARION has a number of advantages compared to the ACT models. First, ACT relies mostly on top-down learning (from given declarative knowledge to procedural knowledge), whereas CLARION can proceed completely bottom-up (from procedural to declarative knowledge) and is able to learn on its own without an external teacher providing instructions of any form. Second, in ACT, both declarative and procedural knowledge are represented in an explicit symbolic form. Thus, ACT does not explain, from a representational viewpoint, the differences in accessibility between the two types of knowledge. CLARION accounts for this difference based on the use of two different forms of representations. The top level of CLARION is symbolic or localist and, thus, naturally accessible/explicit, while the bottom level contains knowledge embedded in a network with distributed representations and is thus inaccessible/implicit. Finally, the reinforcement learning technique in CLARION is effective in dealing with sequential decisions under the condition of limited external feedback such as a payoff/reinforcement signal. The learning methods of ACT models do not work well in this situation.

10.3.3.3.2 *Learning Sequencing Knowledge*

In IDEAL, the CLARION model is extended to the learning of fine-grain sequencing knowledge for adaptive delivery of OLOs.[108–111] Training examples are generated when learners use the learning environment and go through a sequence of OLOs adaptively selected for a course. A learner's learning performance is used as the reinforcement signal. The sequencing model is initialized either randomly or heuristically and is improved when more training examples become available. Different sequencing models can be learned for different types of learners such as female learners and male learners.

For a given topic, a large number of OLOs may be available. IDEAL finds the optimal path through the OLO jungle for individual learners by applying the bottom-up learning approach based on a two-level cognitive model. A rule-based representation is used at the top level to capture declarative knowledge, whereas a back-propagation neural network is used at the bottom level to capture procedural knowledge. Neural network, reinforcement, and symbolic learning methods are applied to obtain the knowledge. A high-level description of the learning algorithm is as follows:

1. Observe the current state x, consisting of the learner's background, learning characteristics, current learning status, and performance data.
2. Compute at the bottom level the value or quality of each of the possible actions $(a_i, i = 1, \Lambda, n)$ associated with the state x: $Q(x, a_1), Q(x, a_2), \Lambda, Q(x, a_n)$, i.e., the quality of selecting each subsequent OLO.
3. Find out all the possible actions $(b_j, j = 1, \Lambda, m)$ at the top level, based on the state x and the existing action rules in place.
4. Choose an appropriate action c from b_j stochastically with Boltzmann distribution based on the weighted sums of the values of a_i (at the bottom level) and b_j (at the top level).
5. Perform the action c, i.e., learn the corresponding OLO, and observe the next state y and (possibly) the reinforcement r. The state y is obtained from x by changing the current learning status and performance data.
6. Update the bottom level in accordance with the *Q-learning* back-propagation algorithm.
7. Update the top level using the rule-extraction-refinement algorithm.
8. Go back to Step 1.

At the bottom level, a Q-value $Q(x, a)$ is an evaluation of the quality of an action a in a given state x. The Q-learning algorithm is a reinforcement learning algorithm for sequential learning tasks.[112] It uses $Q(x, a)$ to estimate the maximum discounted cumulative reinforcement that the agent will receive from the current state x. In terms of both simplicity and performance, Q-learning is the best among similar reinforcement learning methods.[112,113]

In IDEAL, the Q-learning algorithm is implemented in a neural network so that the calculation of Q-values for the current input with respect to all the possible actions is done in a parallel fashion and is thus highly efficient. Figure IV.10.3 shows a four-layered neural network in which the first three layers form a (feedforward or recurrent) back-propagation network for computing Q-values, and the fourth layer with only one node performs stochastic decision making. The output of the third layer indicates the Q-value of each action (represented by an individual node), and the node in the fourth layer determines probabilistically the action to be performed based on the Boltzmann distribution of Q-values:

$$p(a|x) = e^{Q(x,a)/\alpha} \Big/ \sum_i e^{Q(x,a_i)/\alpha}$$

where α controls the degree of randomness (temperature) of the decision-making process.

Applying Q-learning, the training of the back-propagation network is based on minimizing the following error at each step: $error_i = r_i + \beta \max_k Q(y, k) - Q(x, a_i)$, where i is the index of an output node representing the action a_i, y is the new state resulting from action a_i in state x, β is a discount factor that favors the reinforcement received sooner relative to that received later, and r_i is the reinforcement received at the ith step. Based on these error measures, the back-propagation algorithm is applied to adjust internal weights of the neural network (which are randomly initialized before training). This learning process performs both structural credit assignment with back-propagation to assign credit/blame to each element in a state, as well as temporal credit assignment through temporal difference updating to assign credit/blame to the action in a sequence that leads to success or failure. This learning process enables the development of procedural knowledge solely based on the training examples, without *a priori* knowledge.

At the top level of the IDEAL sequence-learning model, declarative knowledge is captured in propositional rules of the form: *current_state* $\Rightarrow$ *action*, where the left-hand side is a conjunction of individual elements each of which refers to a dimension x_i of the input x, and the right-hand side is a selection action a. The rule-extraction-refinement algorithm developed by Sun and Peterson[108] is applied to learn the declarative knowledge (rules) using the information at the bottom level. The declarative knowledge obtained is useful in speeding up the sequence learning process, facilitating the transfer of sequencing knowledge for new learners, and helping communication of this explicit knowledge to humans. The advantage of the bottom level procedural knowledge is that it can be highly efficient once developed.

In IDEAL, a learner is modeled using her background, learning characteristics, current learning status, and performance data. Her performance on a subject will be used as the reinforcement signal during neural network sequence learning. It can be determined based on three factors: Quiz_Score, Study_Score, and Review_Scores.

10.3.4 Adaptive Presentation of Open Learning Objects

Once an OLO is chosen, how to teach, such as how to dynamically construct page contents, can be determined based on the three individual OLO scores. For example, a learner who has poor quiz scores on an OLO and who has not studied the OLO for very long should be treated differently than a learner who has the same quiz score but has spent much more time studying. The second learner should be presented with more background material to improve his comprehension. Furthermore, each time a learner reviews an OLO, it should be taught in a different way than it was the last time. Also, each time a learner is presented an OLO, the presentation should be adapted to his learning style and preferences. Adaptive presentation of OLOs can be enabled by an OLO agent programmed for a reactive or situated behavior using a data-driven approach to provide interactivity based on an XML state sequencing specification in the OLO.

10.3.4.1 Adaptive Content Generation

The OLO content is dynamically generated and adapted to the learner model such that it meets the particular architecture of instruction[86] requirements that are predominantly aligned with a learning theory and with the needs of a particular learner's skill level. Adaptation is ultimately constrained by the designed-in capabilities of the OLO and the learner's actualized model.

The OLO internal composition is described through an extension to the LOM metadata standard[4] that seeks to capture additional elements targeted at the internal structure such as interaction modalities, skill-level adequacy, and multimedia structural relationships. This internal metadata represented in XML provides a standard but extensible framework for active multimedia. It includes reference identifiers to marked-up text, graphics that use the scalable vector graphics (SVG)[114] standard, an

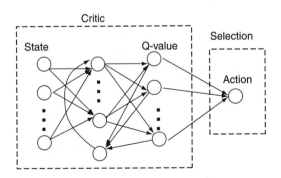

FIGURE IV.10.3   Neural network implementation of Q-learning.

interactive agency specification with information for learner interaction, and typical multimedia components such as images, movie clips, audio clips, etc. The recent introduction of vector graphics description for dynamic content using the scalable vector graphics standard provides an XML-based and compact representation for graphics, which dramatically reduces the transmission time when compared with bit-mapped approaches even when using lossy compression techniques. Several plug-ins and graphics drawing tools, packages, and toolkits for SVG are available such as Adobe,[115] CSIRO,[116] Batik,[117] and WebDraw.[118]

For a novice learner on an OLO who lacks meta-cognitive skills, it is recommended that a behavioral architecture of instruction be adopted. For this architecture, the OLO includes a step-by-step learner-controlled animation using SVG and additional learning style-specific features preselected according to his learning style profile. For an advanced learner who has better meta-cognitive skills, can handle a larger semantic load, and exhibits good short-term memory, the OLO implements the guided exploration architecture of instruction. It includes mark ups that can be used for adaptation features such as explicit navigation controls into the next set of possible OLOs, visual queues as to the recommended choice, a larger grain for the incremental changes introduced by each new state, and learner selection of aural and visual rendering of the contents. More control is vested on the advanced learner than the novice.

The OLO also contains a state sequencing description for learner interaction with the learning material that is interpreted by a client-side agent that also performs backtrack and replay. A sequencing description is an XML fragment that contains the following major elements:

- ⟨animation⟩: This is the root element which contains one or more ⟨state⟩ elements and an attribute identifying the initial state for the animation.
- ⟨state⟩: Each state is expected to represent a stable configuration of the OLO multimedia components as presented to the learner. Each state contains a state name ID attribute and one or more ⟨branch⟩ elements.
- ⟨branch⟩: Associated with each branch is a branch name, a default branch attribute, and next state attributes. It also contains subelements specifying the ⟨condition⟩ under which that branch is taken and the ⟨action⟩ or change to be introduced when taking that branch.
- ⟨condition⟩: A condition specifies one or more ⟨predicate⟩s which must be evaluated and disjunctively combined to determine if the condition is true.
- ⟨predicate⟩: These are expressed in terms of semantic event names, whether an event occurrence is restricted as true or false. If the event variable is true, the event is considered to have happened and to be present within the event stack.
- ⟨action⟩: An action contains attributes that identify the document, the element within that document which are the target of the update, and the type of action. Other attributes specify the generic name of the target element or attribute, which allows us to determine how any update values are to be used.
- ⟨updateValue⟩: This may be empty or hold element values, attributes, or scripts to apply.

10.3.4.2 Adaptive Presentation of Open Learning Objects

Adaptive presentation is enabled by the interpreter portion of the intelligent situated micro-agent that is downloaded as part of the OLO content. The interpreter receives or traps all events generated by the mouse or any other form-input widgets defined in the OLO user interface. It dynamically modifies the OLO DOM tree representation to be rendered inside the browser with plug-ins.[115] Usually, a dedicated JavaScript function is included (a UI listener) for each type of event that can be generated from each widget in the defined UI. These functions record the event into an event stack and invoke the interpreter. The interpreter inspects the stack for new events. Upon finding a new event, it searches each branching condition on the current state as described in the state sequencing description, trying to match the branch predicate. If no matches are found, the event is not recorded into the interaction trace nor removed from the event stack. If the event and any other preceding events from the stack match a branch condition in the state sequencing description, the new state corresponding to the branch condition becomes the current state, and the events are removed from the event stack and stored in the interaction trace.

State transition is accomplished by pushing the current state and branch IDs into the state stack as part of the interaction trace and jumping to the next state indicated by the branch. Any incremental updates specified by the transition are implemented within the DOM tree representation, which causes the browser and any plug-ins to update the rendered view of the OLO DOM tree. Once the incremental changes are inserted, the new state is marked as the current state, and the interpreter waits until activated again by a new event.

The interpreter is also responsible for resuming the presentation of a suspended OLO. In this case, the downloaded OLO will include a non-empty state stack. By rehearsing the sequence of states and branches in the interaction trace, the interpreter is able to resume the presentation of the OLO in the same state as when it was suspended.

10.4 Implementation Techniques

IDEAL is being implemented using prevalent Web and intelligent agent technologies to achieve maximal portability, flexibility, and scalability.

10.4.1 Multi-Agent Software System

Several characteristics specific to Web-based learning make multi-agent systems attractive. First, the learners are widely distributed and the number of potential participants is large. This renders static and centralized systems inadequate. A distributed multi-agent system with personalized agents for each learner is more appropriate. Second, the background, learning characteristics, and goals of learners are different and change over time. Learning material and teaching methodologies are also dynamic in nature. Third, learners may take more than one course at a time. Coordinating the learning on different subjects for each learner enriches the learning experience. Finally, learners tend to get together to discuss study topics and share knowledge. Smooth communications, including visualizing and sharing common contexts, need to be supported. Hence, the multi-agent approach has attracted great attention in Web-based education.[119,120]

IDEAL contains different specialized intelligent agents that support personalized adaptive learning and community-based knowledge sharing. Each learner is assigned a unique personal agent that manages the learner's personal profile, including background, learning characteristics, interests, courses enrolled in, etc. The personal agent can talk to other agents in the system through various communication channels. A course is composed of a collection of OLOs supported by a course agent. The course agent searches, selects, organizes, and delivers the OLOs of the course to the learners. It collects learning statistics from OLOs and applies machine learning methods to construct sequencing models for different types of learners that are used in adaptively selecting and organizing the most appropriate sequence of OLOs for individual learners. The course agent also acts as a mediator for communication among learners in the course. Multiple course agents may exist on distributed sites to provide better efficiency, flexibility, and availability. An OLO management agent indexes, categorizes, and manages OLOs and their relationships. It returns OLOs that match queries submitted by course agents. It can talk to other OLO database agents on the Web and receive updated information about new OLOs.

IDEAL also supports community-based knowledge sharing. In a learning community, instructors, and learners work together systematically toward shared academic goals. Collaboration is stressed, and competition is de-emphasized. By taking an active learning approach, the instructor's primary role shifts from delivering content to setting up learning environments and serving as coach, expert guide, and role model for learners. The learner's role changes as well, from relatively passive observer of teaching and consumer of information to active co-constructor of knowledge and understanding.[121–123] There are three issues in supporting a learning community: how to support various communication channels including one-to-one, one-to-many, and many-to-many channels; how to find other people that share similar interests; and how to visualize and share common contexts. Active support using agent technology based on interaction between software agents and between humans and software agents is promising in addressing these issues.[124] IDEAL supports a learning community through personal agents and course agents. Personal agents acquire a learner's profile and help them by gathering, exchanging, and viewing information from the community. Course agents provide shared information, knowledge, or contexts within the learning community and act as mediators for communication among people. They collect various usage statistics for analyzing the activity patterns, which are, in turn, used to improve the environment settings and update communication guidance and principles. They enforce community rules to provide a more equitable learning environment for learners of different background and for both male and female learners.[125–127]

10.4.2   Open Learning Object Management and Delivery

Currently, the WWW client-side technology is dominated by browsers that have entrenched and imposed their interfaces as a kind of universal user interface. Newer competing technologies based on peer-to-peer (P2P) networking,[128] direct launching Web applications,[129,130] and Web services[131,132] have the potential to subvert this state of affairs in the long term but do not provide a mature solution in the short term. The most popular browsers have been extended through plug-ins to allow many forms of multimedia. In addition, the standardization of client-side scripting in the form of a vendor independent ECMAScript (JavaScript)[133] and the W3C Document Object Model (DOM)[134] has made possible the creation of truly interactive multimedia documents capable of being delivered through the WWW and interactively used by learners through existing browsers equipped with appropriate plug-ins.

On the Web-server side, the creation of dynamic Web content has evolved from the use of the common gateway interface (CGI) technology[135] to the creation of Web applications using component technologies such as Java Beans and its derivatives such as Java Servlets,[136] Java Server Pages,[137] JDOM,[138] etc. These have introduced considerable flexibility and performance enhancements to the development of e-learning applications by allowing individual learner per-session tracking and context sharing for servlets within a servlet container while executing as independent lightweight processes or threads.

OLOs are implemented as interactive structured multimedia and adaptively delivered by a client–server system implemented mostly with XML SVG and ECMAScript on the client side and using servlets and persistent XML databases on the server side. The implementation relies on IEEE and IMS standards such as LOM and LIP.

Figure IV.10.4 illustrates the framework for adaptive delivery of OLOs on the Web. The OLO framework makes the transfer of educational material between different software systems transparent to the learner and as easy as possible. It allows software agents to reach out to the Web to read and make sense of online OLO listings.

OLOs support interactive learning, which combines the visual presentation of course material and executable components. OLOs contain simple predefined code for customized presentation, online interaction, sub-topic sequencing, animation, performance evaluation, etc. Based on learner models obtained from the OLO server, interactive learning tailors the learning material to the needs of individual learners. The performance assessment component collects various kinds of statistics of the learning process, such as how long the learner studies each OLO, his/her performance on the exercises and quizzes, and comments. This information is uploaded to the server when the current session is finished, and used to update the learner model.

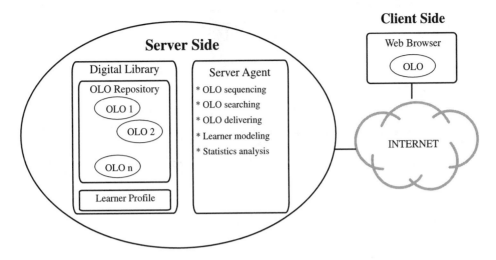

FIGURE IV.10.4   Framework for adaptive delivery of OLOs.

For OLO management, a learner-centered digital library management system, LOVE (learning object virtual exchange), of educational material (e.g., demonstrations, projects, homework, presentations, and examinations) is developed. Other authors and instructors are allowed to submit their educational materials to LOVE as a cooperative that manages intellectual property rights (for authors), material reviewing, and maintenance. LOVE is an important component of IDEAL for reviewers and curators, instructors and authors, and learners. While instructors contribute and share their courseware, learners select any topic for their learning purposes. In addition to user contributed educational material, LOVE is interoperable with distributed search engines of large NSF-supported digital libraries and application software tools to automatically obtain educational material available on the Web.

10.4.3   OLO Content Adaptation and Learner Interaction

Figure IV.10.5 sketches the general process for OLO content adaptation. The open nature of OLOs is supported through the adoption of a data-driven architecture that relies on XML for marking up the object contents. XSL[139] style sheets are used to guide an XSLT[140] engine to generate an adapted version of the OLO. A course agent controls this adaptation process.

It is desirable to provide a standardized and open mechanism for capturing and making accessible outside an OLO the results of interaction between a learner and the OLO. This learner interaction is dependent on the particular user interface that the OLO presents to the learner, and it is contextualized by the OLO itself. Thus, we cannot determine in advance the ultimate form of such interaction trace in general, but we can define a basic mechanism that is sufficiently general by defining a basic markup language that can be extended and specialized for particular groups of related OLO that use specific UI devices and widgets.

The interaction trace can be seen as OLO-independent primitive events and OLO-specific semantic events interpreted in the context of the OLO. Primitive interaction events include clicking a mouse button at some coordinates or pressing a keyboard key, as shown in Table IV.10.5. OLO-specific semantic events include the transition to a new OLO-state as a result of interacting with a particular UI widget. They correspond to the states that the situated micro-agent goes through as the learner interacts with the OLO. These states usually specify dynamic changes or modifications to the multimedia content of the OLO. For example, for an OLO whose multimedia is an SVG graph, each state specifies incremental changes to be introduced to the DOM tree representation of the SVG graph dynamically rendered using the SVG viewer plug-in as a reaction to the learner input.

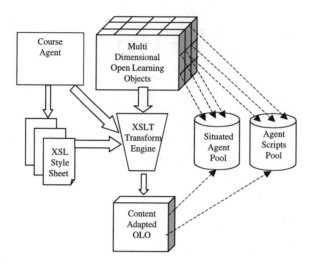

FIGURE IV.10.5   OLO content adaptation process and situated micro-agents.

TABLE IV.10.5   OLO-Independent Primitive Interaction Events

Primitive Event Name	Atributes	Description
Mouse button click	Type: #primitive	Always primitive
	Time:	Time since loading of OLO in milliseconds
	Button:	One of: [left, center, right]
	Coordinates:	X, Y coordinates relative to the visible frame upper-left corner, increasing right and down, measured in pixels
Mouse drag	Type: #primitive	Always primitive
	Time:	Time since loading of OLO in milliseconds
	Coordinates	X, Y coordinates relative to the container frame upper-left corner, increasing right and down, measured in pixels
Key press	Type: #primitive	Always primitive
	Time:	Time since loading of OLO in milliseconds
	Key character:	UTF8 coded character on key pressed
	Key code: [option]	Optional: Key code in hexadecimal when available

Given that each interaction trace is to be associated with an OLO, all time measurements are relative to the time when the OLO is first loaded and made visible to the learner. The OLO loading event is saved as the first primitive interaction trace event, and it is associated with the initial state of the OLO. When the OLO is suspended, a suspend event is added to the trace with the time of suspension. Similarly, a resume event is generated to account for this event when the OLO is restarted. The state reference and time-stamp are stored within the OLO-specific interaction events.

As the learner generates primitive interaction events, they are interpreted by the situated micro-agent engine (interpreter), which may cause a state transition. The primitive interaction events with any new state reached as a result of those events are sent to the OLO server-side agent as a permanent record of the learner interaction with the OLO. On the server side, the interaction trace is persistently stored as part of the learner information record[2] in the XML database. The interaction trace is saved upon completion of the OLO or whenever the learner suspends his interaction with the OLO. This allows restarting the OLO in the same configuration as when it was suspended.

Every primitive event and state transition semantic event is time-stamped using absolute time with a resolution of milliseconds. Because time stamping occurs on the client side, there is very little latency between the actual event and the reported time-stamp. The time stamping of both the state transition semantic events and the learner generated primitive events allows for very accurate determination of the reaction time of learners to the changes associated with the OLO state transition. This reaction time can be used for various purposes ranging from behavior-oriented measurements of reaction delays to correlation with other more cognitive traits such as semantic density, short-term memory, and prior knowledge.

Because both learner-initiated events and OLO-initiated events are time-stamped, it is possible to perform detailed correlation analysis between these. The learner interaction trace information can be data-mined to identify patterns that indicate important features of a group of learners or a set (i.e., a course) of OLOs. Traces from a group of learners can be statistically aggregated to identify problems with particular portions of an OLO, especially when the learners might be taking excessive time to react to some portions of the OLO. This could indicate that such portions need to be reviewed by the authors and improved as deemed necessary. This is a key feedback facility to support continuous OLO improvement. Trace data can also be used to verify the readiness level or deficiencies of a group of learners. This could be used in hybrid e-learning and traditional learning environments[34] where the OLO is used as homework to ready the learners for the in-classroom discussion of exceptional or more difficult material. For individual learners, the trace is used to assess their performance by measuring the total time taken to complete the OLO and the time taken to review.

10.5   Concluding Remarks

IDEAL is a scalable e-learning environment that provides comprehensive and integrated e-learning solutions tailored to each learner's needs, allows each learner to create a personalized path through learning based on that learner's characteristics, background, and learning goals, and contains a large collection of dynamic online learning material, far beyond the existing linear courseware objects.

A prototype of IDEAL together with a sample set of OLOs on Web and agent technologies is currently implemented using prevalent Web and intelligent agent technologies.[7–12,141–143] The IDEAL prototype runs on top of a scalable, reliable, and high-performance distributed environment using object-oriented technologies. The implementation consists of the server side and the client side connected through the Internet. The server side consists of the intelligent agents, the student profiles, and the OLOs that can be dynamically sent to the client and dynamically updated. Prototypes of the agents include interface agents, personal agents, teaching agents, and course agents. They are implemented on top of distributed object-oriented software environments including Java RMI, JATLite,[144] and JavaSpace.[145] The software agents communicate with each other in KQML-style[146] XML SOAP[131] messages through a variety of communication channels, including peer-to-peer, multicasting, and broadcasting. Real-time communication among users is implemented in the forms of chat room, white board, and streaming audio and video.

The client side consists of a browser that has support for XML and Java applets. Applets are used for dynamic processing on the client side, thus reducing the load on the server as well as on the network. The contents in XML are presented either using XSL (extended stylesheet language) or Java applets depending on the level of processing that needs to be done on the client side. The presentation of OLOs is adaptive. The XML file is stored on the Web server. According to the student's request, the Web server responds appropriately, dynamically generating an XSL style sheet using the learner's preferences to render the presentation customized to the learner.

In summary, the Web is an ideal media for large scale, efficient education. It supports one-on-one interaction and eliminates the time and distance constraints. To maximize the benefit of Web-based education, a learning environment should be scalable, efficient, flexible, and intelligent. There are a number of aspects that are essential to building such a reliable learning system. They include security on the client side to prevent the student from printing off the test/quiz and others, effective OLO search techniques, adaptive presentation of OLOs, effective student modeling to model student performance, and a dynamic question set for each quiz/exam every time it is retaken by the student. To further increase the intelligence of IDEAL, we still need to do more research.

Acknowledgments

This research was supported in part by the National Science Foundation under grants DUE-9980375 and EIA-0086230.

References

1. Berners-Lee, T., Hendler, J., and Lassila, O., The semantic Web, a new form of Web content that is meaningful to computers will unleash a revolution of new possibilities, *Sci. Am.*, 2001, http://www.sciam.com/2001/0501issue/0501berners-lee.html.

2. IMS, Learner Information Packaging Information Model Specification, Public Draft Specification, version 1.0, 2000, http://www.imsproject.org/profiles/index.html.

3. IMS, Question and Test Interoperability Specification, Public Draft Specification, version 1.1, 2000, http://www.imsproject.org/question/index.html.

4. IMS, Learning Object Metadata Specification, Public Draft Specification, version 1.2, 2000, http://www.imsproject.org/metadata/index.html.

5. Shank, R., *Virtual Learning: A Revolutionary Approach to Building a Highly Skilled Workforce*, McGraw-Hill, New York, NY, 1997.

6. Brusilovsky, P., Adaptive and intelligent technologies for Web-based education, *Kunstliche Intelligenz*, 4, 19, 1999.

7. Shang, Y. and Shi, H., IDEAL: an integrated distributed environment for asynchronous learning, in *Proceedings of the ACM Workshop on Distributed Communities on the Web (Lecture Notes in Computer Science*, 1830), 182, 2000.

8. Shang, Y., Shi, H., and Chen, S., Agent technology in computer science and engineering curriculum, Proceedings of the 5th ACM Annual Conference on Innovation and Technology in Computer Science Education, Helsinki, Finland, 120, 2000.

9. Shi, H., Shang, Y., and Chen, S., A multi-agent system for computer science education, Proceedings of the 5th ACM Annual Conference on Innovation and Technology in Computer Science Education, Helsinki, Finland, 1, 2000.

10. Shi, H., Shang, Y., and Chen, S., Smart instructional component based course content organization and delivery, Proceedings of the 6th ACM International Conference on Innovation and Technology in Computer Science Education, Canterbury, UK, June 2001.

11. Shang Y., Shi, H., and Chen, S., An intelligent distributed environment for active learning, Proceedings of the 12th World Wide Web Conference, Hong Kong, May, 308, 2001.

12. Shang, Y., Shi, H., and Chen, S., An intelligent distributed environment for active learning, *J. Educ. Resour. Comput.*, 1(2), 1, 2001.

13. Stern, M.K. and Woolf, B.P., Curriculum sequencing in a web-based tutor, in *Intelligence Tutoring System (Proceedings of the 4th International Conference ITS '98)*, Goettl, B.P. et al., Eds., Springer, 584, 1998.

14. Weber, G., Individual selection of examples in an intelligent learning environment, *J. Artif. Intelligence Educ.*, 7(1), 3, 1996.

15. Barr, A., Beard, M., and Atkinson, R.C., The computer as tutorial laboratory: the Stanford BIP project, *Int. J. Man Machine Studies*, 8(5), 567, 1976.

16. Weber, G. and Specht, M., User modeling and adaptive navigation support in www-based tutoring systems, in *User Modeling*, Jameson, A., Paris, C., and Tasso, C., Eds., Springer-Verlag, Heidelberg, 289, 1997.

17. Specht, M. et al., AST: adaptive www-courseware for statistics, in *Proceedings of the Workshop on Adaptive Systems and User Modeling on the World Wide Web at the 6th International Conference on User Modeling, UM97*, Brusilovsky, P., Fink, J., and Kay, J., Eds., 91, 1997.

18. Specht, M. and Oppermann, R., ACE—adaptive courseware environment, *New Rev. Hypermedia Multimedia*, 4, 141, 1998.

19. Henze, N. et al., Adaptive hyperbooks for constructivist teaching, *Kunstliche Intelligenz*, 4, 26, 1999.

20. Brusilovsky, P., Eklund, J., and Schwarz, E., Web-based education for all: a tool for developing adaptive courseware, *Comput. Networks ISDN Syst.*, 30, 291, 1998.

21. Brusilovsky, P., Ritter, S., and Schwarz, E., Distributed intelligent tutoring on the Web, in *Artificial Intelligence in Education: Knowledge and Media in Learning Systems*, Boulay, B.D. and Mizoguchi, R., Eds., IOS, Amsterdam, 482, 1997.

22. Nakabayashi, K. et al., Architecture of an intelligent tutoring system on the WWW, in *Artificial Intelligence in Education: Knowledge and Media in Learning Systems*, Boulay, B.D. and Mizoguchi, R., Eds., IOS, Amsterdam, 39, 1997.

23. Anjaneyulu, K., Concept level modeling on the WWW, in *Proceedings of the Workshop on Intelligent Educational Systems on the World Wide Web at AI-ED '97, 8th World Conference on Artificial Intelligence in Education*, Brusilovsky, P., Nakabayashi, K., and Ritter, S., Eds., 26, 1997.

24. Brusilovsky, P., Schwarz, E., and Weber, G., ELM-ART: an intelligent tutoring system on World Wide Web, in *Intelligent Tutoring Systems. Lecture Notes in Computer Science*, 1086, Frasson, C., Gauthier, G., and Lesgold, A., Eds., Springer-Verlag, Berlin, 261, 1996.

25. Anderson, J.R. and Reiser, B., The LISP tutor, *Byte*, 10(4), 159, 1985.

26. Beck, J.E. and Woolf, B.P., Using a learning agent with a student model, in *Intelligence Tutoring System (Proceedings of the 4th International Conference ITS '98)*, Goettl, B.P. et al., Eds., Springer, 6, 1998.

27. Brusilovsky, P., Methods and techniques of adaptive hypermedia, *User Modeling and User-Adapted Interaction*, 6(2–3), 87, 1996.

28. Jameson, A., Numerical uncertainty management in user and student modeling: an overview of systems and issues, *User Modeling and User-Adapted Interaction*, 5, 193, 1996.

29. Kay, J. and Kummerfeld, B., User models for customized hypertext, in *Intelligent Hypertext: Advanced Techniques for the World Wide Web (Lecture Notes in Computer Science, 1326)*, Nicholas, C. and Mayfield, J., Eds., Springer-Verlag, Berlin, 1997.

30. De Bra, P. and Calvi, L., AHA: an open adaptive hypermedia architecture, *New Rev. Hypermedia Multimedia*, 4, 115, 1998.

31. Hohl, H., Bocker, H.D., and Gunzenhauser, R., Hypadapter: an adaptive hypertext system for exploratory learning and programming, *User Modeling and User-Adapted Interaction*, 6(2–3), 131, 1996.

32. Hoppe, U., Use of multiple student modeling to parameterize group learning, in *Proceedings of AI-ED '95, 7th World Conference on Artificial Intelligence in Education*, Greer, J., Ed., 234, 1995.

33. Oda, T., Satoh, H., and Watanabe, S., Searching deadlocked Web learners by measuring similarity of learning activities, Proceedings of the WWW-Based Tutoring Workshop at the 4th International Conference on Intelligent Tutoring Systems (ITS '98), San Antonio, TX, 1998.

34. Novak, G.M., Patterson, E.T., and Gavrin, A.D., Eds., *Just-In-Time Teaching: Blending Active Learning With Web Technology*, Prentice Hall, Upper Saddle River, NJ, 1999.

35. IUPUI-JITT, http://webphysics.iupui.edu/jitt/jitt.html.

36. Hartman, V.F., Teaching and learning style preferences: transitions through technology, *VCCA J.*, 9(2), 18, 1995.

37. Bonwell, C., Building a supportive climate for active learning, *Natl. Teaching Learning Forum*, 6(1), 4, 1996.

38. Richards, L.G., Promoting active learning with cases and instructional modules, *J. Eng. Educ.*, 84(4), 375, 1995.

39. Rubin, L. and Hebert, C., Model for active learning: collaborative peer teaching, *College Teaching*, 46(1), 26, 1998.

40. Kafai, Y. and Resnik, M., *Constructionism in Practice: Designing, Thinking, and Learning in a Digital World*, Lawrence Erlbaum Associates, Hillsdale, NJ, 1996.

41. Jonassen, D., Evaluating constructivist learning, *Educ. Technol.*, 36(9), 28, 1991.

42. Buron, C., Grinder, M., and Ross, R., Tying it all together: creating self-contained, animated, interactive, Web-based resources for computer science education, *SIGCSE Bull.*, 31(1), 7, 1999.

43. Carver, C., Howard, R., and Lane, W., Enhancing student learning through hypermedia courseware and incorporation of student learning styles, *IEEE Trans. Educ.*, 42(1), 33, 1999.

44. Chou, C., Developing hypertext-based learning courseware for computer networks: the macro and micro stages, *IEEE Trans. Educ.*, 42(1), 39, 1999.

45. Davidovic, A. and Trichina, E., Open learning environment and instruction system (OLEIS), *SIGCSE Bull.*, 30(3), 69, 1998.

46. Latchman, H. et al., Information technology enhanced learning in distance and conventional education, *IEEE Trans. Educ.*, 42(4), 247, 1999.

47. Ehrman, M., The role of personality type in adult language learning: an ongoing investigation, in *Language Aptitude Reconsidered*, Parry, T. and Stansfield, C., Eds., Prentice Hall, Upper Saddle River, NJ, 126, 1990.

48. Jung, C.G., *Psychological Types*, Princeton University Press, Princeton, NJ, 1921.

49. Myers, I.B. and Myers, P.B., *Gifts Differing: Understanding Personality Type*, Consulting Psychologists Press, Palo Alto, CA, 1993.

50. Keefe, J.W., Ed., *Student Learning Styles and Brain Behavior*, National Association of Secondary School Principals, Reston, VA, 1982.

51. Harshman and Paivio, Paradoxical sex differences in self-reported imagery, Canadian *J. Psychol.*, 41, 287, 1987.

52. Zakay, D., Block, R.A., and Tsal, Y., Prospective duration estimation and performance, in *Attention and Performance XVII: Cognitive Regulation of Performance: Interaction of Theory and Application*, Gopher, D. and Koriat, A., Eds., MIT Press, Cambridge, MA, 557, 1999.

53. Bodi, S., Critical thinking and bibliographic instruction: the relationship, *J. Academic Librarianship*, 14(3), 150, 1988.

54. Cranton, P., *Working with Adult Learners*, Wall and Emerson, Toronto, 1992.

55. McNeer, E., Learning theories and library instruction, *J. Academic Librarianship*, 17(5), 294, 1991.

56. Carol, D., Bridging the gap: the contributions of individual women to the development of distance education, in *Toward New Horizons for Women in Distance Education: International Perspectives*, Faith, K., Ed., Routledge, London, 1988.

57. MacKeracher, D., Women as learners, in *The Craft of Teaching Adults*, Barer-Stein, T. and Draper, J.A., Eds., Krieger Publishing Company, Malabar, FL, 1994.

58. Blum, K.D., Gender differences in asynchronous learning in higher education: learning styles, participation barriers and communication patterns, *JALN*, 3(1), 1999.

59. Spender, D., *Men's Studies Modified: The Impact of Feminism on Academic Disciplines*, Pergamon Press, Oxford, 1981.

60. Vanfossen, B., Gender differences in communication, in ITROW's Women and Expression Conference, Institute for Teaching and Research on Women, Towson State University, Maryland, 1996.

61. Burge, E. and Lenskyj, H., Women studying in distance education: a case study, *Am. J. Distance Educ.*, 4(2), 1990.

62. Hensel, N., Realizing gender equality in higher education: the need to integrate work/family issues, *ASHE/ERIC Higher Educ. Rep.*, 2, 1991.

63. Tisdel, E.J., Poststructural feminist pedagogies: the possibilities and limitations of feminist emancipatory adult learning theory and practice, *Adult Educ. Q.*, 48(3), 1998.

64. Bailey, S.M., Shortcoming girls and boys, *Educ. Leadership*, 53(8), 1996.

65. Cross, K.P., *Adults as Learners*, Jossey-Bass Publishing, London, 1981.

66. LTSC, IEEE learning technology standards committee Website, http://ltsc.ieee.org/.

67. IMS, IMS global learning consortium Website, http://imsproject.org/.

68. Advanced Distributed Learning Network, SCORM, Shareable Content Object Reference Model, http://www.adlnet.org/.

69. MacroMedia Inc., eLearning Studio, http://www.macromedia.com/software/elearningstudio/.

70. Adobe, eBook Reader, http://www.adobe.com/products/ebookreader/main.html.

71. CISCO, Interactive Mentor, http://www.cisco.com/warp/public/710/cim/index.html.

72. Reigeluth, C.M., Ed., *Instructional Design Theories and Models: A New Paradigm of Instructional Theory*, Lawrence Erlbaum Associates, Hillsdale, NJ, 1999.

73. Reigeluth, C.M., The elaboration theory: guidance for scope and sequence decisions, in *Instructional Design Theories and Models: A New Paradigm of Instructional Theory*, Reigeluth, C.M., Ed., Lawrence Erlbaum Associates, Hillsdale, NJ, 5, 1999.

74. Gibbons, A.S. et al., Work models: still beyond instructional objectives, *Machine Mediated Learning*, 5(3–4), 221, 1995.

75. van Merriënboer, J.J., *Training Complex Cognitive Skills: A Four-Component Instructional Design Model for Technical Training*, Educational Technology Publications, Englewood Cliffs, NJ, 1997.

76. Wilson, B. and Cole, P., A critical review of elaboration theory, *Educ. Technol. Res. Dev.*, 40(3), 63, 1992.

77. Gibbons, A.S. and Fairweather, P.G., *Computer-Based Instruction: Design and Development*, Educational Technology Publications, Englewood Cliffs, NJ, 1998.

78. Burton, R.R., Brown, J.S., and Fischer, G., Skiing as a model of instruction, in *Everyday Cognition: Its Development in Social Context*, Rogoff, B. and Lave, J., Eds., Harvard University Press, Cambridge MA, 139, 1984.

79. Dunn, R., *How to Implement and Supervise a Learning Style Program*, Association for Supervision and Curriculum Development, ASCD, Alexandria, VA, 1996.

80. Silver, H.S., Strong, R.W., and Perini, M.J., *So Each May Learn: Integrating Learning Styles and Multiple Intelligences*, Association for Supervision and Curriculum Development, ASCD, Alexandria VA, 2000.

81. Felder, R.M. and Silverman, L.K., Index of learning styles (ILS) http://www2.ncsu.edu, /unity/lockers/users/f/felder/public/ILSpage.html.

82. Pham, P.M. Learning styles, http://payson.tulane.edu/ppham/Learning/lstlyes.html.

83. Felder R.M. and Soloman, B.A., Index of learning styles questionnaire, http://www2.ncsu.edu/unity/lockers/users/f/felder/public/ILSdir/ilsweb.html.

84. W3C Consortium, User agent accessibility guidelines, 1.0 Working Draft, June 22, 2001, http://www.w3.org/TR/2001/WD-UAAG10-20010622/.

85. Clark, R.C., Four architectures of instruction, *Performance Improvement J.*, 39(10), 2000.

86. Clark, R.C., *Building Expertise: Cognitive Methods for Training and Performance Improvement*, International Society for Performance Improvement, Washington, D.C., 1998.

87. Murray, W., A practical approach to Bayesian student modeling, in *Intelligent Tutoring Systems (Proceedings of the 4th International Conference, ITS '98)*, Goetth, B. et al., Eds., Springer, 424, 1998.

88. Brusilovsky, P., Student model centered architecture for intelligent learning environment, Proceedings of the 4th International Conference on User Modeling, 1994.

89. Petrushin, V. and Sinista, K., Using probabilistic reasoning techniques for learner modeling, *Proc. World Conf. AI Educ.*, 418, 1993.

90. Villano, M., Probabilistic student models: Bayesian belief networks and knowledge space theory, in *Intelligent Tutoring System (Proceedings of the 2nd International Conference ITS '92)*, Springer, 491, 1992.

91. Bauer, M., A Dempster–Shafer approach to modeling agent references for plan recognition, *User Modeling and User-Adapted Interactions*, 5, 317, 1996.

92. Hawkes, L., Derry, S., and Rundersteiner, E., Individualized tutoring using an intelligent fuzzy temporal relational database, *Int. J. Man Machine Studies*, 33, 409, 1990.

93. Anderson, J. et al., Cognitive tutors: lessons learning, *J. Learning Sci.*, 4(2), 167, 1995.

94. McArthor, D. et al., Skill-oriented task sequencing in an intelligent tutor for basic algebra, *Instructional Sci.*, 17(4), 281, 1988.

95. Anderson, J.R., *Language, Memory, and Thought*, Lawrence Erlbaum Associates, Hillsdale, NJ, 1976.

96. Anderson, J.R., *The Architecture of Cognition*, Harvard University Press, Cambridge, MA, 1983.

97. Anderson, J.R., *Rules of the Mind*, Lawrence Erlbaum Associates, Hillsdale, NJ, 1993.

98. Damasio, A., *Decartes' Error*, Grosset/Putnam, New York, 1994.

99. Smolensky, P., On the proper treatment of connectionism, *Behav. Brain Sci.*, 11(1), 1, 1988.

100. Anderson, J.R., Acquisition of cognitive skill, *Psychol. Rev.*, 89, 369, 1982.

101. Rosenbloom, P., Laird, J., and Newell, A., *The SOAR Papers: Research on Integrated Intelligence*, MIT Press, Cambridge, MA, 1993.

102. Jones, R. and VanLehn, K., Acquisition of children's addition strategies: a model of impasse-free knowledge-level learning, *Machine Learning*, 16, 11, 1994.

103. Berry D. and Broadbent, D., Interactive tasks and the implicit-explicit distinction, *Br. J. Psychology.*, 79, 251, 1988.

104. Lewicki, P., Hill, T., and Czyzewska, M., Nonconscious acquisition of information, *Am. Psychol.*, 47, 796, 1992.

105. Stanley, W. et al., Insight without awareness: on the interaction of verbalization, instruction and practice in a simulated process control task, *Q. J. Exp. Psychol.*, 41A(3), 553, 1989.

106. Sun, R., Learning, action, and consciousness: a hybrid approach towards modeling consciousness, *Neural Networks*, 10(7), 1317, 1997.

107. Sun, R., Merrill, E., and Peterson, T., From implicit skills to explicit knowledge: a bottom-up model of skill learning, *Cognitive Sci.*, 2000.

108. Sun, R. and Peterson, T., Autonomous learning of sequential tasks: experiments and analyses, *IEEE Trans. Neural Networks*, 9(6), 1217, 1998.

109. Gelfand, J., Handelman, D., and Lane, S., Integrating knowledge-based systems and neural networks for robotic skill acquisition, in *Proceedings of the IJCAI*, Morgan Kaufmann, San Mateo, CA, 193, 1989.

110. Maclin, R. and Shavlik, J., Incorporating advice into agents that learn from reinforcements, in *Proceedings of AAAI-94*, Morgan Kaufmann, San Mateo, CA, 694, 1994.

111. Sun, R. and Bookman, L., Eds., *Computational Architectures Integrating Neural and Symbolic Processes*, Kluwer Academic Publishers, Norwell, MA, 1994.

112. Watkins, C., Learning with delayed rewards, Ph.D. thesis, Cambridge University, Cambridge, UK, 1989.

113. Lin, L.J., Self-improving reactive agents based on reinforcement learning, planning, and teaching, *Machine Learning*, 8(3/4), 293, 1992.

114. W3C, Scalable Vector Graphics (SVG) 1.0 specifications, http://www.w3.org/TR/SVG/.

115. Adobe SVG Viewer, http://www.adobe.com/svg/viewer/install/.

116. CSIRO SVG Tool Kit, http://sis.cmis.csiro.au/svg/.

117. Bathik SVG ToolKit, http://xml.apache.org/batik/.

118. JASC Software Inc., WebDraw, http://www.jasc.com/webdraw.asp.

119. Barnett, L. et al., Design and implementation of an interactive tutorial framework, *SIGCSE Bull.*, 30(1), 87, 1998.

120. Weiss, G., Ed., *Multiagent Systems: A Modern Approach to Distributed Artificial Intelligence*, MIT Press, Cambridge, MA, 1999.

121. Angelo, T.A., The campus as learning community: seven promising shifts and seven powerful levers, *AAHE Bull.*, 49(9), 3, 1997.

122. Cross, K.P., Why learning communities? Why now? *About Campus*, 3(3), 4, 1998.

123. Tinto, V., Universities as learning organizations, *About Campus*, 1(6), 2, 1997.

124. Hattori, F. et al., Socialware: multiagent systems for supporting network communities, *Commun. ACM*, 42(3), 55, 1999.

125. Key, M.R., *Male/Female Language*, The Scarecrow Press, Metuchen, NJ, 1975.

126. Spender, D., *Invisible Women*, The Women's Press, London, 1982.

127. Stalker, J., Women and adult education: rethinking androcentric research, *Adult Educ. Q.*, 46(2), 98, 1996.

128. Oram, A., Ed., *Peer-to-Peer: Harnessing the Benefits of a Disruptive Technology*, O'Reilly & Associates, 2001.

129. SUN, Java Web Start, http://java.sun.com/products/javawebstart.

130. JCP, Java network launching protocol, Java Community Process, http://www.jcp.org/.

131. W3C, Simple object access protocol (SOAP) 1.1, http://www.w3.org/TR/SOAP.

132. W3C, Web services description language (WSDL) 1.1, http://www.w3.org/TR/wsdl.

133. ECMA, Standard ECMA-262 ECMAScript language specifications 3rd ed., December 1999, http://www.ecma.ch/ecma1/stand/ecma-262.htm.

134. W3C, Document object model (DOM) technical reports, 2nd ed., http://www.w3.org/DOM/DOMTR.

135. UIUC-NCSA, The common gateway interface, http://hoohoo.ncsa.uiuc.edu/cgi/overview.html.

136. SUN, Java servlet technology: the power behind the server, http://java.sun.com/products/servlet/.

137. SUN, Java server pages: dynamically generated Web content, http://java.sun.com/products/jsp/.

138. JDOM, http://www.jdom.org/.

139. W3C, Extensible stylesheet language (XSL) 1.0 specifications, http://www.w3.org/TR/xsl/.

140. W3C, XSL Transformations (XSLT) Version 1.1, http://www.w3.org/TR/xslt11/.

141. Shang, Y. and Shi, H., A Web-based multi-agent system for interpreting medical images, *World Wide Web*, 2(4), 209, 1999.

142. Chen, S. et al., Personalizing digital libraries for learners, in *Proceedings of the 12th International Conference on Database and Expert Systems Applications (DEXA'2001)*, Munich, Germany, September 2001.

143. Shang, Y., Sapp, C., and Shi, H., An intelligent Web representative, *Information*, 3(2), 2000.

144. JATLite, http://java.stanford.edu/java_agent/html.

145. Freeman, E., Hupfer, S., and Arnold, K., *JavaSpaces Principles, Patterns, and Practice*, Addison-Wesley, Reading, MA, 1999.
146. Finin, T. and Fritzson, R., KQML—a language and protocol for knowledge and information exchange, Proceedings of the 13th International Workshop on Distributed Artificial Intelligence, 126, 1994.

11

The System Perspective of an Intelligent Tutoring System Based on the Inquiry Teaching Approach

L.H. Wong
Institute of Systems Science

Chai Quek
Nanyang Technological University

Abstract. This chapter describes a research effort that has been motivated by a keen desire to bridge the gap between educational and computational science research. It consists of two portions. The first is reinterpreting the intelligent tutoring systems (ITS) architecture in general, and the TAP architecture (an ITS architecture for the inquiry teaching approach that we have developed) in particular, in terms of control theory. The second is the proposal of a new version of TAP architecture whose instructional planning mechanism strictly adheres to the definition of adaptive control.

Inquiry teaching is a dialog-based teaching style that trains the student in the systematic reasoning process. In building a TAP (tutoring agenda planner), a generic framework has been constructed for developing inquiry teaching software. TAP also adopted Peachey & McCalla's dynamic planning techniques[4] to complement Collins & Stevens' inquiry teaching method.[36]

The first part of this chapter focuses on the reinterpretation of the general ITS architecture in terms of control theory. This generic architecture is educationally sound as it adequately describes a teacher's intelligence in handling instruction. Nonetheless, we argue that the architecture is *ad hoc* from the viewpoint of systems structure because it is loosely specified by perceptive words in the relevant literature. The universal control theory does not only provide a means to validate the underlying structure of ITS architecture (and TAP in particular). It also serves as a starting point for the ITS field to explore the concrete and well-established control techniques from which ITS researchers could borrow to improve the design of their systems.

The research effort also paves the way for the tackling of the second issue. A refined TAP architecture that adheres to the strict specifications of adaptive control systems is proposed. In the context of control theory, an adaptive system must be capable of adjusting its control rules in order to improve its performance. The adaptive TAP architecture has demonstrated the possibility of developing ITSs that are not only able to criticize the student's understandings of the domain, but are also able to critique and adapt their own planning rules. Such a performance-driven adaptive planning scheme has the potential of developing into a new direction in ITS research.

11.1   Introduction

Launched in the early 1970s, the paradigm of intelligent computer-assisted instructions (ICAIs) or intelligent tutoring systems (ITSs) represented a methodological shift with the addition of intelligence to computer-aided learning. Systems that fall within this paradigm have a common feature of understanding their domain sufficiently to provide an approximate quality of guidance and instruction similar to a human teacher (see Wenger[1] for an extensive survey of the paradigm).

Almost all of the modern ITSs are modelled after a general architecture of ITS (hence, it has become a *de facto* structure of ITS), which was proposed by several groups of researchers in the mid 1970s (please refer to Section 11.3 for a detailed description of the architecture). The model is educationally sound as it adequately describes a teacher's (more specifically, a one-to-one tutor) competence and intelligence in handling instruction. Nonetheless, we argue that the architecture is *ad hoc* from the viewpoints of system structure and system engineering, simply because the architecture is loosely specified by perceptive words in the relevant literature.

One may argue that this validation problem does not seem to be critical at the moment since the effectiveness of any given ITS could be convincing once a proper evaluation has been taken and the result is satisfactory. Incorporating system theory or control theory (a sophisticated interpretation of system theory) in ITS design is nonetheless a new area that is worth being explored due to the former's universality and sophistication.

Hence, this research aims to reinterpret ITS in terms of system or control theory. Due to the time constraint, it is not intended to give a complete unification of both theories, but rather to provide a foundation for further investigations of this matter since we believe that there are plenty of concepts and techniques in control theory that the ITS community can learn from.

To demonstrate this, we will first reinterpret a typical ITS that we have built, namely, TAP (stands for tutoring agenda planner, which is an ITS that is capable of performing inquiry teaching). After that, we will attempt to introduce one of the control techniques, adaptive control, to TAP.

An adaptive control system is one that provides a means of periodically measuring the system's performance in relation to a given set of criteria as well as a means of automatically modifying the system's adjustable parameters on the basis of the performance assessment. This implies that such a system must react or adapt itself to changes in its environment.

In the past, several groups of ITS researchers claimed that their systems had this adaptive (instructional) planning capability.[2–4] Their claims are valid in the context of AI planning in a loose sense. On the other hand, a stricter definition has been given to adaptive control, and some formal techniques for developing such a system have emerged. The major difference between adaptive control and adaptive planning (flexible planning) is that the former takes "system performance" (again, the definition is different from the context of ITS) into account when system adaptation has taken place.

Hence, the last part of the project proposes a new TAP architecture with the adaptive instructional mechanism that is modelled after a strict and formal definition in adaptive control theory. This will serve as a demonstration of incorporating the latter to ITSs in a more general sense, in order to enhance the instructional planner's capability to cope with dynamic situations.

11.2 Control Theory Fundamentals

11.2.1 Control Systems

A system is a collection of matter, parts, or components that are included inside a specified boundary. This boundary is the separation of the system with surrounding environments that interfere/interact with its operation. The system becomes dynamic when one or more aspects of the system change with time. The effects that originate from outside the system and directly act on it but are unchanged by changes within the system are known as inputs to the system. Thus, system inputs (or reference inputs) are independent of changes in the system state and are produced by external factors or (command) inputs.

A control system is typically produced by connecting various components in a specific configuration to achieve a desired response or output. The system controls the variable output to the desired value by applying the proper input or controlling signal to the system input terminals. The input-output relationship of the system represents the cause and effect relationship of the system. It mathematically represents the process in the system by which the input signal through system parameters controls the output signal to produce the desired output.[5]

11.2.2 Block Diagram

A block diagram is a short-hand symbol of a system that provides pictorial representation of the cause and effect relationship between the input and output of the system (e.g., a real-life example in Figure IV.11.1). The interior of the block usually contains a description or the name of the element of the symbol for the mathematical operation to be performed on the input to obtain the output. The arrows represent the direction of unilateral flow of information or signal, which means that signal can only flow in the direction of the arrow. The operations of addition and subtraction are represented by circles in place of blocks and are known as summing points. The summing points have arrows with plus or minus signs that indicate the summing point to be additive, subtractive, or both. These operations are illustrated in Figure IV.11.2.

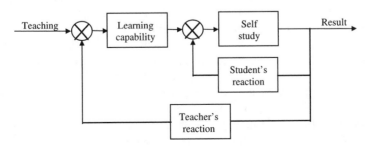

FIGURE IV.11.1 Block diagram of a teaching–learning strategy system. (From Aggarwal, K.K., *Control System Analysis and Designs*, Khanna Publishers, New Delhi, India, 1981. With permission.)

FIGURE IV.11.2   Block diagram of sensing devices.

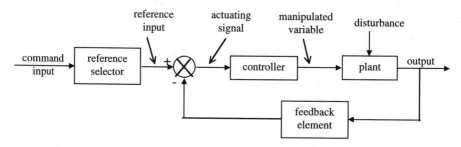

FIGURE IV.11.3   Block diagram of an open-loop control system.

reference input   actuating signal   manipulated variable   disturbance

command input → reference selector → + ⊗ − → controller → plant → output

feedback element

FIGURE IV.11.4   Block diagram of a closed-loop control system.

The use of a block diagram provides a simple means by which the functional relationship of the various components may be shown and reveals the operation of and the functional relationship between various points in the system more readily than observation of the physical system itself. Therefore, it may reveal the similarity of apparently unrelated systems.[6]

11.2.3   Open Loop vs. Closed Loop

Control systems can be broadly classified into two types depending upon whether the controlled variable affects the actual input to the control system or not, that is, whether some kind of a feedback is used or not. When feedback is not used, the system is called an open-loop system; when used, it is called a closed-loop system or feedback control system.

The open-loop systems are the simplest and most economical type of control systems. However, these are generally inaccurate and unreliable and, as such, are not preferred. In an open-loop system, an input known as the reference input is applied directly to the controller, and an output known as the controlled output is obtained. In an open-loop system, the input to the controller is in no way affected by the value of the controlled output. For example, traffic control by means of signals operated on a time basis is an open-loop system. Such a system is schematically shown in Figure IV.11.3.

In a closed-loop control system, the corrective signal that drives the controller is derived from some kind of comparison between the input variable and the control variable. Thus, in general, when a controller, which itself drives a plant, is driven by some functions of a difference of the reference input and a fraction of controlled output (usually this difference is termed as error or actuating signal), the overall system is known as a closed-loop system. This is shown in Figure IV.11.4. Definitions of the terminology found in Figures IV.11.3 and IV.11.4 are available in the next section.

An advantage of the closed-loop control system is that the use of feedback makes the system response relatively insensitive to external disturbances and internal variations in system parameters.

It is thus possible to use relatively inaccurate and inexpensive components to obtain the accurate control of a given plant, which is impossible in the open-loop case.[6]

11.2.4   Terminology

The rest of the terminology necessary to describe control systems is defined as follows (adapted from Ogata[6] and D'Azzo and Houpis[7]). More formal and detailed definitions can be found in the proposed standards of the IEEE.[8]

1. **Plants.** A plant is a piece of equipment to perform a particular operation. In most books on control systems, any physical object to be controlled is called a plant.
2. **Processes.** The Merriam-Webster dictionary defines a process to be an artificial or voluntary, progressively continuing operation that consists of a series of controlled actions or movements systematically directed toward a particular result or end. In most books on control systems, any operation to be controlled is called a process.
3. **Disturbances.** A disturbance is a signal that tends to adversely affect the value of the output of a system. If a disturbance is generated within the system, it is called internal, while an external disturbance is generated outside the system and is considered an input to the system.
4. **Command input.** This is the motivating input signal to the system and is independent of the output of the system.
5. **Reference selector (reference input element).** This is the unit that establishes the value of the reference input. The reference selector is calibrated in terms of the desired value of the system output.
6. **Reference input.** The reference input is the reference signal produced by the reference selector. It is the actual signal input to the control system.
7. **Actuating signal or error.** This is the signal that is the difference between the reference input and the feedback signal. It actuates the control unit in order to maintain the output at the desired value.
8. **Controller.** The controller produces the manipulated variable from the actuating signal.
9. **Manipulated variable.** The manipulated variable is that quantity obtained from the controllers that is applied to the controlled system.
10. **Output (controlled variable).** This is the quantity that must be maintained at a prescribed value.
11. **Feedback element.** The feedback element is the unit that provides the means for feeding-back the output quantity, or a function of the output, in order to compare it against the reference input.

11.2.5   Gain Scheduling Control

Gain scheduling is a technique that moves one step further from ordinary open-loop control (or constant gain control, since the control rules remain unchanged), as illustrated in Figure IV.11.5. A control engineer that adopts such an approach in his/her control system first identifies possible states of the plant in terms of the combinations of the relevant parameters and variables.

The suitable gain (control rules) of each state can then be prespecified. This forms a look-up table. When the plant is running, an appropriate logic for detecting the operating point and choosing the corresponding gain from the table will be activated. The controller adjustment is thus achieved.

One of the disadvantages of gain scheduling is that the adjustment mechanism of the controller gains is precomputed off-line and, therefore, provides no feedback to compensate for incorrect schedules. Unpredictable changes in the plant dynamics may lead to deterioration of performance or even to complete failure.[15]

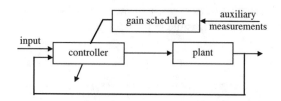

FIGURE IV.11.5   Gain scheduling.

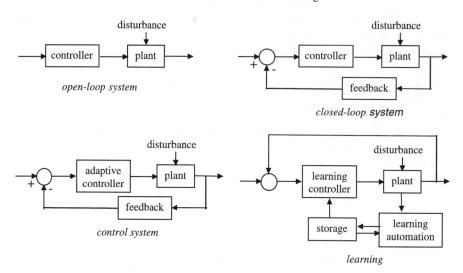

FIGURE IV.11.6   Four levels of sophistication in a control system.

11.2.6   Multi-Level Control

For a complex system, each sub-problem is provided with its own controller, designed according to a local criterion and a local model. These are the first-level controllers. The next higher, that is, second-level controller, then coordinates the operations of the first-level controllers, so that the objectives of the overall system are best served. Complex management systems are normally organized with this form of control arrangement known as multi-level control.[9,10]

11.2.7   Levels of Sophistication in Control Systems

We have presented a simple classification of control systems in the previous section where two classes of control systems, open-loop and closed-loop systems, were introduced. Further evolution of control theory in the last five decades has made such classification inadequate. Gibson[11] divides the space of all control systems into four basic hierarchical levels: (1) open loop; (2) closed loop; (3) adaptive loop; and (4) learning loop (see Figure IV.11.6). As we have already described the first two levels, we now introduce the next two classes.

With the advent of high performance aircraft missiles and space vehicles, dynamic performance was found to vary drastically. Such systems needed parameter adjustments to maintain the desired performance. This added another stage of sophistication where control systems were required to adapt themselves, that is, self-adjust or self-modify (not just the controlled output that closed-loop systems can achieve, but also the control rules), in accordance with the changing environmental conditions. Such a system is called an adaptive control system.[12,13]

The next higher stage of sophistication is achieved in what is known as learning systems. Such systems are designed to recognize familiar stimuli and patterns in an unknown situation. The major difference between an adaptive system and a learning system is that, while the former can cope with present situations, the latter is capable of recognizing new situations on the basis of its past experience, give decisions, and act accordingly.[11]

11.3 Systems View of ITS

This section attempts to interpret the underlying mechanism of ITS in terms of control theory, thereby giving a more formal treatment to the general ITS architecture.

11.3.1 TCAI vs. Open-Loop Control

Traditional Computer-Assisted Instruction (TCAI) is often characterized as branching, scripted courseware with fairly simple and straightforward objectives—exposing students to a body of information or practicing simple skills. As Peachey and McCalla have stated,[4] TCAI tends to have a rigid predetermined course structure that leads the student through a series of lessons. The course designer must explicitly predict and program all possible paths through the lessons.

Consider a branching drill-and-practice program that takes different students on different instructional paths (which are predetermined and encoded by TCAI programmers) through the same lesson: a student who answers A to a multiple-choice question will go in one direction, while the student who answers B goes in another. This is because branching lessons reserve one path for a particular understanding the students may have about the material. However, after the various paths have been completed, all students return to the main stream of instruction, where another part of the subject is taught and where the branching process begins again.[14]

In other words, drill-and-practice TCAI essentially functions as an electronic workbook (with marking, but not diagnosing, capability), not to mention another popular paradigm of TCAI that functions as an electronic textbook and concentrates on pure presentation of course contents (probably with fanciful multimedia capability). The rigidity and inflexibility is obvious. Hence, TCAI is, in general, an open-loop control system. Figure IV.11.7 presents the control view of a TCAI in terms of a block diagram.

The student is the plant to be controlled, since all sorts of CAI programs (including TCAI and ITS) aim to keep the students on the right learning path(s) in order to eventually achieve the ultimate teaching goal(s).

Tutor (or tutoring module) is an *ad hoc* term in this chapter for TCAI since no official term was fixed in the field (while in the context of ITS, the tutoring module is a rather well-established, if not standardized, term for software modules in ITSs). The tutor in a TCAI is the software module that decides what and how to teach next. Based on the definition of TCAI, such a module could be as simple as a scripted branching mechanism (if such a module is more sophisticated and intelligent in a given CAI, then the CAI could be an ITS instead). Hence, tutor is considered as a controller that generates a manipulated variable and feeds it to the plant (student).

In TCAI, the manipulated variable refers to the instructions in the form of pure presentation (plain text or multimedia), drill-and-practice, game, simulation (probably less interactive than in ITS), etc. The plant (student) is supposed to either passively absorb the content (in the cases of presentation and simulation) or actively respond to the tasks (in the cases of drill-and-practice and game).

FIGURE IV.11.7   TCAI as an open-loop system.

A minor difficulty in this mapping process is to identify an element in TCAI that corresponds with the command/reference input (or simply the input) for open-loop systems. Conceptually, the input of a general CAI refers to, for example, the teacher or student's specifications of any combination of the following:

- System and/or instructional preferences
- Prior knowledge of the student
- Target knowledge to be learned by the student
- Time limit
- Other relevant options

Nevertheless, the availability of such input for the open-loop system being discussed varies from one TCAI to another as it depends on the flexibility and functionality provided by each individual TCAI. In fact, most TCAIs always start at a fixed point and probably go through a fixed path without consulting the teacher and/or student to understand their need in advance. From a systems point of view, the input of such TCAIs could just be a signal to start running the software (analogous to the power on/off switch of a factory plant and its controller).

The output of the plant (student) is equivalent to:

- The student's partial or overall result (a macroscopic view, where the effectiveness of the entire instructional session is of one's interest)
- Every single response (could be an answer to a question or an indication of "finished reading a tutorial article"; a "microscopic" view)

Both views can be taken as the means to assess the performance of the controller (tutor). Table IV.11.1 summarizes the mapping of elements in control systems to elements in ITS.

11.3.2 Does the Feedback Mechanism Exist in TCAI?

Based on the preceding discussion, one may question the decision to map TCAIs with open-loop systems, simply because a weak feedback mechanism seems to exist in certain TCAIs. In a drill-and-practice TCAI, for instance, the tutoring module attempts to keep the instructional session right on the predetermined main path. The student's incorrect answers trigger some other predetermined branches to remedy the problems. After this, the instruction would be brought back to the main path. In the context of control theory, however, such a mechanism is considered as the simplified version of gain scheduling rather than feedback.

Recall from the previous section that a control engineer who adopts the gain scheduling technique in designing a system first identifies possible states of the plant and then precomputes the suitable gain (or control rule set) for each of the states, which forms a look-up table. Note that the block diagram in Figure IV.11.5 represents a closed-loop system where both the feedback path and gain scheduling mechanisms are present.

TABLE IV.11.1 Mapping of Control Concepts to TCAI Concepts

Open-Loop System	TCAI
Controller	Tutor
Plant	Student
Manipulated signal	Instruction
Plant output	Student's response

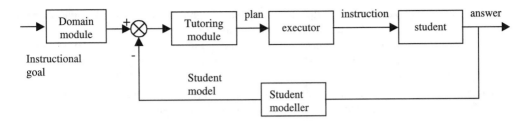

FIGURE IV.11.8   ITS as a closed-loop system.

Hence, a student's responses or answers to a drill-and-practice TCAI are, in fact, the prespecified states for a gain scheduler rather than feedback signals. The key word here is prespecified. As Ioannou and Sun[15] have stated, one disadvantage of gain scheduling is that the adjustment mechanism of the controller gains is precomputed off-line and, therefore, provides no feedback to compensate for incorrect schedules. Unpredictable changes (unexpected responses or answers in the context of TCAI) in the plant dynamics (student's responses) may lead to deterioration of performance or even to complete failure. For most TCAIs, the user interface is implemented in a way that the student's input of incorrect answers that are out of a prespecified range or not among the multiple-choice answer set will trigger the system to prompt the student for answers within the range.

Both early and recent surveys of TCAIs[16–20] suggest ambiguous or inconclusive results. Sophisticated meta-studies by Kulik and associates[21–23] indicate that TCAI success is limited to clearly demarcated situations like short-term studies and lower-level cognitive objectives. This sounds similar to common critiques against open-loop systems such as inaccuracy, unreliability, sensitivity to unexpected external disturbances, and inability to cope with complex plants.

In short, we argue that it is justifiable to characterize general TCAIs as open-loop systems. The feedback mechanism, being the decisive feature that distinguishes closed-loop systems from open-loop systems in the context of control theory, is incidentally one of the major strengths of ICAIs or ITSs over general TCAIs. This will be studied in detail in the next section.

11.3.3   ITS vs. Closed-Loop Control

Figure IV.11.8 depicts the system's view of a typical ITS (or ICAI) architecture. At first glance, one might have an impression that the system's view is similar to that of TCAI with an additional feedback path (although the path itself constitutes a major improvement of ITS over TCAI). That is, one can map student to plant, tutoring module to controller, or student's responses and answers to output. Indeed, the ITS field differs from the fundamental concepts of TCAI and is analogous to closed-loop control.

Nevertheless, it is important to compare two control systems by examining their block diagrams— a block diagram reveals the functional and causal relationships between major modules of a system. It is equally important to study the functionality of each of the modules and their respective roles in the entire system, especially the controller and the feedback element, which play important roles in characterizing a control system.

11.3.4   Feedback Path

One of the major differences between Figure IV.11.7 and IV.11.8 is the feedback path. In a typical ITS, a student's answers and/or responses are diagnosed and the student model is subsequently updated. The student model then serves as the basis for the tutoring module to revise its teaching plan, which will affect/modify the instruction to be presented to the student. Hence, the diagnoser and the student modeller are analogous to the feedback element that will process the output of the plant (student's answers) and convert it to a special format of signal that will be used by the controller (tutoring module) by varying its gain (instructional replanning).

The above description suits both overlay and bug catalog models of student modelling. For the latter, bugs of the student identified by SM are represented as part of the special format of signals that is sent to the controller.

The incorporation of the feedback mechanism in ITS also brings in actuating signal or error, that is, the difference between the expert model (domain knowledge) and the student model, as the input to the controller (tutoring module).

11.3.5   Intelligent Modules in ITS

The second main difference is the intelligence that exists in various ITS modules that makes the paradigm superior to TCAI. The feedback element (diagnoser and student modeller) could either be a look-up table or an intelligent agent that concentrates on revealing the plant's (student's) internal states (knowledge states) and/or variations from ideal output (that is, bugs). A more sophisticated feedback element could even identify the sources/reasons of such variations. The controller (tutoring module) is capable of planning and replanning the instructions online (which is what the tutoring module of TCAI lacks). It often incorporates general AI planning techniques.

An interesting observation is that the primary output (manipulated signal) of the controller (tutoring module) in TCAI is usually the user-interface presentation of the questions or the text or multimedia for delivering the domain knowledge, which could be directly fed into the plant (presented to the student). In the controller of ITS, however, the primary output is an explicit instructional plan (for example, to specify combinations of teaching goal + domain knowledge to be taught + delivery method), which is not supposed to be disclosed to the student. Usually, a final control element (FCE), e.g., an executor, is needed to transform every plan to the corresponding user-interface presentation and feed it into the plant (present to the student). Hence, Figure IV.11.8 could be further refined where an extra block (executor module) is added between the controller and the plant in order to transform the current plan to a format that is understandable by and useful for the plant (student).

Next, the command input of the system could be instructional goals (this is true for most of the ITSs, but there are some exceptional cases which will be investigated later) which are prespecified by the designer(s) or chosen by teachers or students before each instructional session starts. The specification of instructional goal(s) provides the controller/tutoring module (and, subsequently, the plant/student) a clear ultimate target (or a group of targets) to achieve. Sophisticated controllers (tutoring modules in ITSs) are able to determine and then concentrate on an instructional path leading to the target(s), while setting irrelevant elements aside.

Some ITSs do not require explicit instructional goal(s) to be specified. Instead, they aim to cover their domain knowledge completely. This indeed could be taken as an implicit instructional goal. Hence, it is appropriate to map instructional goal(s) to the command input of the systems view of generic ITS architecture.

11.3.6   Domain Module

The third main difference is an additional block in the block diagram that represents an ideal model of the plant (domain module or expert model in the context of ITS). This module does not exist in general TCAI because developers of the older paradigm rigidly encode the domain knowledge as course presentations and/or questions (and their answers). Hence, a TCAI memorizes the pedagogical decisions in the tutoring module that implicitly represent the domain knowledge, whereas an ITS understands the knowledge as the module is explicitly encoded as a stand-alone entity through a knowledge engineering process. Table IV.11.2 summarizes the mapping of elements in control systems to elements in ITS.

TABLE IV.11.2   Mapping of Control Concepts to ITS Concepts

Closed-Loop System	ITS
Controller	Tutoring module
Final control element	Executor/user-interface
Plant	Student
Reference model	Domain module
Feedback element	Diagnoser and Student modeller
Command input	Instructional goal(s)
Manipulated signal	Instructional plan
Feedback signal	Student model
Plant output	Student's response

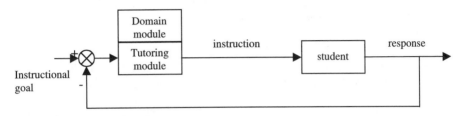

FIGURE IV.11.9   Systems view of SCHOLAR.

11.4   Case Study I: SCHOLAR

SCHOLAR[24,25] can be considered historically as the first ITS project. Being the system that launched the field, it is interesting to investigate its underlying structure. SCHOLAR is a mixed-initiative tutor that teaches factual knowledge of South American geography. A mixed-initiative tutor engages the student in a two-way conversation. It asks the student a series of questions. The student is allowed to interrupt the dialog flow by posing a request (e.g., asking another question or asking to change the topic). Once the tutor has responded to the student's request, it refocuses on the previously suspended issue. SCHOLAR also allows the student to request for switching to question-and-answer (Q/A) mode, where the tutor will stop raising new problems and concentrate on answering the student's questions.

Essentially, the SCHOLAR system is comprised of four basic components (domain, tutoring, student modelling, and communication), which are also found in most of its descendants. The functional relationships between the modules are depicted by Figure IV.11.9 (the communication module is omitted as it is insignificant to the control structure).

The notable part of the structure is that the output of the student model (which is supposed to be the feedback element in general ITS structure) is directly connected to the actuating summing point (before the controller, that is, the tutoring module) instead of going through a student modeller. The reason is that although Carbonell proposed the revolutionary idea of (global) student modelling for his system (which is later referred to as overlay model by the ITS field),[24,25] he did not explain how the student model could be referred to by the tutoring module in instructional planning.

Instead, SCHOLAR relies only on the student's individual responses in deciding the next move. It simply compares the student's every answer with the model answer and gives comments on the answer, then generates the next question. If the student poses a question instead, the tutor will access the domain module (which represents facts of South American geography in the form of a semantic network), answers the student, and resumes the discussion of the tutor's last question. Furthermore,

SCHOLAR generates questions in a random manner, which gives SCHOLAR's dialogs a somewhat disconnected flavor.

Nevertheless, what makes SCHOLAR more advanced than typical TCAI is the feedback mechanism. It does not simply resemble the open-loop gain-scheduling construct of TCAIs. General TCAIs can only handle a fixed, small amount of states (that is, the variety of student's responses) at each instructional point (for example, such a system will not accept an answer that is not among the answers provided by multiple-choice questions). SCHOLAR's NLP and mixed-initiative capability and the retrievable domain module make the system able to not only accept a larger variety of answers to each question, but also allows the student to pose questions. The developer of SCHOLAR cannot simply form a look-up table for each situation (like designing a gain scheduler) as the acceptable states are virtually limitless (in actual fact, though, they are bounded by the size and coverage of the domain knowledge base). On the other hand, the idea of instructional goals was not incorporated into SCHOLAR. Thus, there is no obvious (online) input for its control structure.

Through the analysis, it is obvious that the ITS field, at the time SCHOLAR was developed, was still in its formation stage whereby many brilliant bits and pieces of new ideas were pending integration into the whole picture of the new paradigm.

Other early ITSs that exemplify the similar structures are WHY,[26] SOPHIE,[27–30] and STEAMER[31] (localized or no student modelling), etc. Among these, WHY is one of the earliest ITSs that have incorporated the idea of instructional goals in their pedagogy design. The rest of the systems are a more open-ended learning environment (that is, they are not pure ITSs).

Meanwhile, certain ITS research groups started to pay more attention on diagnosis and student modelling (e.g., INTEGRATION;[32,33] WUSOR-II[34,35]), which produced the earliest batch of ITS with a similar structure to Figure IV.11.8.

11.5   Case Study II: TAP-2

The TAP project seeks to construct a shell for inquiry teaching software by transferring Collins and Stevens' cognitive theory for inquiry teaching[36] (a human theory) into a computational model. Inquiry teaching, a dialog-based teaching style, is characterized by forcing students to actively engage in articulating theories and principles that are critical for an in-depth understanding of a domain. TAP (tutoring agenda planner), the proposed architecture, is constructed within the context of ITS. Further details of this project are available in Wong, Quek, and Looi.[37]

TAP-2[38] is an ITS shell for delivering minimal inquiry teaching sessions, which has adopted Peachey and McCalla's globalized curriculum planning technique[4] to complement Collins and Stevens' architecture[36] that concentrates on the localized delivery planning of inquiry teaching. An executor with various teaching strategies and a dialog manager complements the integrated planner. Furthermore, it provides the mechanism for switching between inquiry and other supplementary teaching styles within a tutoring session, since inquiry teaching itself is not necessarily suitable for teaching all kinds of knowledge (it is good for teaching concepts and procedures, but not facts). This also tallies with Elsom-Cook's argument[39] stating that using multiple pedagogic strategies (that is, teaching styles) can provide a very powerful learning environment.

11.5.1   Collins and Stevens' Theory of Inquiry Teaching

To formalize the inquiry teaching style, Collins and Stevens[36] propose a cognitive theory that consists of:

- The goals and subgoals of teachers
- The strategies used to realize different goals and subgoals
- The control structure for selecting and pursuing different goals and subgoals

Teachers typically pursue several subgoals simultaneously. Each goal is associated with a set of strategies for selecting cases, asking questions, and giving comments. In pursuing goals simultaneously, teachers maintain an agenda that allows them to deliver the various goals.[26,40]

There are two top-level goals that teachers in inquiry dialogs pursue:

- Teaching students particular rules or theories
- Teaching students how to derive rules or theories

There are several subgoals with each of these top-level goals. The identified goals and subgoals are shown below.

1. Teach a general rule or theory.

   - Debug incorrect hypothesis.
   - Teach how to make predictions in novel cases.

2. Teach how to derive a general rule or theory.

   - Teach what question to ask.
   - Teach how to formulate a rule or theory.
   - Teach what is the nature of a rule or theory.
   - Teach how to test a rule or theory.
   - Teach students to verbalize and defend rules or theories.

In a later paper,[41] Collins and Stevens attempted to analyze the inquiry strategies of the "very best teachers" for whom they could obtain films or transcripts. Looking at the fine structure of the dialogs, they noted recurring patterns of strategies in selecting cases, asking questions, and making comments. They characterized the individual strategies in terms of condition–action pairs or productions. These are classified into the four following categories, namely,

- Case Selection Strategies　　　　(16 productions, denoted as CSS)
- Entrapment Strategies　　　　(12 productions, denoted as ENS)
- Hypothesis Identification Strategies　　　　(17 productions, denoted as IS)
- Hypothesis Evaluation Strategies　　　　(14 productions, denoted as ES)

The control structure that the teacher uses to sequence different goals and subgoals consists of four basic parts:

- A set of strategies for selecting cases
- A student model
- An agenda
- A set of priority rules for adding goals and subgoals to the agenda

Given a set of top-level goals, the teacher selects cases that optimize the ability of the student to master those goals. Subsequently, the teacher begins by questioning the student about the cases and the rules related to them. The answers reveal what the student does and does not know. This will modify the student model. Whenever specific bugs in the student's theory or reasoning processes are identified, subgoals are created to correct the bug.

However, the cognitive theory does not address the issue of the ordering of the top-level teaching goals. It concentrates on local or intra-goal planning but does not include planning at the global curriculum level. In this respect, the theory is incomplete and, hence, has to be supplemented by other global planning techniques, for example, the architecture proposed by Peachey and McCalla.[4]

11.5.2   Peachey and McCalla's Curriculum Planner

Peachey and McCalla[4] noted that an underlying weakness of ITS is the absence of global knowledge about the course being taught. Domain knowledge is stored in the form of bits and pieces. Hence, they proposed that planning techniques be used to create large, individualized courses that could handle a broader subject area.

The architecture consists of five components: a domain knowledge database, a student model, a collection of teaching operators, a planner, and a plan executor. The planner develops a teaching plan that is tailored to a particular student being taught. The executor then uses the teaching plan to guide the student through the course.

A STRIPS-like planning technique is used to plan the global strategies using the local information: teaching operators.[42] The teaching operators are stored independently in the domain knowledge base. At the beginning, the planner retrieves the teaching operators and links them together, according to the prerequisite (pre-condition) relationships, to form a curriculum network (a directed graph). Next, the planner attempts to find a path in the network that will lead to the accomplishment of the ultimate goal(s). This represents the eventual concepts or skills to be taught.

The executor starts teaching the sequence of actions and diagnosing the student's answers. If a teaching action fails to make the student understand the expected effect(s), the selected path is broken and a new path has to be identified. Finally, if all possible paths are broken, the student is considered to have failed to acquire the ultimate knowledge of the course.

11.5.3   TAP-2 Architecture

Figure IV.11.10 depicts the software architecture of TAP-2. A complete TAP-2-based ITS consists of two major parts: the TAP kernel and the domain-dependent module. The TAP kernel is essentially a domain-independent ITS shell, specifically for inquiry-centered dialog. It consists of three software modules, a control module, a planner, and a student modeller, and maintains three databases: an agenda, a tutoring history (TH), and a student model (SM).

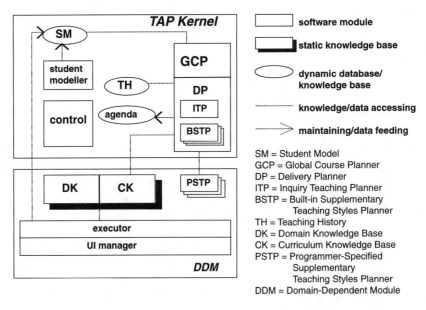

FIGURE IV.11.10   Architecture of TAP-2.

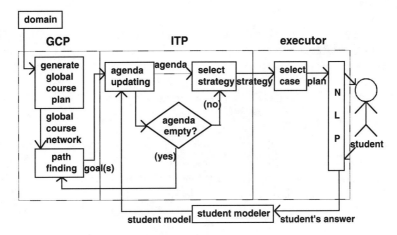

FIGURE IV.11.11   Program flow of TAP-2.

In particular, the planner combines Peachey and McCalla's global curriculum planner (GCP) and a delivery planner (DP). It retrieves the top-level goals specified by the domain and, hence, produces and maintains the agenda. The DP is further subdivided into several sub-planners to facilitate switching between different teaching environments (these are instances of built-in supplementary teaching styles planner or BSTP). Currently, there are only three teaching sub-planners specified in the DP: (1) inquiry teaching (sub-) planner (ITP); (2) assessment sub-planner; and (3) expository sub-planner. The agenda stores the goals to be pursued in a certain order. The SM, maintained by the student modeller, reflects what the student understands at any time.

The domain-dependent module (DDM) is the domain-dependent part of the system. The domain knowledge base in TAP-1 is broken down into two interrelated knowledge bases:

- A domain knowledge base (DK) that stores the knowledge of the subject domain
- A curriculum knowledge base (CK) that provides course information and course specifications needed for various steps in the planning process (e.g., local information for GCP, ultimate goal specifications, preset SM, etc.)

Other software modules in the DDM are an executor and a user-interface manager. The domain programmer could also develop a programmer-specified supplementary teaching styles planner to introduce more teaching environments to the system.

The general program flow of TAP-2 is as follows (refer to Figure IV.11.11):

1. The program state is initialized. The preset SM specified by the CK is loaded into the active SM (refer to the later sessions).

[2–4: GCP]

2. From the list of ultimate goals, GCP selects a goal for current focus.
3. GCP selects the shortest unblocked path leading to the ultimate goal.
4. GCP selects a goal from the eligible goals in the selected path and passes the goal to DP.

[5–6: DP]

5. DP breaks the top-level goal(s) into several subgoals and adds them onto the agenda.
6. Given the first item of the agenda (a goal/subgoal), DP switches the planning task to the relevant sub-planner.
   Assuming that DP has selected ITP, the system proceeds to execute Step 7.

[7–8: ITP]

(In case DP has selected another teaching style, this portion is to be replaced by the planning process of the respective sub-planner.)

7. ITP selects strategy (strategies) based on the combination of goal + subgoal.
8. ITP selects suitable case(s) for the current plan. The goal + subgoal + strategy + case combination is recorded in the TH and passed to the executor.

[9–10: executor/UI manager]

9. The executor interprets the delivery plan and starts the execution. In the meantime, the executor frequently passes control to the UI manager for user interfacing.
10. The UI manager compiles the student's response and passes this to the executor for diagnosis, which in turn passes the diagnosis to the student modeller.

[11: student modeller]

11. The student modeller updates the SM.

[12: DP]

12. If the student modeller reveals bug(s) of the students, the system creates subgoals to correct it (them). These debugging subgoals will be ordered based on some priority rules and added on top of the agenda. In some cases, the pursuit of debugging subgoals is delayed.
13. If a debugging subgoal has failed eventually and it causes the current path in the global curriculum plan to be blocked, then loop back to Step 4 to select an alternative path; otherwise, loop back to Step 6 to execute the next item on the agenda.
14. Loop back to Step 2 until the ultimate goals are all achieved or no more paths leading to unachieved ultimate goals can be chosen.

11.5.4   Systems View of TAP-2: A Multi-Level Closed-Loop Control System

Recall that the program flowchart of TAP-2 as depicted in Figure IV.11.11 has provided readers a first glance at the control structure of the framework. Indeed, the flowchart is fairly brief and was constructed as a system engineering treatment to both frameworks. Figure IV.11.12 depicts the summarized control structure of general ITSs that incorporates both curriculum planning and delivery planning capabilities. This shows a clear distinction between curriculum and delivery planning, as highlighted by Brecht (Wasson),[43] that at least two feedback paths should be involved. Some examples of the prior architectures that adopt such a distinction are Meno-tutor,[44,45] TUPITS,[46] COACH,[47] TAPS,[48] PEPE,[43] Murray's BB1-based instructional planner,[49–51] and FITS.[52,53] All these architectures were built, expanded, and adapted from the control structure, as depicted in Figure IV.11.12.

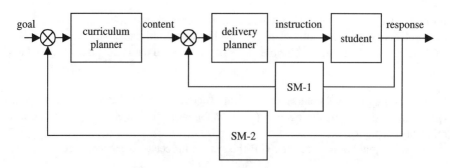

FIGURE IV.11.12   Systems view of curriculum and delivery planning.

Not surprisingly, this summarized diagram suits TAP-2, as the delivery planner is treated as a black box. The more detailed block diagram of TAP-2 in Figure IV.11.13 shows its relatively complex delivery planning mechanism. In this figure and the rest of the figures in this section, $\{\langle a \rangle\}$ denotes a set of elements $\langle a \rangle$. In order to elaborate the functional and causal relationships between the software modules, certain modules are decomposed into smaller units.

In the context of control theory, the outer loop is called servo control (or trajectory), which concentrates on driving the plant (student) to the objective (ultimate goals) at the fastest rate possible. The inner loop is termed regulatory control, whose task is to restrict the plant's output (student's response) within an acceptable range (achieve the subgoal under a given goal). The combination of the macroscopic servo control and microscopic regulatory control is commonly found in many industrial controllers and is an instance of a multi-level control mechanism.

11.5.5 Curriculum Planning

The flow in Figure IV.11.13 starts with the input signal, represented by a set of ultimate goals (as specified in the curriculum knowledge base or CK). The signal is fed into the first controller, the path selector (which is the first function of the global curriculum planner or GCP, denoted as GCP-1 in the diagram). The path selector accesses the prerequisite relationship list (that is, a list of pre (Goal-1, Goal-2), where Goal-1 is the prerequisite of Goal-2) of the domain knowledge base (DK) (this portion of DK is denoted as DK-1) to determine an optimal path leading to each ultimate goal.

The path is then dispatched to the second controller, the goal selector (which is the second function of GCP, denoted as GCP-2 in the diagram). The goal selector then accesses the goal library (list of goals and their attached subgoals) of DK (denoted as DK-2), picks one goal from the eligible goals, and produces goal + subgoal_set.

11.5.6 Switching between Teaching Styles vs. Gain Scheduling

As specified in the TAP-2 architecture, domain programmers are allowed to specify non-inquiry subgoals that must be handled by their relevant sub-planners (sub-controllers), such as expository, assessment, general Socratic dialog, simulation, etc. After the delivery planning of each subgoal is done, the output of the corresponding sub-planner is fed to the same executor (analogous to the final control element or FCE in control theory) for execution, and then the instruction is delivered to the student. This is not a strict specification as different copies of FCEs with different functionality could be constructed for each of the sub-planners.

This construct is, indeed, analogous to the gain scheduling mechanism of control systems. The system's states which are required by the gain scheduling system to decide the switching of gains are current subgoals. In the context of TAP-2 (and some other multiple-teaching-style systems, e.g., DOMINIE,[39,54]) teaching styles are analogous to gains. This is because the same input (for example, a piece of knowledge to be taught) that is fed into different sub-planners of different teaching styles will produce different kinds of instructions. As for the case of control systems, the same input, which is fed to different controllers with different gains, would produce different output signals. The whole idea is that each subgoal has its corresponding sub-planner, thus the selection of a new subgoal (change of the state) might cause the switching of sub-planners (change of the controller gain). In gain scheduling, there is a finite amount of prespecified states that will trigger the change of the controller gain, just as there is a finite set of prespecified subgoals in TAP-2.

11.5.7 ITP

Figure IV.11.14 depicts the internal control structure of ITP in TAP-2. In TAP-2, the selection of strategy and case(s) are concatenated together. Therefore, DK-3 (case library) and ITP-2 (strategy

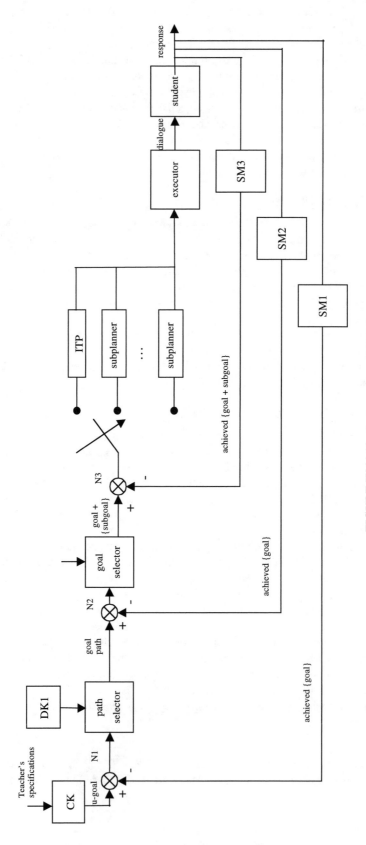

FIGURE IV.11.13 Systems view of TAP-2.

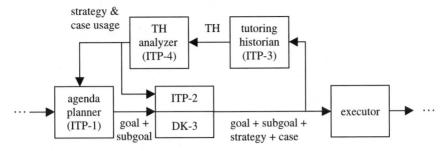

FIGURE IV.11.14   Systems view of ITP in TAP-2.

and case selector) are lumped together in Figure IV.11.14. The interaction between DK-3 and ITP-2 is represented here as a partitioned black box, but, in fact, the algorithm involved (refer to Figure 4.12 in Wong[55]) is rather complex.

The tutoring historian path has an additional element called TH analyzer. By referring to TH (the tutoring history), this element essentially summarizes the usage of cases and passes the statistics to ITP-2 for the case selection process ("drill a case to death," if possible). In the ITS architectural view of the TAP-2 system (refer to Figure IV.11.10), this task is handled by the ITP. Hence, the analyzer is denoted as ITP-4. The output of the analyzer is eventually fed back to the ITP-2.

11.5.8   Student Modelling as Feedback Paths

The system involves three major feedback paths (student modelling). The innermost path captures the plant's output and passes it to the feedback element (student modeller, denoted as SM-3), which in turn updates the student model and passes the set of achieved goal + subgoal combinations to the summing point between DK-2 and ITP-1. The difference between the output of DK-2 (set of current pursuing goal + subgoal combinations) and this feedback signal (set of achieved combinations) is sent to the gain scheduling switch for replanning.

Whereas the innermost path is the feedback path of delivery planning, the two outer paths are those of curriculum planning. The feedback element in the middle path is denoted as SM-2, whose output is a set of achieved goals (not a combinations of goal + subgoal). The goal set subtracts the eligible goal list, and the difference (that is, unachieved goals) is fed into the goal selector (GCP-2).

The feedback element (SM-1) of the outermost path sends the same set of achieved goals (same as the set of goals generated by SM-2) to the summing point between DK-1 and GCP-2 (denoted as N2) only when the currently selected path is blocked. The effect is that the achieved ultimate goals will be removed from the ultimate goal list that is supposed to be fed into GCP-2. The rest of the achieved non-ultimate goals will be ignored.

Note that the distinction between SM-1, SM-2, and SM-3 is not the same as GCP-1 and GCP-2; ITP-1, ITP-2, and ITP-3; or DK-1, DK-2, and DK-3. GCP-1 and GCP-2 are different components of GCP (path selector and goal selector), and so are ITP-1, -2, and -3 and DK-1, -2, and -3. On the other hand, SM-1, -2, and -3 all refer to the entire student modeller, as the sole student modeller in the framework could provide the information required by the three feedback paths (either achieved goal set or achieved set of goal + subgoal combinations). Nevertheless, the system would not come across such a situation where all the feedback paths are simultaneously activated (because the paths represent three separate planning steps that should not be activated in parallel). The mutual exclusion of the three loops is represented as shown in Figure IV.11.13, where three separate branches from the student's output are drawn (in contrast with Figure IV.11.11, where there is only one arrow coming from the student and pointing to the only student modeller block), each of which is pointed to a different SM block.

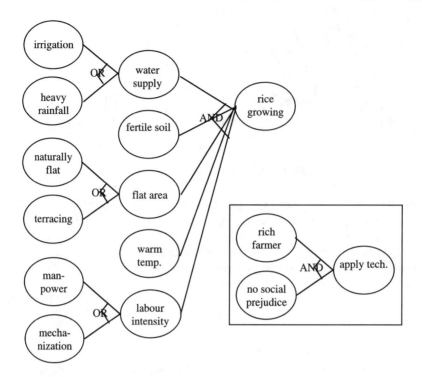

FIGURE IV.11.15   The AND–OR graph representation of domain knowledge of PADI-2.

11.5.9   Domain Case Study: PADI-2

In order to experiment on the functionality of TAP-2, an expanded rice growing domain, PADI-2, has been implemented and integrated with the TAP-2 kernel. The domain has two dependent variables that will be taught separately: the capability of a place to grow rice and the capability of a place to apply agricultural technology (e.g., irrigation, terracing, mechanization, etc.). Figure IV.11.15 depicts the causal network of the domain. A dialog dissection between PADI-2 and a student (and the change of the internal states of the system) is given in Figure IV.11.16.

11.6   Adaptive Planning in ITS—Using A-TAP as the Case Study

11.6.1   Adaptive Control Systems

A system that is designed to have, at all times, a very small sensitivity to variations in plant characteristics so that the complex, overall system remains unchanged is, in a sense, an adaptive system. This, in turn, implies that an adaptive control system is basically a feedback control system that automatically achieves a desired response in the presence of extreme changes in the controlled system's parameters and/or major external disturbances.

Such a system must, therefore, automatically measure the characteristics of the output of the control system and of the process under control and, on the basis of these measurements, adjust the overall system towards some previously defined optimum conditions or characteristics. It should be noted here that:

- The adaptive loop must be closed via the comparison of the actual performance with the desired standard.
- The signal computed by the adaptive controller must be used for parameter adjustments.

Dialog	Goal & Subgoal	Strategy & Case	Student Model
(PART 1)			
T: Consider Yangtze Plain and Nigeria. Do you know why Yangtze Plain produces rice while Nigeria does not? S: Because Nigeria is short of fertile soil. T: So you think fertile soil is a necessary factor for rice growing. S: Yes.	g_FW (fresh water supply as a necessary factor for rice growing) + sg3 (from a hypothesis	st1 (compare positive & negative exemplars)-positive: Yangtze Plain; negative: Nigeria	(blank)
T: However, Nigeria actually has fertile soil. Think about an other possibility. S: That is because Nigeria lacks rainfall. T: So you think heavy rainfall is a necessary factor for rice growing. S: Yes.			g_FS (fertile soil as a necessary factor for rice growing)
T: OK. Back to the case of Yangtze Plain. Suppose Yangtze Plain doesn't have heavy rainfall, do you think it still grows rice? S: Yes. T: Why? S: Because Yangtze Plain has irrigation facilities.	g_FW + sg1 (debug a wrong hypothesis)	st4 (use back previous positive case: Yangtze Plain)	g_FS, bug (g_FW)
T: So you think either heavy rainfall or irrigation facilities supplies water for rice. S: Yes.			g_FS, g_FW
(PART 2)			
T: How about Christmas Island? Do you think there is a lot of rice? S: No. T: Why do you say so? S: That is because Christmas Island lacks flat terrain. T: So you think flat terrain is a necessary factor for rice growing. S: Yes.	g_FA (flat area as a necessary factor for rice growing) + sg3	st2 (Select a positive case - Christmas Island)	g_FS, g_FW
T: How about Japan? Do you think there is a lot of rice? S: No. T: Actually Japan really grows rice, despite being mountainous. Do you know why? S: Because Japan has technology for terracing land.	g_FA, sg1	st3 (Select a contradictory case: Japan)	g_FS, g_FW (bug(g_FA)
T: So you mean either flat terrain or terracing area is a rice growing factor. S: Yes.			g_FS, g_FW, g_FA

FIGURE IV.11.16   A dissection of dialog between PADI and a student (T = tutor; S = student).

On the other hand, as an adaptive system, self-organizing features must exist. If the adjustment of the system parameters is done only by direct measurement of the environment, the system is not adaptive.[6]

It may, therefore, be seen that such a definition demands that three main functions be performed. They are:

- Identification of dynamic characteristics of the plant
- Decision-making based on the identification of the plant
- Modification or actuation based on the decision made

For the purpose of identification, it may be necessary to formulate an index of performance by which the level of quality of the system performance may be established.[56] The very basis of adaptive control rests on the premise that there are some conditions of operation or performance for the system that are better than others. A performance index is a number that indicates the goodness of system performance. A control system is considered optimal if the values of the parameters are chosen so that the selected performance index is minimum or maximum.[6] For example, if the difference between the ideal output and the actual output of a system (error) is selected as the performance index, the objective is to minimize the performance index.

11.6.2   Adaptive Planning in ITS

We argue that new ITSs that are modelled after adaptive control theory are not only able to criticize the student's understandings to the domain, but are also able to criticize the effectiveness of their own instructional planning rules by referring to the periodically measured performance index. Their planning rules no longer remain static, but are adaptive and more responsive to individual students' cognitive strengths and weaknesses (for example, preference of learning styles).

We believe that this performance-driven adaptive planning scheme has the potential to become a new and exciting direction for ITS research (and perhaps for pure educationists who concentrate on developing instructional theories). Once a more generic treatment of this proposal is established, a whole set of adaptive control techniques would be opened for the ITS field to explore.

11.6.3   Adaptive Control in Existing ITSs

Self-improving ITSs could be achieved in both domain and tutoring modules. Since the self-improving capability of the domain module (e.g., INTEGRATION;[32,33] GEO[57,58]) makes little or no impact on the overall control structure of an ITS, we will concentrate on the study of a self-improving tutoring module.

In this session, two existing ITSs that are claimed by their developers or field surveys to have self-tuning, self-improving, or adaptive capabilities in instructional planning will be analyzed in the context of adaptive control theory. This is to give the readers a better idea of how adaptive control can be achieved in ITS.

11.6.3.1   O'Shea's Self-Improving Tutor

In his doctoral dissertation, O'Shea[3,59] describes a tutor "as a first attempt at giving a tutoring system some ability to set up experiments using variations of its strategies and to adopt those that seem to produce the best results."

The domain is the solution of quadratic equations of the form $x^2 + c = bx$, whose answers can be obtained by clever guesses and with the help of a few rules. The tutor presents example problems on the basis of which students are to discover and master the rules. The tutor considers three sources of information during instructional planning: (1) task difficulty matrix (that relates specific features in a problem to teaching goals); (2) student model (a set of hypotheses about the student's current mastery of each of the few rules he/she must learn); (3) teaching strategies (a set of production rules).

The self-improvement facilities deal only with the teaching strategies and tune parameters that govern trade-offs between conflicting goals in the teaching task. Other parameters that can be modified include the frequency of encouraging remarks, the number of guesses allowed before the system gives out the solution, and other limits on the tutor's patience. Finally, the system can monitor the effort it expends in optimizing its performance by varying the time it spends on two teaching tasks: forming a student model by testing diagnostic hypotheses, and generating examples by matching problem features with pedagogical requirements.

For improvements to be possible, not only is it necessary to have an explicit and modular representation of the strategies, but the tutorial objectives must also be clearly defined. The instructional

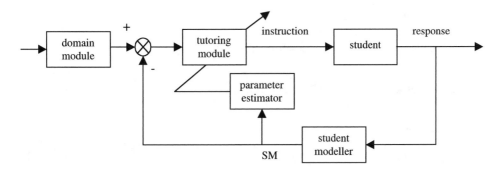

FIGURE IV.11.17   Systems view of O'Shea's Quadratic Tutor.

experiments embodied in the variations of the system's parameters revolve around four distinct tutorial goals:

1. To increase the number of students completing the session
2. To improve their scores on the post-test
3. To decrease the time taken by students to learn all four rules and their combinations
4. To decrease the computer time used in the process

The foundation for the experimentation process is a database of assertions which O'Shea calls a *theory of instruction* for the domain. Each assertion in the database indicates the expected results of a specific experimental modification performed on the production rules. Results are expressed in terms of possible or definite effects on teaching performance with respect to one goal.

After choosing one of the four goals, the system selects all the assertions that propose modifications leading to a possible improvement with respect to that goal; then it executes the changes these assertions suggest. Statistical tests evaluate the variations in performance caused by modifications with respect to each of the four goals. The overall improvement of the system is then computed as a linear function of the goals' respective variations. If a significant improvement is achieved, the modifications are adopted.

Figure IV.11.17 depicts the underlying control structure of the tutor. As usual, one can map the controller to its tutoring module, the plant to the student, and the reference model to its domain module. The new module is the parameter estimator where the adjustable parameters of the tutoring module are tuned (as mentioned in the third paragraph of this section) on the basis of certain performance measurements. The tutor measures its performance in terms of the overall improvement of the system that is represented in a linear function of the goals' respective evaluations. It is obvious that this system meets the requirement of adaptive control systems. However, it is not considered as a learning system as it cannot qualitatively generate new behaviors but only tune existing strategies.

11.6.3.2   DOMINIE

DOMINIE[39,54] is a tutoring system to allow the experimentation of multiple teaching strategies in an ITS for teaching how to use computer-based interfaces. The DOMINIE interface represents the goal structure of tasks that can be performed using the interface. An overlay student model is used to represent the student's knowledge as a subset of the total domain knowledge.

The overall philosophy incorporated within DOMINIE is one of decreasing intervention, where the amount of tutorial control is relinquished as the user becomes more knowledgeable. Seven teaching and assessment strategies are available to the system. These include: cognitive apprenticeship, successive refinement, discovery learning, Socratic diagnosis, abstraction, practice, and direct assessment. DOMINIE constantly assesses which of the seven strategies is most appropriate based on these competing factors, namely:

- Achieving a balance between teaching and assessment
- The appropriateness of the strategy to a given area
- The student's prior success with the different strategies
- The student's personal preferences

DOMINIE is a meta-level system that uses meta rules to guide the system to switch between available teaching styles. Two examples of such rules are:

- IF it is suspected that the student has a misconception about x and in the past the student was able to overcome misconceptions when they were pointed out, THEN use a Socratic diagnosis strategy.
- IF the student wishes to browse ahead in the curriculum, THEN use a cognitive apprenticeship strategy.

After a teaching style has been selected, the instructional planner will choose a strategy from the strategy set of that particular teaching style.

Ohlsson[60] argued that one of the requirements for successful ITSs is that they be adaptive, and thus an ITS must have a wide range of instructional strategies. Much ITS literature labels such meta-level tutoring systems as adaptive or adaptable tutoring systems. This is acceptable in a loose sense as the meta rules warrant the switching between various sets of strategies (or fine-tuning of certain parameters in strategies). In the context of adaptive control theory, however, the key idea is performance measurements.

For the case of DOMINIE, all the competing factors (that contribute to the conditions of the meta rules) but the student's prior success with the different strategies have nothing to do with performance, let alone performance measurements (that is, to maximize parameters like success rate, or to minimize parameters like failure rate).

Fortunately, DOMINIE incorporates the following set of meta rules to accomplish its core strategy of decreasing intervention:

- IF the rating of student is very poor
  THEN change strategy to direct_assessment
- IF the rating of student is poor
  THEN change strategy to socratic_diagnosis
- IF the rating of student is average
  THEN change strategy to cognitive_apprenticeship
- IF the rating of student is good
  THEN change strategy to successive_refinement
- IF the rating of student is very good
  THEN change strategy to discovery_learning

Although it was not elaborated in the relevant literature on how the rating of the student is determined, it could be designed in such a way where performance measurement is taken. For example, the system could periodically assess the student and collect the raw scores, or compare the current student model with the previous ones. This performance measure could then be used by the system to decide if a teaching style shift should be made. This could be considered as adaptive tutoring and its aim is to maximize the rating of the student.

Indeed, meta rule construct is a powerful but not necessary or sufficient approach in constructing an adaptive control system. It is not a necessity because the adaptability can also be achieved by adjusting parameters in a fixed set of strategies (for example, O'Shea's quadratic tutor). It is insufficient because performance measurements must also be incorporated to guide the switching of the strategy set.

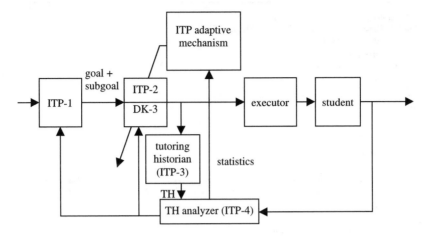

FIGURE IV.11.18   Systems view of ITP in A-TAP.

In the context of control theory, the formal terminology for parameter adjustment is parameter adaptation, whereas that for meta rule construct is self-organization. They represent two major paradigms in adaptive control systems.

11.6.4 Adaptive TAP

In this section, an attempt to extend the TAP-2 architecture from a closed-loop control system to an adaptive control system is made. The hypothetical system, A-TAP (stands for adaptive TAP), as presented in Figure IV.11.18, is meant for illustrating the possibility of adopting an adaptive control mechanism in an ITS. The figure depicts the ITP of A-TAP, the only portion of the latter architecture that is different from TAP-2. On the other hand, A-TAP inherits the two curriculum planning loops and the teaching style switching mechanism in TAP-2. Indeed, the main focus of A-TAP is on how ITP could be made adaptive.

The new architecture essentially has the same control structure and software modules of TAP-2, except for an additional block for performance measurement and its output that can tune ITP-2. Recall that control rules tuning and performance measurement are two key elements of any adaptive control system. Therefore, it is necessary to involve both of them in the new architecture.

The performance measurement block is a module that collects relevant data from various sources, analyzes them, and then derives the performance index. The performance index is subsequently fed into the adaptive system to tune the control rules.

From Figure IV.11.18, one can observe that the input signal of the performance measurement block is TH. Furthermore, the tutoring historian path is more complex than in TAP-2. The new structural design suits the proposed adaptive control mechanism that will be discussed later.

11.6.4.1 Adaptive Planning in A-TAP

In performing adaptive control, there are two common ways to adapt the control rules: switching between prespecified sets of rules (e.g., adaptive meta-systems), and adjusting parameters in the rules (e.g., O'Shea's quadratic tutor[2]). For A-TAP, the 36 dialog session rules (DSR's) that are derived from Collins and Stevens' 59 inquiry strategies are considered control rules. In TAP-2, the (summarized) algorithm in selecting the rules is shown in Figure IV.11.19. Because there are no adjustable parameters in these rules, a viable choice is to adopt the first approach. The purpose of doing so is to improve the student's performance index.

Lists (global; non-volatile memory)
Unused_ Case := all the cases stored in DK [initially]
Used_Cases := (blank) [initially]

STRUCTURE (local)
Rule_List (eligible rules with their suitable cases)

VARIABLE (local)
Boolean   NeedNewCase
PROCEDURE   FireDSRs

NeedNewCase := TRUE
Rule_List := (blank)

FOR all the inquiry rules

    IF the goal+subgoal+rule combination has been used in the dialog before OR
       the combination should not be selected as per domain programmer's specification
         THEN continue with the next iteration of the FOR loop (check next rule)
    ENDIF

    IF the rule requires ONE case
      THEN select a case from the Used_Cases
      IF a case is found
        THEN NeedNewCase := FALSE
      ELSE IF NeedNewCase = TRUE
        THEN select a case from the the Unused_Cases
      ELSE
        THEN continue with the next iteration of the FOR loop
      ENDIF

        IF a case is found in either Used_Cases or Unused_Cases
          THEN add the rule and the case into Rule_List
    ENDIF

    ELSE   (the rule requires TWO cases)
      select the first case from the Used_Cases
      IF   no case in found AND NeedNewCase = FALSE
        continue with the next iteration of the FOR loop
      ELSE IF no case is found
      THEN select the first case from the Unused_Cases
        IF no case is found
          THEN continue with the next iteration of the FOR loop
        ENDIF
      ENDIF

      select the second case form the Used_Cases
      IF the second case is found AND the selected first case is also a used case
        THEN NeedNewCase := FALSE
      ELSE IF no case is found AND NeedNewCase = TRUE
        THEN select the second case from the Unused_Cases
        IF the no case is found
          THEN continue with the next iteration of the FOR loop
        ENDIF
      ELSE IF no case is found   (AND NeedNewCase = FALSE)
        THEN continue with the next iteration of the FOR loop
      ENDIF

    ENDIF

    IF Rule_List has more than one rule
      THEN pick a rule with the most preferred case(s) being attached to
    ENDIF

ENDFOR

END PROCEDURE

FIGURE IV.11.19   DSR firing algorithm.

There are two areas in TAP that can be made adaptive, namely, switching of DSR subcategories and case selection. An algorithm to handle the hypothetical adaptive inquiry planning will be proposed after a description of these two areas is presented.

In the following discussion, an instructional session refers to the execution of a TAP-based tutor from beginning to end, whereas a dialog session refers to the execution of a plan that is represented by a goal + subgoal + strategy (DSR) + case(s), that is, the execution of a single dialog template.

11.6.4.2 Switching of DSR Subcategories

In A-TAP, we further subcategorize the 36 DSRs based on their nature as:

1. Discuss a case (CSS1-4; IS4-6).
2. Compare two cases (CSS5-8; IS11-14; ES11-14).
3. Discuss a counterexample (CSS9-12).
4. Discuss a hypothetical case (CSS13-16).

In practice, individual students may be more accustomed to one subcategory of DSRs than another. For example, student K may perform better when comparing two cases since a careful study of the similarities and differences between two cases could help him/her generalize the knowledge. On the other hand, student L may become easily confused when more than one case is presented; hence discussing a single case may be more helpful to him/her. In other words, we argue that whether a student's thinking style suits the type of teaching strategies being employed is one of the factors that affects the student's performance.

In order to identify the more appropriate DSR subcategories for a student, the system has to trace the performance of the student for every subcategory. The basic idea is to determine the success rate, in terms of percentage, for each subcategory. The higher the percentage of successful sessions (that is, the student achieves the combination of {goal + subgoal} of each of the sessions) over the number of attempted sessions for a DSR subcategory, the higher the priority the rules that belong to the subcategory will be given. This can be expressed as the following mathematical formula:

$$DSR\_group\_success\_rate = (number\_of\_successful\_attempts\_of\_DSR\_group /$$
$$total\_number\_of\_attempts\_of\_DSR\_group)^*100\%$$

This is considered as a performance measurement because the success rate is maximizable. That is, the system fine-tunes the control rules (switches between different DSR subcategories) in order to maximize the overall success rate. Note that since different students have different preferences in the subcategories, the decision of subcategory switching is localized. That is, it is only meant for the current instructional session and will not be saved and used by subsequent sessions that will probably deal with different students.

Figure IV.11.20 shows the algorithm to handle the adaptation of DSRs. This simplified algorithm has left out some irrelevant planning steps in ITP. This is because the main objective of presenting Figure IV.11.20 is to elaborate on the subcategory switching. The full algorithm will be presented subsequently.

The performance measurement (takes place in the line "compute the success rates of all DSR groups" in Figure IV.11.20) is a summarizing process. At the beginning of an instructional session, there is no relevant value of the performance index to which the system can refer. Therefore, a performance index construction process is performed. For the first twelve iterations (refer to "IF $n < 12$"), ITP alternates among the four DSR groups in selecting DSRs. After three rounds ($3 \times 4 = 12$ iterations), the system constructs the initial performance index. Then, the system can make use of the performance index to prioritize the DSR groups. Usually, a DSR will be chosen from the DSR group with the highest success rate (percentage). However, if none of the DSRs in the DSR group is suitable for the current plan, or if all the DSRs have been used to deliver the same combination of

```
CONSTANT LIST DSR_group1 := {  (list of DSR s)  }

CONSTANT LIST DSR_group2 := {  (list of DSR s)  }
CONSTANT LIST DSR_group3 := {  (list of DSR s)  }
CONSTANT LIST DSR_group4 := {  (list of DSR s)  }
VARIABLE Current_group
VARIABLE current_DSR
VARIABLE
case_list
STATIC VARIABLE
Integer  n := 0 (initial value)
        (STATIC means memory will hold the value of n when the system exits the procedure)


PROCEDURE SelectStrategy
INPUT partial_plan {current_goal, current_subgoal}

IF n < 12
        Current_group := {DSR_group1, DSR_group2, DSR_group3, DSR_group4} [n mod 4]
ELSE
        Compute the success rates of all DSR groups
        WHHLE true
                Current_group := DSR group with the next highest success rate
                IF there is a suitable DSR in Current_group that could form an unselected combination of
                        goal+subgoal+DSR
                                Exit WHILE loop
                ENDIF
        END WHILE
ENDIF
n := n+1

Choose a suitable current_DSR from Current_group
Return current_DSR

END PROCEDURE
```

FIGURE IV.11.20   Algorithm for selecting DSRs.

goal + subgoal, then the DSR group with the next highest success rate will be chosen. This is the initialization phase, similar to self-organizing control.

Note that there is no fixed number of iterations needed to construct the initial performance index (of course, it is advisable to have a number that is a multiple of the number of rule sets). It is a trade-off between more iterations for better accuracy of the performance index and less iterations for starting the adaptive control earlier.

The algorithm fixes the performance measurement point at the beginning of every single dialog session. Again, this could be flexible. The system could also measure the performance at the end of a fixed number of dialog sessions (for example, 5 sessions). This flexibility is particularly important when the performance measurement becomes a significant computational overhead on the system.

In the case where the student keeps providing incorrect answers to repeated attempts at goal + subgoal combination (with different DSRs), the success rate for the current DSR group could drop. Therefore, other previously unflavorable or lower-ranked groups are likely to be given another chance to perform the teaching task. As in general adaptive control systems, even if the plant (student) gives a satisfactory performance, the controller might occasionally experiment with adjusted rules or different rule sets to seek further improvement in the performance.

11.6.4.3 Case Selection

TAP-2 adopts Collins'[61] recommendation of "drilling a case to death" as the most important criterion in selecting cases. This is based on the notion that the student should be freed from studying unfamiliar details of cases and instead concentrate on picking up the inquiry skills.

However, the first few cases being chosen are probably those cases with the highest preference ratings that are specified by the domain programmers. Consider the following scenario: a domain programmer who designs a geography tutor for Singapore students assigns the highest preference rating to a case called Jurong (Singapore's largest industrial area). Most Singapore students should have little problem dealing with this case. Then an exchange student from a foreign country tries the tutor. She is likely to commit plenty of case bugs corresponding to this case since she may know very little about or have never heard of Jurong. Similarly, it is not advisable to keep discussing Isle of Man (an autonomous territory of the United Kingdom) with most Singapore students.

TAP-2 ignores all case bugs committed by the student and assumes that the student is supposed to become more familiar with more frequently used cases. The assumption might be valid if the student has reasonably good memory. Even so, if new factors are discussed for the same old case, there is no guarantee that the student will be able to produce less case bugs.

In A-TAP, the intention of minimizing case bugs is addressed. The basic idea is to relax the first two of the following set of case selection rules adopted by TAP.

MCS1: Select the most recently used case(s), if any (to ensure minimal switching of cases).
MCS2: Select the most frequently used case(s), if any.
MCS3: Select the previously used case(s).
MCS4: Select the more common/popular case(s) (among all the unused cases).

The approach could be summarized as shown in the algorithm of Figure IV.11.21. Again, this is just a simplified algorithm for demonstration purposes. In the algorithm, suitable case means the case that is suitable for the input plan goal + subgoal + DSR.

During an inquiry teaching session, the tutor repeatedly asks the student about the situations of the selected cases, for instance, "Do you know if A grows rice?" "Does B have heavy rainfall?" The *rate_of_case_bugs* is the percentage of the student's wrong answers to such questions.

At the beginning, each case is given three trials without being checked against its *rate_of_case_bugs* (see the WHILE loop conditions). After that, only these cases with *rate_of_case_bugs* of less than 50% will get through. For those cases that have higher *rate_of_case_bugs*, their *rate_of_case_bugs* will be reduced by 5% (analogous to forgetting factors in self-organization in adaptive control theory) when they are deselected. Otherwise, they will not have the chance to be chosen again. In this case, the case with the next highest preference rating will be checked. If the trend goes on until no more suitable case can be chosen, then the most preferred case will be again selected regardless of its *rate_of_case_bugs*.

The three figures involved in the algorithm, 3 (maximum number of trials for each new case), 50% (upper bound of *rate_of_case_bugs* for a case to be selected), and 5% (each time the *rate_of_case_bugs* of a deselected case is reduced by this percentage), are again adjustable. To adjust the maximum number of trials for each case is a trade-off between obtaining a more precise initial *rate_of_case_bugs* (after more trials) and starting the adaptation of the MCS rules earlier. To adjust the upper bound of *rate_of_case_bugs* for a case to be selected is a trade-off between giving a more preferred case greater chances and trying alternative cases with lower preference ratings. To adjust the percentage of the *rate_of_case_bugs* of a deselected case is a trade-off between letting the more preferred cases be reselected earlier and giving other less preferred cases greater chances of being attempted.

Therefore, the *rate_of_case_bugs* often does not reflect what its label suggests. The forgetting factors mechanism gives the more preferred cases with higher *rate_of_case_bugs* some chances

---

```
VARIABLE
CASE selected_case

PROCEDURE SelectCase
IINPUT goal+subgoal+DSR

Execute MCS1-MCS4 to obtain a selected_case

WHILE number_of_times_being_selected(selected_case) > 3 AND
        rate_of_case_bugs(selected_case) > 50%                    (this is a performance criterion)
                rate_of_case_bugs(selected_case) = rate_of_case_bugs(selected_case) - 5%
                selected_case := next most preformance suitable case (by checking preference_rating)
                IF on more suitable case meets the performance criterion
                        selected_ case := the most preferred suitable case       (re-selected)
                        EXIT WHILE loop
                ENDIF
ENDWHILE

Return selected_case

END PROCEDURE
```

FIGURE IV.11.21   Algorithm for the adaptive case selection process.

of reselection. For example, a case that has a 49% $rate\_of\_case\_bugs$ suddenly jumps to 63%. It will then be suspended for at least three rounds until the $rate\_of\_case\_bugs$ drops to a value that is less than 50%.

Obviously, the three figures must be carefully determined in order to address the two policies of case selection: (1) minimize case switching (as suggested by Collins[61]); (2) minimize case bugs (our proposal). The two policies might become contradictory when the most frequently selected case triggers a lot of case bugs from the student. In this situation, we propose a more adventurous attitude in trying out new cases, but this must be well controlled to prevent excessive case switching. The other possibility is to do the checking of $rate\_of\_case\_bugs$ once every n (say, 3 to 5) dialog sessions rather than in every session. This will also reduce the frequency of case switching.

What does this mechanism have to do with adaptive control? The relaxation of MCS1 and MCS2 could be considered an effort of qualitatively adapting the case selection rules. The adaptation is based on the periodically updated $rate\_of\_case\_bugs$ of the currently selected case, which is taken as a performance index. The aim is to minimize the overall $rate\_of\_case\_bugs$ of the entire instructional session.

11.6.4.4   An Adaptive Version of the DSR Firing Algorithm

In the previous section, the adaptive versions of strategy and case selection have been discussed separately. Since A-TAP inherits the coupling of strategy and case selection from TAP-2, an algorithm based on the DSR firing algorithm in Figure IV.11.19 is constructed. Figure IV.11.22 depicts the new algorithm (bold denotes the new changes).

Note that the lists of *Unused_Cases* and *Used_Cases*, and the boolean NeedNewCase, are removed from the revised algorithm. This will be handled by the procedure *SelectCase* in Figure IV.11.21. Cases with $rate\_of\_case\_bugs$ are definitely used.

The major change in the new algorithm is to wrap a WHILE loop around the original algorithm, thereby allowing DSR switching to be performed. The first IF condition in the WHILE loop appears

STRUCTURE (local)
Rule_List (eligible rules with their suitable cases)

PROCEDURE FireDSRs

NeedNewCase :=TRUE
Rule_List := (blank)

Compute the success rates of all DSR groups
WHILE   current rule has not been selected AND there are still some suitable rules unchecked
    IF one or more DSR_groups have been selected as Current_group for less then 3 times
        Current_group := the DSR_group that is less selected
  ELSE
        Current_group := DSR group with the next highest success rate
        FOR all the DSR s in **Current_group**
            IF the goal + subgoal + rule combination bas been used in the dialog before OR the
                combination should not be selected as per domain programmer s specification
                Continue with the next iteration of the FOR loop (check next rule)
            ENDIF

            Check other conditions (i.e., NOT conditions about cases)
            IF other conditions are unsatisfied
                Continue with the next iteration of the FOR loop
            ENDIF

            IF the rule requires ONE case
                Selected case :=SelectCase(goal + subgoal + DSR)
                        ---executing algorithm in Figure 6.7
                IF a case is found
                    Add the rule and the case into Rule_LIST
ELSE
                  Continue with the next iteration of the FOR loop
                ENDIF
            ELSE   (the rule requires TWO cases)
                Selected case1 :=SelectCase(goal + subgoal + DSR)
                IF no case is found
                  Continue with the next iteration of the FOR loop
                ENDIF

                (At this point, the first case should have already been selected)
                Selected case2 :=SelectCase(goal + subgoal + DSR)
                IF no case is found
                  Continue with the next iteration of the FOR loop
                ENDIF

                Add the rule and cases to Rule_LIST
            ENDIF
            IF Rule_List has more than one rule
                Pick a rule with the most preferred case(s) being attached to
            ENDIF
         ENDFOR
ENDWHILE
END PROCEDURE

FIGURE IV.11.22   Adapted DSR firing algorithm.

different from the one specified in Figure IV.11.20, but they share the same idea. In the new algorithm, it is still desirable to give each DSR group a try of at least 3 times (such that the initial success rate for each group could be determined) before the adaptive mechanism is activated. This is the initialization phase.

The following explains the interaction between the performance measuring block and the tutoring historian, as specified in Figure IV.11.18. The tutoring historian obtains the tutoring plan of goal + subgoal + DSR + case(s) (the output of ITP-2) and the student's responses, and couples them together as a new TH entry. The student's responses that are of interest to the system consist of two elements: (1) whether the goal + subgoal is achieved; (2) the correctness of the student's answers to questions regarding the cases. The TH is then accessed by the performance measuring element when the needs arise. For DSR selection, the performance measuring element computes the success rate of each DSR group. For case selection, it determines the rate of case bugs for each suitable case. The outcome is then fed into ITP-2 for adaptive tuning of the planning rules. The aim here is to maximize the overall success rate of dialog sessions (execution of goal + subgoal combinations) and minimize the overall rate of case bugs. Meanwhile, TH is directly accessed by ITP-2 in checking whether the selected goal + subgoal + DSR has been selected before.

11.6.4.5 Discussion

The adaptive approach for A-TAP is an implicit control method, that is, it is largely based on the result of trial-and-error. To construct an explicit control mechanism within A-TAP is out of the scope of this project since it requires concrete instructional theories and/or extensive empirical studies to induce meta-rules. The tuned rules in A-TAP are not intended to be stored for future use as the tutor in different instructional sessions may handle differently students who would have different background knowledge, learning styles, and preferences.

Furthermore, the approach is adventurous in nature as it could experiment with alternatives that are previously unflavorable whenever the opportunity arises. In certain aspects, it conflicts with Collins and Stevens' original theory (for instance, the case selection policy as discussed above).[36] Nonetheless, since performance measurement of the student does exist, the mechanism is actually performing self-criticism on its own planning rules that are modelled after the original theory.

The effectiveness of the adaptive planning process relies on the length of the instructional session (in terms of the amount of dialog sessions). The more dialog sessions the system goes through, the more closely the system can determine the suitable DSRs and cases of the student. For TAP-based tutors with relatively small domain knowledge (hence, a small amount of top-level goals), the power of the adaptive planning mechanism maybe negligible. This is because the inductive process of performance measurement would only be able to rely on a small number of dialog sessions.

Although not evaluated by empirical approaches, we believe that the hypothetical architecture of A-TAP has convincingly demonstrated the possibility of incorporating an adaptive planning mechanism in any new or existing ITSs. We also believe that, apart from the planning rules and case selections as demonstrated in A-TAP, there are many other areas of instructional planning that could be made adaptive. Therefore, it is our hope that a new ITS design methodology based on adaptive or even learning control theory could be explored in the future. Hence, more ITSs with the capability of tuning their own planning rules could be developed.

11.7   Conclusion

The reinterpretation of ITS in general, and TAP in particular, in terms of control systems was motivated by the desire of validating the educationally sound, but structurally *ad hoc*, architectures.

ITS is a multi-disciplinary field that involves both educationists and computer scientists. In a pragmatic view, an ITS software could prove its value by incorporating an education or learning theory (usually a human theory) and going through a carefully planned evaluation. That is, unlike other fields of computer science, the ITS field almost solely relies on humans (educationists and students) to recognize the value of its products.

However, the field apparently does not intend to address the issue that system scientists might be concerned with if they are asked to evaluate the architecture of general ITS: Is it structurally and systematically sound? The generic architecture was merely described by natural language without going through systematic analysis; hence, it was questionable to be considered as a formal system within the context of systems theory. In other words, as compared to other fields of computer science, ITS is a software paradigm that is closer to human nature and perceptions than being mathematical and systematic.

Perhaps our attempt to give general ITS architecture and TAP architectures a more stringent systems view is a theoretical effort. Whether the result has direct impact on the ITS field is still unknown (but is worth being explored). It nonetheless unveils the underlying control structure of typical ITSs that will help ITS researchers better understand the system structure of the paradigm. This is probably analogous to the effort of an ITS to diagnose a student's bugs to reveal his/her underlying misconceptions rather than simply recording the incorrect answers in the student model.

A more obvious contribution of the unification effort is that it serves as a starting point for the ITS field to learn from the field of control theory. As a universal theory with much longer history and diversification as well as sophisticated techniques, control theory is definitely an area that ITS researchers can learn from. For instance, the gain scheduling technique of control theory has been extensively used in ITSs, including TAP, as revealed by this chapter. Another example is to design an instructional planner by incorporating adaptive control theory and techniques.

Implemented based on the typical rule-based expert system technology, TAP offers dynamic instructional plan, but the planning rules are fixed. Inquiry teaching is a sophisticated and dynamic teaching style that is difficult to be fully simulated by TAP. One possible way to improve on this architecture is to redesign the instructional planning components in TAP to make it more adaptable to individual students' performance and preferences.

Therefore, we propose the adaptive control approach in the context of control theory to be incorporated within TAP. Adaptive control theory is a more mature, well-defined, sophisticated, and, more importantly, universal system theory that many other areas including ITS could learn and benefit from. Its universality has been demonstrated by the successful reinterpretation of Quadratic Tutor and DOMINIE's adaptive capability in terms of the theory.

Therefore, our proposal gives a new dimension to instructional planning in ITS. A strict definition and set of criteria are provided to construct an ITS with adaptive control capability. New ITSs that are modelled after the adaptive control theory are not only able to criticize the student's understandings to the domain, but are also able to criticize the effectiveness of their own planning rules by referring to the periodically measured performance index. Their planning rules no longer remain static, but are adaptive and more responsive to individual students' cognitive strengths and weaknesses (for example, preference of learning styles). We believe that this performance-driven adaptive planning scheme has the potential to become a new and exciting direction for ITS research (and perhaps for pure educationists who concentrate on developing instructional theories).

The future direction of this project is to work out a more generic treatment of this novel idea in ITS in order to guide future ITS researchers to incorporate the approach in constructing their new ITSs or improving their existing systems. This is not an easy task as such an approach could be planner-dependent in different ITSs. For instance, Quadratic Tutor and DOMINIE employ the approach in different ways—the former is capable of adjusting the parameters of its planning rules, while the latter switches between different sets of planning rules. Both of these mechanisms are very common in adaptive control systems. In adaptive control theory, the former is known as parameter adaptation and the latter is known as self-organization. A generic treatment of this idea through a thorough study of adaptive control theory would provide guidelines to assist future ITS researchers to incorporate the right mechanism in the right planner. Meanwhile, more sophisticated techniques for adaptive control systems could be investigated to see if they could be adopted in ITS design.

Another aspect that one could examine is explicit adaptive control in ITSs. This requires an explicit representation of rules adjusting conditions that are based on a well-established education theory, instead of the trial-and-error nature of implicit adaptive control in the above-mentioned systems. This task will be very theoretical as it actually requires the research and development of new instructional theories.

Finally, a feasibility study could be done for employing adaptive control in curriculum planning, that is, making curriculum planning rules adjustable. Again, such a self-tuning mechanism must suit a specification of adaptive controllers: self-tuning must be based on the periodically measured performance index.

References

1. Wenger, E., *Artificial Intelligence and Tutoring Systems*, Morgan Kaufmann, Los Altos, CA, 1987.
2. O'Shea, T., *Self-Improving Teaching Systems: An Application of Artificial Intelligence to Computer-Aided Instruction*, Birkhauser, Verlag, Basel, 1979.
3. Goldstein, I.P., The genetic graph: a representation for the evolution of procedural knowledge, in *Intelligent Tutoring Systems*, Sleeman, D.H. and Brown, J.S., Eds., Academic Press, London, 1982.
4. Peachey, D.R. and McCalla, G.I., Using planning techniques in intelligent tutoring systems, *Int. J. Man Machine Studies*, 24, 79 1986.
5. Verma, S.N., *Automatic Control Systems*, Khana Publishers, New Delhi, India, 1979.
6. Ogata, K., *Modern Control Engineering*, Prentice-Hall, New Delhi, India, 1970.
7. D'Azzo, J.J. and Houpis, C.H., *Feedback Control System Analysis and Synthesis*, 2nd ed., McGraw-Hill, New York, 1966.
8. IEEE, Terminology for Automatic Control, ASA-C85.1, 1963.
9. Mesarovic, M.D. et al., Advances in multilevel control, Proceedings of the Tokyo Symposium on Systems Engineering for Control System Design, Tokyo, Japan, August 1965.
10. Kulikowski, R., Optimal control of aggregated multi-level system, Proceedings of the Third Congress, I.F.A.C., London, 1966.
11. Gibson, J.E., *Nonlinear Automatic Control*, McGraw-Hill, New York, 1963.
12. Wright-Patterson Air Force Base, Proceedings of the Self Adaptive Flight Control Systems Symposium, WADC TR 59-49, Dayton, Ohio, March 1959.
13. Blakelock, J., *Automatic Control of Aircraft and Missiles*, John Wiley & Sons, New York, 1965.
14. Keller, A., *When Machines Teach: Designing Computer Courseware*, Harper and Row, New York, 1987.
15. Ioannou, P.A. and Sun, J., *Robust Adaptive Control*, Prentice Hall, Upper Saddle River, N.J, 1996.
16. Rockart, J.F. and Morton, M.S., *Computers and the Learning Process in Higher Education*, McGraw-Hill, New York, 1975.
17. Brebner, A. et al., Teaching elementary reading by CMI and CAI, *Association for Educational Data Systems: Convention Proceedings*, 1980.
18. Dence, M., Toward defining a role for CAI: a review, *Educ. Technol.*, 20(11), 50, 1980.
19. Hallsworth, H.J. and Brebner, A., Computer-Assisted Instruction in Schools: Achievements, Present Developments and Projections for the Future, Calgary: Faculty of Education, Computer Applications Unit.
20. Gershman, J. and Sakamota, E., Computer-assisted remediation and evaluation: a CAI project for Ontario Secondary Schools, *Educ. Technol.*, 21(3), 40, 1980.

21. Kulik, J.A., Kulik, C.L.C., and Cohen, P., Effectiveness of computer-based college teaching: a meta-analysis of finding, *Rev. Educ. Res.*, 50(4), 525, 1980.
22. Kulik, J.A. and Bangert-Drowns, R.L., Effectiveness of technology in precollege mathematics and science teaching, *J. Educ. Syst.*, 12(2), 137, 1983.
23. Kulik, J.A., Bangert, R.L., and Williams, G.W., Effects of computer-based teaching on secondary school students, *J. Educ. Psychol.*, 75(1), 19, 1983.
24. Carbonell, J.R., Mixed-initiative man-computer instructional dialogues, Ph.D. dissertation, Massachusetts Institute of Technology, Cambridge, MA, 1970.
25. Carbonell, J.R., AI in CAI: an artificial intelligence approach to computer-assisted instruction, *IEEE Trans. Man Machine Syst.*, 11(4), 190, 1970.
26. Stevens, A.L. and Collins, A., The goal structure of a Socratic tutor, Proceedings of the National ACM Conference, Association for Computing Machinery, New York, 256.
27. Brown, J.S. Burton, R.R., and Bell, A.G., SOPHIE: a step towards a reactive learning environment, *Int. J. Man Machine Studies*, 7, 675, 1975.
28. Brown, J.S. and Burton, R.R., Multiple representation of knowledge of tutorial reasoning, in *Representation and Understanding: Studies in Cognitive Science*, Bobrow, D. and Collins, A., Eds., Academic Press, New York, 1975.
29. Brown, J.S., Uses of AI and advanced computer technology in education, in *Computers and Communications: Implications for Education*, Seidel, R.J. and Rubin, M., Eds., Academic Press, New York, 1977.
30. Brown, J.S., Burton, R.R., and de Kleer, J., Pedagogical, natural language, and knowledge engineering techniques in SOPHIE I, II, and III, in *Intelligent Tutoring Systems*, Sleeman, D.H. and Brown, J.S., Eds., Academic Press, London, UK, 1982.
31. Stevens, A.L. and Roberts, B., Quantitative and qualitative simulation in computer-based training, *J. Comput. Based Instruction*, 10(1), 16, 1983.
32. Kimball, R., Self-optimizing computer-assisted tutoring: theory and practice, Technical Report 206, Psychology and Education Series, Institute of Mathematical Studies in the Social Sciences, Stanford University, Stanford, CA, 1973.
33. Kimball, R., A self-improving tutor for symbolic integration, in *Intelligent Tutoring Systems*, Sleeman, D. and Brown, J.S., Eds., Academic Press, London, UK, 1982.
34. Carr, B., Wusor-II: a computer-aided instruction program with student modelling capabilities, AI Lab Memo 417 (Logo Memo 45), Massachusetts Institute of Technology, Cambridge, MA, 1977.
35. Carr, B. and Goldstein, I.P., Overlays: a theory of modelling for computer-aided instruction, AI Lab Memo 406 (Logo Memo 40), Massachusetts Institute of Technology, Cambridge, MA, 1977.
36. Collins, A. and Stevens, A.L., A cognitive theory for inquiry teaching, in *Teaching Knowledge and Intelligent Tutoring*, Goodyear, P., Ed., Ablex, Norwood, NJ, 203, 1991.
37. Wong L.H., Quek C., and Looi, C.K., TAP: a software architecture for an inquiry dialogue based tutoring system, *IEEE Trans. Syst. Man Cybernetics*, 28(3), 315, 1998.
38. Wong L.H., Quek C., and Looi, C.K., TAP-2: a framework for an inquiry dialogue based tutoring system, *Int. J. Artif. Intelligence Educ. (IJAIED)*, 9(1–2), 88, 1998.
39. Elsom-Cook, M., Using multiple teaching strategies in an ITS, Proceedings of Intelligent Tutoring Systems (ITS-88), Montreal, Canada, 286, 1988.
40. Collins, A., Warnock, E.H., and Passafiume, J.J., Analysis and synthesis of tutorial dialogues, in *The Psychology of Learning and Motivation*, Bower, G.H., Ed., Academic Press, New York, 1975.
41. Collins, A. and Stevens, A.L., Goals and strategies for inquiry teachers, in *Advances in Instructional Psychology II*, Glaser, R., Ed., Erlbaum, Hillsdale, N.J., 1982.
42. Fikes, R.E. and Nilsson, N.J., STRIPS: a new approach to the application of theorem proving to problem solving, *Artif. Intelligence*, 2, 189, 1971.

43. Brecht (Wasson), B.J., Determining the focus of instruction: content planning for intelligent tutoring systems, Ph.D. dissertation, University of Saskatchewan, Saskatoon, Canada, 1990.

44. Woolf, B. and McDonalds, D.D., Context dependent transitions in tutoring discourse, Proceedings of the Fifth Annual Conference of the American Association for Artificial Intelligence, Austin, TX, 355, 1984.

45. Woolf, B. and McDonalds, D.D., Building a computer tutor: design issues, *IEEE Comput.*, 17(9), 61, 1984.

46. Woolf, B. et al., Knowledge primitives for tutoring systems, Proceedings of Intelligent Tutoring Systems (ITS-88), Montreal, Canada, 491, 1988.

47. Winkels, R., Breuker, J., and Sandberg, J., Didactic discourse in intelligent help systems, Proceedings of Intelligent Tutoring Systems (ITS-88), Montreal, Canada, 279, 1988.

48. Derry, S.J., Hawkes, L.W., and Ziegler, U., A plan-based opportunistic architecture for intelligent tutoring, Proceedings of the International Conference on Intelligent Tutoring Systems (ITS-88), Montreal, Canada, 116, 1988.

49. Murray, W.R., Control for intelligent tutoring systems: a blackboard-based dynamic instructional planner, Proceedings of the 4th International Conference on Artificial Intelligence and Education (AI-ED '89), Amsterdam, 150, 1989.

50. Murray, W.R., A blackboard-based dynamic instructional planner, Research Report No. R-6376, Artificial Intelligence Center, FMC Corporation, Santa Clara, CA, 1990.

51. Murray, W.R., A blackboard-based dynamic instructional planner, Proceedings of the Eighth National Conference on Artificial Intelligence, Boston, MA, 434, 1990.

52. Mizoguchi, R. and Ikeda, M., A generic framework for ITS and its evaluation, in *Advanced Research on Computers in Education*, Lewis, R. and Otsuki, S., Eds., North-Holland, Amsterdam, 63, 1991.

53. Ikeda, M. and Mizoguchi, R., FITS: a framework for ITS—a computational model of tutoring, *J. Artif. Intelligence Educ.*, 5(3), 319, 1994.

54. Spensley, F. et al., Using multiple teaching strategies in an ITS, in *Intelligent Tutoring Systems: At the Crossroads of Artificial Intelligence and Education*, Frasson, C. and Gauthier, G., Eds., Ablex, Norwood, NJ, 1990.

55. Wong, L.H., A structured system's view of an intelligent tutoring system based on inquiry teaching approach, Ph.D. thesis, Nanyang Technological University, Intelligent Systems Laboratory, Singapore, 1998.

56. Aggarwal, K.K., *Control System Analysis and Designs*, Khanna Publishers, New Delhi, India, 1981.

57. Duchastel, P. and Imbeau, J., (1986), Instructible ICAI, Doc. 86-07 du Laboratoire d'Intelligence Artificielle en Education, Université Laval, Quebec, Canada, 1986.

58. Duchastel, P., Interfacing learning, Poster session, Computer–Human Interaction Conference, Seattle, WA, 1990.

59. O'Shea, T., A self-improving quadratic tutor, *Int. J. Man Machine Studies*, 11, 97, 1979.

60. Ohlsson, S., Some principles of intelligent tutoring, in *AI and Education*, Lawler, R. and Yazdani, M., Eds., Ablex, Norwood, NJ, 1987.

61. Collins, A., personal conversation, 1995.

12

An Intelligent Educational Metadata Repository

Nick Bassiliades
Aristotle University of Thessaloniki

Fotis Kokkoras
Aristotle University of Thessaloniki

Ioannis Vlahavas
Aristotle University of Thessaloniki

Dimitrios Sampson
Informatics and Telematics Institute
Thessaloniki, Greece

Abstract. Recently, several standardization efforts for e-Learning technologies have given rise to various specifications for educational metadata, that is, data describing all the "entities" involved in an educational procedure. The internal details of systems that utilize these metadata are still an open issue since these efforts are primarily dealing with "what" and not "how." In this chapter, under the light of these emerging standardization efforts, we present X-DEVICE, an intelligent XML repository system for educational metadata. X-DEVICE can be used as the intelligent back-end of a WWW portal on which "learning objects" are supplied by educational service providers and accessed by learners according to their individual profiles and educational needs. X-DEVICE transforms the widely adopted XML binding for educational metadata into a flexible, object-oriented representation and uses intelligent second-order logic querying facilities to provide advanced, personalized functionality. Furthermore, a case study is presented in which learning object metadata and learner's profile metadata are combined under certain X-DEVICE rules in order to dynamically infer customized courses for the learner.

12.1   Introduction

Educational applications are among the most promising uses of the World Wide Web. There is already a large amount of instructional material on-line, most of it in the form of static multimedia HTML documents, some of which are enriched with a level of interactivity via Java-based technologies. Unfortunately, the majority of these approaches has been built on general-purpose, noneducational standards and fails to utilize the Web's great potential for distributing educational resources that are easily located and interoperate with each other.

Although computers have been used in education for almost 20 years, in the form of computer-based training (CBT) and computer-assisted instruction (CAI), the level of sophistication of such systems was rather primitive, since they were primarily dealing with the sequencing of the instructional material. In contrast to this, research in information technology-assisted education reached sophisticated levels during the 1990s, taking into account issues like pedagogy, individual learner, and interface, apart from the basic educational material organization. Following the recent "e-" trend, these approaches are just beginning to appear on the Internet. The reason for this late adoption is mainly the substantial effort required to bring them onto the Web since all of them were designed without the Web in mind.

On the other hand, it is obvious that our society has already moved away from the "once for life" educational model. The complexity and continuous evolution of modern enterprises' activities require continuous training of their personnel. The networked community enables the management and enhancement of knowledge in a centralized yet personal way, while keeping track and merging new intellectual resources into that process.

The above requirements and advances have led us to the "lifelong learning" concept. The idea is to integrate the WWW technology with a novel, dynamic, and adaptive educational model for continuous learning. The result will be a learning environment that will enable the individual learner to acquire knowledge just in time, anytime and anywhere, tailored to his/her personal learning needs.

There is an ongoing, large-scale effort to improve the interoperability over the Internet. The eXtended Mark-Up Language (XML[43]) is massively used to define the semantics of the information required to achieve the *Semantic Web* idea.[44] In the domain of education, these efforts deal primarily with the definition of educational metadata, that is, data describing all the "entities" involved in an educational process. The internal details of systems that utilize these metadata are still an open issue, since these efforts are primarily dealing with "what" and not "how." Although in an early stage of development, these standardization efforts will enable the various educational resource developers to create autonomous, on-line educational material that will be used by multiple tutorials, will operate independently of any single tutorial, and will be adjusted to each learner's individual needs.

In this chapter, we present X-DEVICE,[7] an intelligent XML repository system for educational metadata. X-DEVICE can be used as the intelligent back-end of a WWW portal on which "learning objects" are supplied by educational service providers and accessed by learners according to their individual profiles and educational needs. X-DEVICE transforms the widely adopted XML binding for educational metadata into a flexible, object-oriented representation and uses intelligent second-order logic querying facilities to provide advanced, personalized functionality. Furthermore, a case study is presented in which learning object metadata and learner's profile metadata are combined under certain X-DEVICE rules in order to dynamically infer customized courses for the learner.

The outline of this chapter is as follows: Section IV.12.2 refers to the standardization efforts in the education and describes the current trends in intelligent approaches in e-Learning; Section IV.12.3 overviews the management of XML data and documents; Section IV.12.4 describes the storage model of X-DEVICE, i.e., it describes how XML data are mapped onto the object data model of X-DEVICE; Section IV.12.5 presents the X-DEVICE deductive rule language for querying XML data through several examples on querying educational metadata objects. Section IV.12.6 presents an e-Learning case study, which combines, in an intelligent way, metadata from different information models in order to dynamically generate a course adapted to an individual learner. Finally, Section IV.12.7 concludes this chapter and discusses future work. There are five appendices to this chapter that contain the DTDs for the various educational metadata used throughout this chapter, their equivalent X-DEVICE object schemata, a sample XML document with learner information, its X-DEVICE representation, and finally, the syntax of the X-DEVICE rule language.

12.2   Preparing the Base for Advanced E-Learning

The term e-Learning is used to describe a wide range of efforts to provide educational material on the Web. These efforts include a diversity of approaches, ranging from static HTML pages with multimedia material to sophisticated interactive educational applications accessible on-line. The state-of-the-art in this latter category is the *adaptive and intelligent Web-based educational systems* (AIWES). As the term indicates, such approaches have their roots in the fields of intelligent tutoring systems (ITS) and adaptive hypermedia systems (AHS). Actually, most of the current implementations are Web-enabled adaptations of earlier stand-alone systems. The main features of AIWES are:[8]

Adaptive Curriculum Sequencing: The material that will be presented to the learner is selected according to his learning request that is initially stated to the system. The sequencing refers to various levels of granularity; e.g., which concept (topic, lesson) should be presented next, which is the next task (exercise, problem) the user should deal with, etc. Either type of sequencing is performed based on the learner's model, that is, the perception that the system has about the learner's current knowledge status and goals.

Problem Solving Support: We can identify three levels of support. In *Intelligent analysis of learner's solution*, the system waits for the final solution and responds with the errors the student has made. In *Interactive Problem Solving Support*, the system is continuously monitoring the learner and is capable of giving hints, signaling errors, or even auto-executing the next step of an exercise. Finally, in *Example-Based Problem Solving Support*, the system is able to suggest relevant problems that the user has successfully dealt with in the past.

Adaptive Presentation: This feature refers to the system's ability to adapt the content of the supplied curriculum to the learner's preferences.

Student Model Matching: This is a unique feature in AIWES. It allows the categorization of the learners to classes with similar educational characteristics and the use of this information for collaborative problem solving support and intelligent class monitoring.

None of the currently available e-Learning systems delivers the advanced functionality described earlier. Most of the current e-Learning approaches are limited to simple hyperlinks between content pages and "portal pages" which organize a set of related links. This happens because most of the existing educational material can be easily published on-line thanks to MIME data format standards. This is very important since educational material is expensive to create in terms of cost and time. This reusability, however, does not exist at the content, tutorial, or pedagogical levels.

One interesting approach on e-Learning is described by Murray,[34] where a framework called model for distributed curriculum (MDC) is introduced. MDC uses a topic server architecture to allow a Web-based tutorial to include a specification for another tutorial where the best fit to this specification will automatically be found at run time. A specific reasoning mechanism toward this functionality is not presented though.

Another approach in e-Learning, which takes into account personalization aspects, is the DGC.[42] This is a tool that generates individual courses according to the learner's goals and previous knowledge and dynamically adapts the course according to the learner's success in acquiring knowledge. DGC uses "concept structures" as a roadmap to generate the plan of the course.

The lack of widely adopted methods for searching the Web by content makes it difficult for an instructor or learner to find educational material on the Web that addresses particular learning and pedagogical goals. In addition, the lack of standards prevents the interoperability of educational resources. Toward this direction and under the aegis of the IEEE Learning Technology Standards Committee (LTSC),[20] several working groups are developing technical standards, recommended practices and guides for software components, tools, technologies, and design methods that facilitate the development, deployment, maintenance, and interoperation of computer implementations of educational components and systems. Two of the most important LTSC groups are the Learning Object Metadata (LOM) group and the Learner Model (LM) group.

The LOM working group is dealing with the attributes required to adequately describe a learning object,[21] that is, any digital or nondigital mean, which can be used during technology-supported learning. Such attributes include type of object, author, owner, terms of distribution, format, requirements to operate, etc. Moreover, LOM may also include pedagogical attributes, such as teaching or interaction style, mastery level, grade level, and prerequisites. The Learner Model Group is dealing with the specification of the syntax and semantics of attributes that will characterize a learner and his/her knowledge abilities.[22] These will include elements such as knowledge, skills, abilities, learning styles, records, and personal information.

Another working group, whose work is referenced later in this chapter, is the Content Packaging group, which is trying to define a single unit of transmission for the media components (text, graphics, audio, video) and all the supporting material of a learning object. This will enable, among other conveniences, the activation of learning content with a single click of a URL in a browser.

Since no information is included in these standards on how to represent metadata in a machine-readable format, the IMS Global Learning Consortium has developed a representation of LOM in XML.[25] Similar bindings have been developed by IMS for the learner model metadata (LIP[24]) and the content packaging metadata (CP[23]). For technical details, the reader can refer to Appendix A of this chapter.

One of the most ambitious efforts on e-Learning that makes use of educational metadata is the Advanced Distributed Learning initiative.[3] Recently, ADL released the SCORM (Sharable Course-ware Object Reference Model) that attempts to map existing learning models and practices so that common interfaces and data may be defined and standardized across courseware management systems and development tools.

Another approach to utilizing educational metadata in personalized e-Learning is CG-PerLS,[28] a knowledge-based approach for organizing and accessing educational resources. CG-PerLS is a model of a WWW portal for learning objects that encodes the learning technologies metadata in the conceptual graph knowledge representation formalism, and uses related inference techniques to provide advanced, personalized functionality. CGPerLS allows learning resource creators to manifest their material, and allows client-side learners to access these resources in a way tailored to their individual profile and educational needs, and dynamic course generation based on fine- or coarse-grained educational resources.

All the above standardization efforts will require time to mature. More time will be required to build systems to conform to these specifications. The internal details of such systems are an open issue, since these standardization efforts are primarily dealing with "what" and not "how." Meanwhile, every day more and more educational material is becoming available. Therefore, there is an urgent need for methods to efficiently organize what is available today and what will become available in the near future, before the educational resource providers conform to the results of the standardization efforts.

The X-DEVICE, which is presented in the following sections, is an XML-based, intelligent educational metadata repository system. X-DEVICE translates the DTD definitions of the XML binding of the educational metadata, into an object database schema that includes classes and attributes. The XML metadata are translated in objects of the database and stored in an underlying object-oriented database. In its current state, X-DEVICE can handle LOM, LIP, and CM metadata and perform reasoning over them using second-order logic querying facilities. It can be used as the intelligent back-end of a WWW portal on which "learning objects" are supplied by educational service providers and accessed by learners according to their individual profiles and educational needs.

12.3 Managing XML Data

XML is the currently proposed standard for structured or even semistructured information exchange over the Internet.[43] However, the maintenance of this information is equally important. Integrating, sharing, reusing, and evolving information captured from XML documents are essential for building long-lasting applications of industrial strength.

There exist two major approaches to manage and query XML documents. The first uses special purpose query engines and repositories for semistructured data.[9,17,30,33,35] These database systems are built from scratch for the specific purpose of storing and querying XML documents. This approach, however, has two potential disadvantages. First, native XML database systems do not harness the sophisticated storage and query capability already provided by existing database systems. Second, these systems do not allow users to query seamlessly across XML documents and other (structured) data stored in database systems.

Traditional data management has an enormous research background that should be utilized. The second approach to XML data management is to capture and manage XML data within the data models of either relational,[14,16,39,40] object-relational,[27,41] or object databases.[12,36,38,47] Our system, X-DEVICE, stores XML data in the object database ADAM,[18] because XML documents have by nature a hierarchical structure that better fits the object model. Also, references between or within documents play an important role and are a perfect match for the notion of object in the object model. This better matching between the object and document models can also be seen in the number of earlier approaches in storing SGML multimedia documents in object databases.[1,2,9,37]

Capturing XML data in traditional DBMSs alone does not suffice for exploiting the facilities of a database system. Effective and efficient querying and publishing of these data on the Web are actually more important since they determine the impact this approach will have on future Web applications. There have been several query language proposals,[2,10–11,13,19,45] for XML data. Furthermore, recently the WWW consortium issued a working draft proposing Xquery,[46] an amalgamation of the ideas present in most of the proposed XML query languages of the literature. Most of them have a functional nature and use path-based syntax. Some of them,[2,11,13] including Xquery,[46] have also borrowed an SQL-like declarative syntax, which is popular among users.

The X-DEVICE system is a deductive object-oriented database that stores XML documents as objects and uses logic-based rules as a query language for XML data.[7] X-DEVICE integrates high-level, declarative rules (namely deductive and production rules) into an active OODB that supports only event-driven rules,[15] built on top of Prolog. This is achieved by translating each high-level rule into one event-driven rule. The condition of the declarative rule compiles down to a set of complex events that is used as a discrimination network that incrementally matches the rule conditions against the database.

X-DEVICE automatically maps XML document DTDs to object schemata, without losing the document's original order of elements. XML elements are represented either as first-class objects or as attributes based on their complexity. The deductive rule language of X-DEVICE uses special operators for specifying complex queries and materialized views over the stored semistructured data. Most of these operators have a second-order syntax (i.e., variables range over class and attribute names), but they are implemented by translating them into first-order rules (i.e., variables can range over class instances and attribute values), so that they can be efficiently executed against the underlying deductive object-oriented database.

The advantages of using a logic-based query language come from their well-understood mathematical properties. The declarative character of these languages also allows the use of advanced optimization techniques. Logic has also been used for querying semistructured documents, in WebLog[29] for HTML documents and in F-Logic/FLORID[31] and XPathLog/LoPix[32] for XML data.

12.4   The X-Device Storage Model

The X-Device system[7] incorporates XML documents with a schema described through a DTD into an object-oriented database. Specifically, X-Device parses XML documents, which describe learning resources and learners' information (including their DTD definitions), and transforms them as follows: DTD definitions are translated into an object database schema that includes classes and attributes, while XML data are translated into objects of the database. Generated classes and objects are stored within the underlying object database.

The mapping between a DTD and the object-oriented data model uses the object types presented in Figure IV.12.1, and is done as follows:

Elements are represented as either object attributes or classes. More specifically:

- If an element has PCDATA content (without any attributes), it is represented as an attribute of the class of its parent element. The name of the attribute is the same as the name of the element and its type is string.
- If an element has either a) children elements, or b) attributes, then it is represented as a class that is an instance of the `xml_seq` metaclass. The attributes of the class include both the attributes of the element and its subelements. The types of the attributes of the class are determined as follows:

  - Simple character (PCDATA) subelements correspond to attributes of string type, except when they also have attribute elements. In this case, the element is represented as a class, the element attributes as class attributes, and the element content as a string attribute called `content`.
  - Element attributes correspond to object attributes of string type.
  - Children elements or subelements that are represented as objects correspond to object reference attributes.

Attributes of elements are represented as object attributes. The types of the attributes are currently only strings or object references, since DTDs do not support data types. Attributes are distinguished from subelements through the `att_lst` meta-attribute.

In Appendix B, the OODB schemata for the educational metadata used throughout this chapter are presented. The DTDs for these educational metadata are shown in Appendix A. More details for the XML-to-object mapping algorithm of X-Device can be found in Bassiliades et al.[7] Appendix C shows an XML document that conforms to the LIP DTD (see Appendix A), and Appendix D shows how this document is represented in X-Device as a set of objects.

There are more issues that a complete mapping scheme needs to address, besides the above mapping rules. First, elements in a DTD can be combined through either sequencing or alternation.

```
class xml_elem
   attributes
        alias         (attribute-xml_elem, set, optional)
        empty         (attribute, set, optional)

class xml_seq
   is_a              xml_elem
   attributes
        elem_ord      (attribute, list, optional)
        att_lst       (attribute, set, optional)

class xml_alt
   is_a              xml_elem
```

FIGURE IV.12.1   X-Device object types for mapping XML document.

Sequencing means that a certain element must include *all* the specified children elements with a specified order. This is handled by the above mapping scheme through the existence of multiple attributes in the class that represents the parent element, each for each child element of the sequence. The order is handled outside the standard OODB model by providing a meta-attribute (`elem_ord`) for the class of the element that specifies the correct ordering of the children elements. This meta-attribute is used for returning results to the user in the form of an XML document, or by the query mechanism.

On the other hand, *alternation* means that *any* of the specified children elements can be included in the parent element. Alternation is also handled outside the standard OODB model by creating a new class for each alternation of elements, which is an instance of the `xml_alt` metaclass, and it is given a system-generated unique name. The attributes of this class are determined by the elements that participate in the alternation. The types of the attributes are determined as in the sequencing case. The structure of an alternation class may seem similar to a sequencing class; however, the behavior of alternation objects is different, because they must have a value for exactly one of the attributes specified in the class (e.g., see Appendix D, instances of class `activity_alt1`).

The alternation class is always encapsulated in a parent element. The parent element class has an attribute with the system-generated name of the alternation class, which should be hidden from the user for querying the class. Therefore, a meta-attribute (`alias`) is provided with the aliases of this system-generated attribute, i.e., the names of the attributes of the alternating class. Mixed content elements are handled similarly to alternation of elements, whereas the plain text elements are represented as string attributes, with the name `content`.

Another issue that must be addressed is the mapping of the occurrence operators for elements, sequences, and alternations. More specifically, these operators are handled as follows:

- The "star"-symbol (*) after a child element causes the corresponding attribute of the parent element class to be declared as an optional, multivalued attribute.
- The "cross"-symbol (+) after a child element causes the corresponding attribute of the parent element class to be declared as a mandatory, multivalued attribute.
- The question mark (?) after a child element causes the corresponding attribute of the parent element class to be declared as an optional, single-valued attribute.
- Finally, the absence of any symbol means that the corresponding attribute should be declared as a mandatory, single-valued attribute.

The order of children element occurrences is important for XML documents; therefore the multivalued attributes are implemented as lists and not as sets.

Empty elements are treated in the framework described above, depending on their internal structure. If an empty element does not have attributes, it is treated as a PCDATA element, i.e., it is mapped onto a string attribute of the parent element class. The only value that this attribute can take is `yes`, if the empty element is present. If the empty element is absent, the corresponding attribute does not have a value. On the other hand, if an empty element has attributes, it is represented by a class. Finally, unstructured elements that have content ANY are not currently treated by X-DEVICE; therefore, we have transformed all the elements with ANY structure in the DTDs of the educational metadata into PCDATA elements.

12.5 The X-DEVICE Deductive Query Language

Users can query the stored XML documents using X-DEVICE, by: a) submitting the query through an HTML form, b) submitting an XML document that encapsulates the X-DEVICE query as a string, or c) entering the query directly in the text-based Prolog environment. In any of the above ways, the X-DEVICE query processor executes the query and transforms the results into an XML document that is returned to the user.

In this section, we give a brief overview of the X-DEVICE deductive rule language. More details about X-DEVICE can be found in Bassiliades et al.[5-7]

12.5.1 First-Order Syntax and Semantics

The syntax for X-DEVICE deductive rules is given in Appendix E. Rules are composed of condition and conclusion. The condition defines a pattern of objects to be matched over the database, and the conclusion is a derived class template that defines the objects that should be in the database when the condition is true. The first rule in Example 1 defines that an object with attributes identifier=ID and title=T exists in class lp_related_resources if there is an object with OID G in class general with attributes identifier=ID, title=T and the string 'Logic Programming' inside the attribute keyword.

Example 1
```
if G@general(identifier:ID,title:T,keyword ∋ 'Logic Programming')
then lp_related_resources(identifier:ID,title:T)

if LP@lp_related_resources(identifier:ID) and
   G@general(identifier=ID) and
   L@lom(general=G,relation ∋ R) and
   R@relation(kind='Requires',resource:RS) and
   RS@resource(identifier:ID1\=ID) and
   G1@general(identifier=ID1,title:T1)
then lp_related_resources(identifier:ID1,title:T1)
```

Class lp_related_resources is a derived class, i.e., a class whose instances are derived from deductive rules. Only one derived class template is allowed at the THEN-part (head) of a deductive rule. However, there can exist many rules with the same derived class at the head. The final set of derived objects is a union of the objects derived by the two rules. For example, the transitive closure of the set of educational resources required from 'Logic Programming' resources is completed with the second (recursive) rule of Example 1, which recursively adds to the result resources that are required by the resources that are in the result already. The second rule is very complicated because it must retrieve the OID of the lom objects that have already been placed in the result, through their text identifier. In Example 13 we will present a more comprehensive version of this set of rules.

The syntax of the basic rule language is first-order. Variables can appear in front of class names (e.g. LP, G), denoting OIDs of instances of the class, and inside the brackets (e.g. ID, T, R), denoting attribute values (i.e., object references and simple values, such as integers, strings, etc.). Variables are instantiated through the ':' operator when the corresponding attribute is single-valued, and the '∋' operator when the corresponding attribute is multi-valued. Since multi-valued attributes are implemented through lists (ordered sequences), the '∋' operator guarantees that the instantiation of variables is done in the predetermined order stored inside the list.

Conditions also can contain comparisons between attribute values, constants, and variables (e.g., kind='Requires'). Negation is also allowed if rules are safe, i.e., variables that appear in the conclusion must also appear at least once inside a non-negated condition. Furthermore, the use of arbitrary Prolog goals inside the condition of rules is allowed in X-DEVICE. In this way the system can be extended with several features, which are outside of the language and therefore cannot be optimized. For example, in the first rule of Example 1 the search for the keyword 'Logic Programming' is case-sensitive. If a case-insensitive search is required instead, then the following rule calls upon a Prolog predicate (either built-in or user-defined) that transforms the keywords in capital case and then compares them to `'LOGIC PROGRAMMING'`.

```
if G@general(identifier:ID,title:T,keyword ∋ K) and
   prolog{upper_case(K,K1), K1=='LOGIC PROGRAMMING'}
then lp_related_resources(identifier:ID,title:T)
```

A query is executed by submitting the set of stratified rules (or logic program) to the system, which translates them into active rules and activates the basic events to detect changes at base data. Data are then forwarded to the rule processor through a discrimination network (much like in a production system fashion). Rules are executed with fixpoint semantics (semi-naive evaluation), i.e., rule processing terminates when no more new derivations can be made. Derived objects are materialized and are either maintained after the query is over or discarded on the user's demand. X-DEVICE also supports production rules, which have at the THEN-part one or more actions expressed in the procedural language of the underlying OODB.

The main advantage of the X-DEVICE system is its extensibility that allows the easy integration of new rule types as well as transparent extensions and improvements of the rule matching and execution phases. The current system implementation includes deductive rules for maintaining derived and aggregate attributes. Among the optimizations of the rule condition matching is the use of an RETE-like discrimination network, extended with reordering of condition elements, for reducing time complexity and virtual-hybrid memories, for reducing space complexity.[4] Furthermore, set-oriented rule execution can be used for minimizing the number of inference cycles (and time) for large datasets.[5]

12.5.2 Extended Language Constructs for Querying XML Data

The deductive rule language of X-DEVICE supports constructs and operators for traversing and querying tree-structured XML data, which are implemented using second-order logic syntax (i.e., variables can range over class and attribute names) that can also be used to integrate schemata of heterogeneous databases.[6]

These XML-aware constructs are translated into a combination of a) a set of first-order logic deductive rules, and/or b) a set of production rules that query the metaclasses of the OODB, instantiate the second-order variables, and dynamically generate first-order deductive rules.

Throughout this section, we will demonstrate the use of X-DEVICE for querying XML data through examples on educational metadata objects. The DTDs for these metadata are shown in Appendix A, while their equivalent X-DEVICE object schemata are shown in Appendix B. More details about the translation of the various XML-aware constructs to the basic first-order rule language can be found elsewhere.[7]

12.5.2.1 Path Expressions

X-DEVICE supports several types of path expressions into rule conditions. The simplest case is when all the steps of the path must be determined. For example, the second rule of Example 1 can be expressed as follows:

Example 2
```
if LP@lp_related_resources(identifier:ID) and
   L@lom(identifier.general=ID, kind.relation='Requires',
      identifier.resource.relation:ID1\=ID) and
   G1@general(identifier=ID1,title:T1)
then lp_related_resources(identifier:ID1,title:T1)
```

The path expressions are composed using dots between the "steps," which are attributes of the interconnected objects, which represent XML document elements. The innermost attribute should be an attribute of "departing" class, i.e., `relation` is an attribute of class `lom`. Moving to the left, attributes belong to classes that represent their predecessor attributes. Notice that we have adopted a right-to-left order of attributes, contrary to the C-like dot notation that is commonly assumed, because we would like to stress out the functional data model origins of the underlying ADAM OODB.[18] Under this interpretation, the chained "dotted" attributes can be seen as function compositions.

Another case in path expressions is when the number of steps in the path is determined, but the exact step name is not. In this case, a variable is used instead of an attribute name. This is demonstrated by the following example, which returns the identifiers of all direct subelements of `lom` objects that have `'Logic Programming'` in their keywords.

***Example* 3**
```
if L@lom(identifier.E:ID,keyword.general ∋ 'Logic Programming')
then identifiers(identifier:ID)
```

Variable `E` is in the place of an attribute name, therefore it is a second-order variable, since it ranges over a set of attributes, and attributes are sets of things (attribute values). Deductive rules that contain second-order variables are always translated into a set of rules whose second-order variable has been instantiated with a constant. This is achieved by generating production rules, which query the metaclasses of the OODB, instantiate the second-order variables, and generate deductive rules with constants instead of second-order variables. More details can be found in Bassiliades et al.[6,7] In this example, two such bindings actually exist for variable `E`, namely elements `general` and `metadata`.

The most interesting case of path expressions is when some part of the path is unknown, regarding both the number and the names of intermediate steps. This is handled in X-DEVICE by using the "star" (*) operator in place of an attribute name. Such path expressions are called "generalized." Example 3 can be rewritten using the star (*) operator as:

***Example* 4**
```
if L@lom(identifier.*:ID,keyword.general ∋ 'Logic Programming')
then identifiers(identifier:ID)
```

The above query has different semantics than those in Example 3 because it involves any subtree with an `identifier` leaf of any length originating from `lom` objects. Actually, the `identifier.*` path is resolved with the following three concrete paths:

```
identifier.general
identifier.metadata
identifier.resource.relation
```

Sometimes, one of the steps of the path expression involves a recursive element, i.e., an element that contains other elements of the same type, as for example the element `definition` of the LIP objects (Appendix A). Example 5 illustrates the use of recursive elements in path expressions. Specifically, the rule in Example 5 retrieves the names of all modules that have been taken by some Mr. John Smith in the first year of the curriculum. Since `definition` is a recursive element, it is not determined at which depth of the tree the required constants will be found, thus, the star (*) symbol after the `definition` element. Notice that once the appropriate level of the definition elements is found (where the `'Curriculum'` constant exists), it is known that `'Module'` definitions are exactly one step further.

Example 5
```
if L@learnerinfo(text.formname.identification='Mr. John Smith',
        definition*.activity:D1) and
   D1@definition(indexid.referential.contenttype='Year1',
        tyvalue.typename='Curriculum', definition ∋ D2) and
   D2@definition(tyvalue.typename='Module',
        indexid.referential.contenttype:ModName)
then modules(module:ModName)
```

12.5.2.2 Ordering Expressions

X-DEVICE supports expressions that query an XML tree based on the ordering of elements. Here we will demonstrate all types of ordering expressions through several examples. Example 6 retrieves the first two modules that Mr. John Smith took during his second year.

Example 6
```
if L@learnerinfo(text.formname.identification='Mr. John Smith',
        definition*.activity:D1) and
   D1@definition(indexid.referential.contenttype='Year2',
        tyvalue.typename='Curriculum', definition ∋=<2 D2) and
   D2@definition(indexid.referential.contenttype:ModName)
then modules(module:ModName)
```

The $\ni_{=<2}$ operator is an absolute numeric ordering expression that returns the first two elements of the corresponding list-attribute. More such ordering expressions exist for every possible position inside a multivalued attribute. The operator $\ni_{=n}$ that returns the n-th element in a sequence has the shortcut notation $\ni_n$. When there are multiple such expressions in a path expression, then there is another shortcut notation, which is demonstrated in Example 7, that retrieves the subject of the second lecture of the third module that Mr. John Smith attended during his second year.

Example 7
```
if L@learnerinfo(text.formname.identification='Mr. John Smith',
        definition*.activity:D1) and
   D1@definition(indexid.referential.contenttype='Year2',
        tyvalue.typename='Curriculum',
        fielddata.definitionfield2.definition3:Lecture) and
then lecture(subject:Lecture)
```

Except for the absolute numeric ordering expressions, there are also relative ordering expressions, which are demonstrated in the following two examples. The first one (Example 8) retrieves the subjects of the lectures that Mr. John Smith attended *between* the lectures of 'Boolean Logic' and 'FET Transistors'. Notice that the query also determines the name of the module, which must be 'Electronics_101'.

Example 8
```
if L@learnerinfo(text.formname.identification='Mr. John Smith',
        definition*.activity:D1) and
   D1@definition(indexid.referential.contenttype='Electronics_101',
        definitionfield ∋ L1) and
   D1@definition(definitionfield ∋ L2) and
   L1@definitionfield(fielddata='Boolean Logic') and
```

```
    L2@definitionfield(fielddata='FET Transistors') and
    D1@definition(definitionfield ∋between(L1,L2) L) and
    L@definitionfield(fielddata:Lecture) and
then lecture(subject:Lecture)
```

The operator `between(L1,L2)` is a relative ordering expression that returns all elements in a sequence after the one with an OID identified by the instantiations of the variable `L1` and before the ones with OID `L2`.

In some cases, the relative ordering expression coexists with an absolute numeric ordering expression, which is called a complex ordering expression. Example 9 demonstrates this by retrieving the subjects of the first two lectures that Mr. John Smith attended *after* the lectures of `'Boolean Logic'` during the `'Electronics_101'` module.

Example 9

```
if L@learnerinfo(text.formname.identification='Mr. John Smith',
        definition*.activity:D1) and
    D1@definition(indexid.referential.contentype='Electronics_101',
        definitionfield ∋ L1) and
    L1@definitionfield(fielddata='Boolean Logic') and
    D1@definition(definitionfield ∋{after(L1),=<2} L) and
    L@definitionfield(fielddata:Lecture) and
then lecture(subject:Lecture)
```

The operator $∋_{\{after(I),=<2\}}$ is a complex ordering expression that consists of the relative ordering expression `after(L1)` followed by the absolute numeric ordering expression `=<2`.

12.5.2.3 Exporting Results

So far, only the querying of existing XML documents through deductive rules has been discussed. However, it is important that the results of a query can be exported as an XML document. This can be performed in X-DEVICE by using some directives around the conclusion of a rule that defines the top-level element of the result document.

When the rule processing procedure terminates, X-DEVICE employs an algorithm that begins with the top-level element designated with one of these directives and navigates recursively all the referenced classes, constructing a result in the form of an XML tree-like document. More details can be found in Bassiliades et al.[7] Example 10 demonstrates how XML documents (and DTDs) are constructed in X-DEVICE for exporting them as results. Actually, Example 10 is the same as Example 5; the only difference being the keyword `xml_result` around the derived class.

Example 10

```
if L@learnerinfo(text.formname.identification='Mr. John Smith',
        definition*.activity:D1) and
    D1@definition(indexid.referential.contentype='Year1',
        tyvalue.typename='Curriculum',definition ∋ D2) and
    D2@definition(tyvalue.typename='Module',
        indexid.referential.contentype:ModName)
then xml_result(modules(module:ModName))
```

The keyword `xml_result` is a directive that indicates to the query processor that the encapsulated derived class (`modules`) is the answer to the query. This is especially important when the query consists of multiple rules. In order to build an XML tree as a query result, the objects that correspond to the elements must be constructed incrementally in a bottom-up fashion, i.e., first the simple elements that are near the leaves of the tree are generated and then combined into more complex elements toward the root of the tree. The above query produces an awkward forest of `modules` elements, with the following DTD:

```
<!DOCTYPE modules [
  <!ELEMENT modules (module)>
  <!ELEMENT module (#PCDATA)>
]>
```

To see why the above is an awkward result, regard the input XML document of Appendix C, which will produce the following answer document:

```
<modules>
  <module>Electronics_101 </module>
</modules>
<modules>
  <module>Maths_101</module>
</modules>
```

A better-looking XML document would be one that encapsulates both `module` elements in the same (root) `modules` element, like the following:

```
<modules>
  <module> Electronics_101</module>
  <module> Maths_101 </module>
</modules>
```

The above XML document, where an element has multiple occurrences of the same subelement type, conforms to the following DTD, instead of the one presented above:

```
<!DOCTYPE modules [
  <!ELEMENT modules (module*)>
  <!ELEMENT module (#PCDATA)>
]>
```

In order to construct the above tree-structured DTD, we should use the `list` construct in the rule conclusion to wrap the `module` elements inside the top-level element `modules`:

```
if L@learnerinfo(text.formname.identification='Mr. John Smith',
        definition*.activity:D1) and
  D1@definition(indexid.referential.contentype='Year1',
        tyvalue.typename='Curriculum',
        tyvalue.typename.definition='Module',
        indexid.referential.contentype.definition:ModName)
then xml_result(modules(module:list(ModName)))
```

The `list(ModName)` construct in the conclusion denotes that the attribute `module` of the derived class `modules` is an attribute whose value is calculated by the aggregate function `list`. This function collects all the instantiations of the variable `ModName` and stores them under a strict order into the multivalued attribute `module`. More details about the implementation of aggregate functions in X-DEVICE can be found in Bassiliades et al.[5]

In order to produce an even more structured DTD with the `module` element not being a mere PCDATA element, but having an internal structure as well, X-DEVICE offers a wrapping construct as a shortcut notation:

```
if L@learnerinfo(text.formname.identification='Mr. John Smith',
        definition*.activity:D1) and
   D1@definition(indexid.referential.contentype='Year1',
        tyvalue.typename='Curriculum',
        tyvalue.typename.definition='Module',
        indexid.referential.contentype.definition:ModName)
then xml_result(modules(module(name:ModName)))
```

This would produce the following slightly more complicated DTD:

```
<!DOCTYPE modules [
 <!ELEMENT modules (module*)>
 <!ELEMENT module (name)
 <!ELEMENT name (#PCDATA)>
]>
```

which corresponds to the following result document:

```
<modules>
 <module>
  <name >Electronics_101</name>
 </module>
 <module>
  <name >Maths_101 </name>
 </module>
</modules>
```

Another directive for constructing XML documents is `xml_sorted`, which is similar to `xml_result` and is used for sorting the elements of the result according to a group of element values specified in the rule head. Example 11 repeats the rule of Example 10, sorting the result according to the module name.

***Example* 11**

```
if L@learnerinfo(text.formname.identification='Mr. John Smith',
        definition*.activity:D1) and
   D1@definition(indexid.referential.contentype='Year1',
        tyvalue.typename='Curriculum',
        tyvalue.typename.definition='Module',
        indexid.referential.contentype.definition:ModName)
then xml_sorted([ModName],modules(module(name:ModName)))
```

12.5.2.4　Alternative Attribute Expressions

The presentation of X-DEVICE so far has assumed that path expressions contain steps that refer only to normal XML elements. However, step expressions can also refer to element attributes, using the ($^\wedge$) symbol as a prefix before the name. Example 12 retrieves all the comments in the English language found anywhere inside the LIP object of Mr. John Smith.

Example 12

```
if L@learnerinfo(text.formname.identification='Mr. John Smith',
          comment.*:C) and
   C@comment(content:Text,^xml_lang='en')
then xml_result(comments(comment:list(Text)))
```

The system attribute is another special attribute that can be used in X-DEVICE. Example 13 demonstrates the use of special attributes. Recall the two rules of Example 1 that retrieve the identifiers and titles of all educational resources required for "teaching" Logic Programming. The first rule collects all the resources that directly contain the `Logic Programming` keyword, while the second rule recursively adds to the result resources by retrieving the OID of the lom objects that have already been placed in the result, through their text identifier. Things would be easier if one could record in the result class lp_related_resources the OID of the recorded lom object, so that the second rule knows where to begin. However, we would not like this lom OID to appear in the final result. This is achieved by prefixing the name of the "hidden" attribute with an exclamation point (!). In this way, it is considered a system attribute that will not appear in the result.

Example 13

```
if L@lom(identifier.general:ID,title.general:T,
          keyword.general ∋ 'Logic Programming')
then xml_result(
          lp_related_resources(!orig_lom:L,identifier:ID,title:T))

if LP@lp_related_resources((!orig_lom:L) and
   L@lom(kind.relation='Requires',
     identifier.resource.relation:ID1) and
   L1@lom(identifier.general=ID1,title.general:T1)
then lp_related_resources(!orig_lom:L1,identifier:ID1,title:T1)
```

Notice that although both rules refer to the same derived class lp_related_resources, only one of them contains the xml_result directive. However, this is not a strict language rule; it does not matter if several rules contain the xml_result or any other result directive, as long as the following constraints are satisfied:

- Only one type of result directive is allowed in the same query.
- Only one derived class is allowed at the result.

12.6 E-Learning Case Study

In this section we present an e-Learning case study, which demonstrates the functionality of the X-DEVICE system as an intelligent educational metadata repository. Specifically, the system (Figure IV.12.2) maintains LOM objects submitted by learning resource providers as well as LIP objects of its registered users (learners). Users submit their learning requests through a Web interface, and the system dynamically generates a Content Package. The latter provides access to the suggested educational resources that cover the learning request.

X-DEVICE uses deductive rules to suggest educational material by intelligently combining information on educational resources (LOM objects) with information about an individual learner (LIP objects). Specifically, we assume that the LIP objects of the database include past learner activities in terms of LOM objects they have "learned." This may sound like an oversimplification, but we can assume that when new learners are added to the system their past activities expressed in various

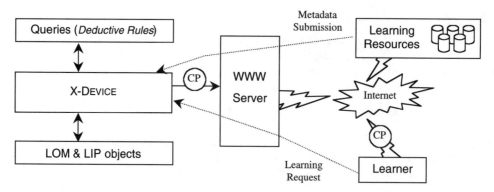

FIGURE IV.12.2   The architecture of the X-DEVICE system.

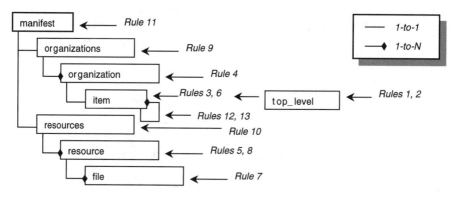

FIGURE IV.12.3   Outline of the case study.

"agreed" formats are "translated" to the format required by our system, i.e., sets of LOM objects that were used during these past activities. Furthermore, we assume that when a learner has a new learning goal (e.g., `Logic Programming`), this is represented in the goal sub-tree of his/her LIP object. The objective of this case study is to construct a content package that includes all the required LOM objects for achieving the new learning goal, except those that have already been "taken" by the learner. The outline of the constructed CP tree and the rules that generate it are shown in Figure IV.12.3.

In order to identify the LOM objects that can achieve the learning goal, we first match the goal directly with the title of the LOM object (Rule 1). If the match fails, then we try to find LOM objects whose keywords are contained within the learning goal (Rule 2), e.g., `Logic`, `Programming`. A more sophisticated approach would use ontologies to better resolve the user's learning goal. The information found with either of the above ways is maintained within the user-defined class `top_level`.

12.6.1   Rule 1

```
if L@learnerinfo(text.formname.identification:LName,
        tyvalue.typename.goal='Education',
        tyvalue.typename.status.goal='New',
        short.description.goal:Subject) and
   L1@lom(title.general=Subject,identifier.general:ID)
then top_level(learner:LName,goal:Subject,lom_id:ID,
   lom_title:Subject,!org_learner:L,!org_lom:L1)
```

12.6.2 Rule 2

```
if L@learnerinfo(text.formname.identification:LName,
        tyvalue.typename.goal='Education',
        tyvalue.typename.status.goal='New',
        short.description.goal:Subject $ K) and
    not T1@tmp_elem1 and
    L1@lom(keyword.general ∋ K,title.general=LTitle,
        identifier.general:ID)
then top_level(learner:LName,goal:Subject,lom_id:ID,
        lom_title:LTitle,!org_learner:L,!org_lom:L1)
```

Notice that the above two rules also keep the OIDs of the original LOM and LIP objects that initiated the whole procedure, in order to use them in subsequent retrieval of information about these objects. This is achieved by using system attributes, which will be hidden away from the final XML result. Also notice the use of the ($) operator, which searches if its right-hand-side argument (string) is a substring of its left-hand-side argument.

Subsequently, the "discovered" top-level LOM objects are added to the content package as organizations and items. Specifically, each top-level LOM object is considered to be an alternative study suggestion for the learning goal and will be represented by a separate organization element. Each of these organization elements will have just one item subelement, which also represents the corresponding top-level LOM object. The LOM objects that are related to these top-level objects will be recursively "hanged" below these top-level item elements.

Rule 3 creates the top-level item elements for each of the top-level LOM objects. Before the item object is created, the rule condition ensures that the same LOM object has *not* been "attended" by the learner in the past, by checking his/her activity records for completed educational activities that their source is LOM and their ID matches the ID of the "discovered" object. Activities can be nested, thus we use the recursive element `activity*`. Furthermore, there may be cases where "discovered" top-level LOM objects are connected through 'Required' relations. In order to avoid repetitions of items in the CP tree, if the same LOM object has already been recorded as an item object, it is not created again. Subsequently, Rule 4 creates an organization object for each top-level LOM object and points to the corresponding item object. Notice that we use a Prolog predicate to generate a unique identifier for the organization object by combining the string of the goal with a number.

12.6.3 Rule 3

```
if TL@top_level(lom_id:ID,lom_title:LTitle,
        !org_learner:L,!org_lom:L1) and
    not L@learnerinfo(tyvalue.typename.activity*='Education',
        tyvalue.typename.status.activity*='Completed',
        source.sourcedid.learningactivityref.activity*='lom',
        id.sourcedid.learningactivityref.activity*=ID) and
    not I@item(^identifier=ID,!org_learner=L)
then item(^identifier:ID,title:LTitle,!org_lom:L1,!org_learner:L)
```

12.6.4 Rule 4

```
if TL@top_level(lom_id:ID,goal:Subject) and
    I@item(^identifier=ID) and
    prolog{generate_id(Subject,OID)}
then organization(^identifier:OID,title:Subject,item:list(I))
```

For each organizational item, a corresponding resource object that holds information about physical resources of the LOM object must be created and linked to the item object. Rule 5 creates the resource object and stores information about the type of the resource. Notice that the OID of the corresponding item object is kept for use later (Rule 6) in cross-referencing the resource from the item object. Notice that Rule 6 is a "derived-attribute rule,"[5] which defines the value that an attribute of an *existing* object should have.

12.6.5   Rule 5

```
if I@item(!org_lom:L) and
   L@lom(format.technical:Type) and
   prolog{generate_id(resource,RID)}
then resource(^identifier:RID,^type:Type,!org_item:I)
```

12.6.6   Rule 6

```
if R@resource(^identifier:RID,!org_item:I)
then I@item(^identifierref:RID)
```

The resource object is not complete until it is linked with the physical file information of the corresponding LOM object, which we assume corresponds to the `location` element of the `technical` sub-tree of the LOM object. However, the `file` subelement of `resource` is an object itself, so it must first be created (Rule 7) and then linked to the `resource` object (Rule 8).

12.6.7   Rule 7

```
if R@resource(!org_item:I) and
   I@item(!org_lom:L) and
   L@lom(location.technical:File)
then file(^href:File,!org_resource:R)
```

12.6.8   Rule 8

```
if F@file(!org_resource:R)
then R@resource(file:list(F))
```

Finally, the `organization` and `resource` objects must be linked together into a `manifest` object. First, all the `organization` objects are linked together into one `organizations` object (Rule 9), according to the DTD of CP (see Appendix A). The same also is done for the `resource` objects (Rule 10). Finally, Rule 11 generates the top-level `manifest` object, generating an identifier from the learner's name. Notice the use of the `xml_result` directive to indicate the top-level element of the result XML document.

12.6.9   Rule 9

```
if O@organization
then organizations(organization:list(O))
```

12.6.10   Rule 10

```
if R@resource
then resources(resource:list(R))
```

12.6.11   Rule 11

```
if TL@top_level(learner:LName) and
   R@resources and
```

```
     O@organizations and
     prolog{generate_id(LName,MID)}
then xml_result(manifest(^identifier:MID,
               organizations:O,resources:R))
```

Now that the result tree has been created, it is time to navigate through the LOM objects that are required by the top-level LOM objects, check if they have been encountered by the learner during his past activities, create the corresponding items, and finally, link them to their parent item/LOM object. Rule 12 creates the items and keeps the OID of their parent item object, which is used by Rule 13 to link the two items. Notice that Rule 12 checks if the item objects to be created have already been created for the same learner. This could happen when multiple LOM objects have common 'Required' LOM objects.

12.6.12 Rule 12

```
if I@item(!org_lom:L,!org_learner:L2) and
   L@lom(kind.relation='Requires',
       identifier.resource.relation:ID1) and
   L1@lom(identifier.general=ID1,title.general:T1) and
   not L2@learnerinfo(tyvalue.typename.activity*='Education',
        tyvalue.typename.status.activity*='Completed',
        source.sourcedid.learningactivityref.activity*='lom',
        id.sourcedid.learningactivityref.activity*=ID1)
   not I1@item(^identifier=ID1,!org_learner=L2)
then item(^identifier:ID1,title:T1,!org_lom:L1,!org_learner:L2,
   !parent_item:I)
```

12.6.13 Rule 13

```
if I1@item(!parent_item:I)
then I@item(item:list(I1))
```

Although the resulting class (manifest) has already been created in Rule 11, it is not necessary to continue deriving objects, because the order of rule execution is determined by stratification and the results will be exported after the logic program reaches a fixpoint. The creation of the resource objects that correspond to the recursively added item objects is taken care of by rules already defined (Rules 5 through 8).

The XML document that will be created by the previous set of rules (Figure IV.12.3) will be packaged along with the corresponding physical resources (e.g., files) and sent to the learner. The DTD of this document is a subset of the full CP DTD (see Appendix A).

```
<!ELEMENT manifest (organizations, resources)>
<!ATTLIST manifest
     identifier ID #REQUIRED>
<!ELEMENT organizations (organization*)>
<!ELEMENT organization (title?, item*)>
<!ATTLIST organization
     identifier ID #REQUIRED>
<!ELEMENT item (title?, item*)>
<!ATTLIST item
```

```
    identifier ID #REQUIRED
    identifierref CDATA #IMPLIED>
<!ELEMENT resources (resource*)>
<!ELEMENT resource (file+)>
<!ATTLIST resource
    identifier ID #REQUIRED
    type CDATA #REQUIRED>
<!ELEMENT title (#PCDATA)>
<!ELEMENT file EMPTY>
<!ATTLIST file
    href CDATA #REQUIRED>
```

The case study presented here does not deal with what the user does in order to study the suggested LOM objects and what happens after each of the suggested items is completed. However, we assume that the studied LOM objects are added to the activity sub-tree of the learner's LIP object, so that the next time the learner has a new learning goal the objects will not be considered again (see Rule 3).

12.7   Conclusions and Future Work

In this chapter, we presented X-DEVICE, an intelligent repository system for educational metadata. X-DEVICE transforms the widely adopted XML binding for LOM, LIP, and CP educational metadata into a flexible, object-oriented representation, and uses intelligent second-order logic querying facilities to provide advanced, personalized learning content access. As demonstrated by the case study, X-DEVICE can be used as the intelligent back-end of a WWW portal on which "learning objects" are supplied by educational service providers and accessed by learners according to their individual profiles and educational needs.

X-DEVICE stores an XML document into an OODB by automatically mapping the schema of the XML document (DTD) to an object schema and XML elements to database objects, treating them according to their structure complexity, without losing the relative order of elements in the original document. Furthermore, X-DEVICE employs a powerful deductive rule query language for expressing queries over the stored XML data. The deductive rule language has certain constructs (such as second-order variables, general path and ordering expressions) for traversing tree-structured data that were implemented by translating them into first-order deductive rules.

In comparing X-DEVICE with other XML query languages (e.g., XQuery), it seems that the high-level, declarative syntax of X-DEVICE allows users to express, everything that XQuery can express in a more compact and comprehensible way, with the powerful addition of fixpoint recursion and second-order variables. Furthermore, users can also express complex XML document views, a fact that can greatly facilitate customizing information for e-learning, as was demonstrated by the case study.

Concerning the functionality of X-DEVICE as an educational portal, we plan to incorporate the learner's assessment results described in the IMS Question & Test Interoperability Specification (QTI[26]). This information can be used to improve the overall knowledge transfer from the system to the learner by preferring to serve him/her with a specific learning object from a set of similar ones, based on the assessment results obtained by learners with similar profiles. This will require substantial research on aspects related to the learner's model.

Furthermore, the suitability of the suggested learning content would be improved by using ontologies to better resolve the end-users' learning request. Further improvement could be achieved by using curriculum description information. In its current state, X-DEVICE assumes a curriculum description defined indirectly by the relations between learning objects. On the other hand, curriculum information contains higher-level pedagogical knowledge and as a result can lead to a better teaching strategy for the suggested educational material.

Acknowledgments

Part of the work presented in this chapter was partially financially supported by the European Commission under the IST No. 12503 project "KOD—Knowledge on Demand" through the Information Society Technologies Programme (IST). Dr. Bassiliades was supported by a scholarship from the Greek *Foundation of State Scholarships* (F.S.S.–I.K.Y.). Dr. Kokkoras was supported by a scholarship from the Informatics and Telematics Institute.

References

1. Abiteboul, S., Cluet, S., Christophides, V., Milo, T., Moerkotte, G., and Siméon, J., "Querying Documents in Object Databases," *Int. J. on Digital Libraries*, 1(1), 5–19, 1997.
2. Abiteboul, S., Quass, D., McHugh, J., Widom, J., and Wiener, J.L., "The Lorel Query Language for Semistructured Data," *Int. Journal on Digital Libraries*, 1(1), 68–88, 1997.
3. ADL, "Sharable Courseware Object Reference Model (SCORM)," Version 1.1, Jan. 2001, (http://www.adlnet.org).
4. Bassiliades, N. and Vlahavas, I., "Processing Production Rules in DEVICE, an Active Knowledge Base System," *Data & Knowledge Engineering*, 24(2), 117–155, 1997.
5. Bassiliades, N., Vlahavas, I., and Elmagarmid, A.K., "E-DEVICE: An extensible active knowledge base system with multiple rule type support," *IEEE TKDE*, 12(5), 824–844, 2000.
6. Bassiliades, N., Vlahavas, I., Elmagarmid, A.K., and Houstis E.N., "InterBaseKB: Integrating a Knowledge Base System with a Multidatabase System for Data Warehousing," *IEEE TKDE*, (to appear) 2002.
7. Bassiliades, N., Vlahavas, I., and Sampson, D., "Using Logic for Querying XML Data," accepted for publication at Web-Powered Databases, D. Taniar, W. Rahayu (eds.), Idea-Group Publishing, USA, 2002.
8. Brusilovsky, P., "Adaptive and Intelligent Technologies for Web-based Education," in C. Rollinger and C. Peylo (Eds.), *Kunstliche Intelligenz* (The German AI Journal) 13(4), 1999.
9. Buneman, P., Davidson, S.B., Hillebrand, G.G., and Suciu, D., "A Query Language and Optimization Techniques for Unstructured Data," *Proc. ACM SIGMOD Conf.*, 505–516, 1996.
10. Buneman, P., Fernandez, M., and Suciu, D., "UnQL: A Query Language and Algebra for Semistructured Data Based on Structural Recursion," *VLDB Journal*, 9(1), 2000.
11. Chamberlin, D., Robie, J., and Florescu, D., "Quilt: an XML Query Language for Heterogeneous Data Sources," *Int. Workshop WebDB*, 53–62, 2000.
12. Chung, T.-S., Park, S., Han, S.-Y., and Kim, H.-J., "Extracting Object-Oriented Database Schemas from XML DTDs Using Inheritance," K. Bauknecht, S.K. Madria, G. Pernul (Eds.), *Proc. 2nd Int. Conf. EC-Web 2001*, Munich, Germany, 2001, LNCS 2115, 49–59.
13. Deutsch, A., Fernandez, M., Florescu, D., Levy, A., and Suciu, D., "A Query Language for XML," *WWW8/Computer Networks*, 31(11–16), 1155–1169, 1999.
14. Deutsch, A., Fernandez, M.F., and Suciu, D., "Storing Semistructured Data with STORED," *ACM SIGMOD Conf.*, 431–442, 1999.
15. Diaz, O. and Jaime, A., "EXACT: An Extensible Approach to Active Object-Oriented Databases," *VLDB Journal*, 6(4), 282–295, 1997.

16. Florescu, D. and Kossmann, D., "Storing and Querying XML Data using an RDMBS," *IEEE Data Eng. Bulletin*, 22(3), 27–34, 1999.

17. Goldman, R. and Widom, J., "DataGuides: Enabling Query Formulation and Optimization in Semistructured Databases," *Proc. Int. Conf. VLDB*, 436–445, 1997.

18. Gray, P.M.D., Kulkarni, K.G., and Paton, N.W., *Object-Oriented Databases, A Semantic Data Model Approach*, Prentice Hall, London, 1992.

19. Hosoya, H. and Pierce, B., "XDuce: A Typed XML Processing Language," *Int. Workshop WebDB*, 111–116, 2000.

20. IEEE Learning Technology Standards Committee (LTSC), http://ltsc.ieee.org/.

21. IEEE P1484.12/D6.0, "Draft Standard for Learning Object Metadata," Feb. 2001.

22. IEEE P1484.2/D7, "Draft Standard for Learning Technology — Public and Private Information (PAPI) for Learners (PAPI Learner)," Nov. 2000.

23. IMS Global Learning Consortium, Content Packaging (CP) Specification, Version 1.1.2, Final Specification, http://www.imsglobal.org/content/packaging/index.html.

24. IMS Global Learning Consortium, Learner Information Package (LIP) Specification, Version 1.0, Final Specification, Mar. 2001, http://www.imsglobal.org/profiles/index.html.

25. IMS Global Learning Consortium, Learning Resource Meta-data (LOM) Specification, Version 1.2, Final Specification, May 2001, http://www.imsglobal.org/metadata/index.html.

26. IMS Global Learning Consortium, Question & Test Interoperability Specification, Version 1.1, Feb. 2001, http://www.imsproject.com/question.

27. Klettke, M. and Meyer, H., "XML and Object-Relational Database Systems—Enhancing Structural Mappings Based on Statistics," *Int. Workshop WebDB*, 63–68, 2000.

28. Kokkoras, F., Sampson, D.G., and Vlahavas, I., "CG-PerLS: Conceptual Graphs for Personalized Learning Systems," *8th Panhellenic Conf. on Informatics*, Cyprus, Nov. 2001.

29. Lakshmanan, L.V.S., Sadri, F., and Subramanian, I.N., "A Declarative Language for Querying and Restructuring the WEB." RIDE-NDS, 12–21, 1996.

30. Lucie Xyleme, "A Dynamic Warehouse for XML Data of the Web," *IEEE Data Engineering Bulletin*, 24(2), 40–47, June 2001.

31. Ludäscher, B., Himmeröder, R., Lausen, G., and May, W., Christian Schlepphorst, "Managing Semistructured Data with FLORID: A Deductive Object-Oriented Perspective," *Information Systems*, 23(8), 589–613, 1998.

32. May, W., "XPathLog: A Declarative, Native XML Data Manipulation Language," IDEAS 2001: 123–128.

33. McHugh, J., Abiteboul, S., Goldman, R., Quass, D., and Widom, J., Lore: "A Database Management System for Semistructured Data," *ACM SIGMOD Record*, 26(3), 54–66, 1997.

34. Murray, T., "A Model for Distributed Curriculum on the WWW," *Journal of Interactive Media in Education*, 5 (1998) (http://www-jime.open.ac.uk/98/5).

35. Naughton, J. et al., "The Niagara Internet Query System," *IEEE Data Engineering Bulletin*, 24(2), 27–33, June 2001.

36. Nishioka, S. and Onizuka, M., "Mapping XML to Object Relational Model," *Proc. Int. Conf. on Internet Computing*, 171–177, 2001.

37. Ozsu, M.T., Iglinski, P., Szafron, D., El-Medani, S., and Junghanns, M., "An Object-Oriented SGML/HyTime Compliant Multimedia Database Management System," presented at ACM Multimedia, Seattle, WA, USA, 1997.

38. Renner, A., "XML Data and Object Databases: A Perfect Couple?," *Proc. Int. Conf. on Data Engineering*, 143–148, 2001.

39. Schmidt, A., Kersten, M.L., Windhouwer, M., and Waas, F., "Efficient Relational Storage and Retrieval of XML Documents," *Int. Workshop WebDB*, 47–52, 2000.

40. Shanmugasundaram, J., Tufte, K., Zhang, C., He, G., DeWitt, D.J., and Naughton, J.F., "Relational Databases for Querying XML Documents: Limitations and Opportunities," *Int Conf. VLDB*, 302–314, 1999.

41. Shimura, T., Yoshikawa, M., and Uemura, S., "Storage and Retrieval of XML Documents Using Object-Relational Databases," *Proc. Int. Conf. on Database and Expert Systems Applications*, Florence, Italy, 206–217, 1999.
42. Vassileva, J., "Dynamic Course Generation on the WWW," *8th World Conference on AI in Education (AI-ED '97)*, Knowledge and Media in Learning Systems, Kobe, Japan, 1997.
43. W3 Consortium, "Extensible Markup Language (XML) 1.0 (2nd Edition)," Recommendation, Oct. 2000, http://www.w3.org/TR/REC-xml.
44. W3 Consortium, "Semantic Web Activity Statement," http://www.w3.org/2001/sw/Activity.
45. W3 Consortium, "XML Path Language (XPath) Ver. 1.0," Recommendation, Nov. 1999, http://www.w3.org/TR/xpath.
46. W3 Consortium, "XQuery 1.0: An XML Query Language," Working Draft, June 2001, http://www.w3.org/TR/xquery.
47. Yeh, C.-L., "A Logic Programming Approach to Supporting the Entries of XML Documents in an Object Database," *Int. Workshop PADL*, 278–292, 2000.

Appendix A.   DTDs for LOM, LIP, and CP Objects

This appendix contains the DTDs for the LOM (Learning Object Metadata),[25] LIP (Learner Information Packaging),[24] and CP (Content Packaging)[23] objects that are used throughout this chapter. Notice that, due to space limitations, only a part of the LIP DTD that is used in the chapter is actually shown. Furthermore, some of the DTDs have been simplified for presentation purposes.

LOM

```
<!ELEMENT lom (general?, lifecycle?, metametadata?, technical?, educational?,
rights?, relation*, annotation*, classification*)>
<!ELEMENT general (identifier?, title?, catalogentry*, language*, description*,
keyword*, coverage*, structure?, aggregationlevel?, extension?)>
<!ELEMENT catalogentry (catalog, entry, extension?)>
<!ELEMENT lifecycle (version?, status?, contribute*, extension?)>
<!ELEMENT contribute (role, centity*, date?, extension?)>
<!ELEMENT date (datetime?, description?)>
<!ELEMENT metametadata (identifier?, catalogentry*, contribute*, metadatascheme*,
language?, extension?)>
<!ELEMENT technical (format*, size?, location*, requirement*, installationremarks?,
otherplatformrequirements?, duration?, extension?)>
<!ELEMENT requirement (type?, name?, minimumversion?, maximumversion?, extension?)>
<!ELEMENT duration (datetime?, description?)>
<!ELEMENT educational (interactivitytype?, learningresourcetype*,
interactivitylevel?, semanticdensity?, intendedenduserrole*, context*,
typicalagerange*, difficulty?, typicallearningtime?, description?, language*,
extension?)>
<!ELEMENT typicallearningtime (datetime?, description?)>
<!ELEMENT rights (cost?, copyrightandotherrestrictions?, description?, extension?)>
<!ELEMENT copyrightandotherrestrictions (source, value)>
<!ELEMENT relation (kind?, resource?, extension?)>
<!ELEMENT resource (identifier?, description?, catalogentry*, extension?)>
<!ELEMENT annotation (person?, date?, description, extension?)>
<!ELEMENT classification (purpose?, taxonpath*, description?, keyword*,
extension?)>
```

```
<!ELEMENT taxonpath (source?, taxon?)>       <!ELEMENT keyword (#PCDATA)>
<!ELEMENT taxon (id?, entry?, taxon?)>       <!ELEMENT coverage (#PCDATA)>
<!ELEMENT structure (source, value)>         <!ELEMENT source (#PCDATA)>
<!ELEMENT type (source, value)>              <!ELEMENT value (#PCDATA)>
<!ELEMENT name (source, value)>              <!ELEMENT version (#PCDATA)>
<!ELEMENT aggregationlevel (source, value)>  <!ELEMENT installationremarks
                                                 (#PCDATA)>
<!ELEMENT status (source, value)>            <!ELEMENT otherplatformrequirements
                                                 (#PCDATA)>
<!ELEMENT role (source, value)>              <!ELEMENT typicalagerange (#PCDATA)>
<!ELEMENT interactivitytype (source, value)> <!ELEMENT location (#PCDATA)>
<!ELEMENT learningresourcetype (source,      <!ATTLIST location
value)>                                          type (URI | TEXT) #IMPLIED>
<!ELEMENT interactivitylevel (source, value)><!ELEMENT identifier (#PCDATA)>
<!ELEMENT semanticdensity (source, value)>   <!ELEMENT extension (#PCDATA)>
<!ELEMENT intendedenduserrole (source,
value)>                                      <!ELEMENT catalog (#PCDATA)>
<!ELEMENT context (source, value)>           <!ELEMENT language (#PCDATA)>
<!ELEMENT difficulty (source, value)>        <!ELEMENT vcard (#PCDATA)>
<!ELEMENT cost (source, value)>              <!ELEMENT datetime (#PCDATA)>
<!ELEMENT kind (source, value)>              <!ELEMENT metadatascheme (#PCDATA)>
<!ELEMENT purpose (source, value)>           <!ELEMENT format (#PCDATA)>
<!ELEMENT centity (vcard)>                   <!ELEMENT size (#PCDATA)>
<!ELEMENT person (vcard)>                    <!ELEMENT minimumversion (#PCDATA)>
```

```
<!ELEMENT title (#PCDATA)>              <!ELEMENT maximumversion (#PCDATA)>
<!ELEMENT entry (#PCDATA)>              <!ELEMENT id (#PCDATA)>
<!ELEMENT description (#PCDATA)>
```

LIP

```
<!ELEMENT learnerinformation (comment?, contenttype?, (identification | goal | qcl |
activity | competency | transcript | accessibility | interest | affiliation |
securitykey | relationship | ext_learnerinfo)*)>
<!ATTLIST learnerinformation
    xml:lang CDATA "en">
<!ELEMENT identification (comment?, contenttype?, (formname | name | address |
contactinfo | demographics | agent)*, ext_identification?)>
<!ELEMENT goal (typename?, comment?, contenttype?, date*, priority?, status?,
description?, goal*, ext_goal?)>
<!ELEMENT qcl (typename?, comment?, contenttype?, title?, organization?,
registrationno?, level?, date*, description?, ext_qcl?)>
<!ELEMENT activity (typename?, comment?, contenttype?, date*, status?, units?,
(learningactivityref | definition | product | testimonial | evaluation)*,
description?, activity*, ext_activity?)>
<!ELEMENT name (typename?, comment?, contenttype?, partname*)>
<!ELEMENT description (short | long | full)+>
<!ELEMENT full (comment?, media+)>
<!ELEMENT contenttype (comment?, (referential | temporal | privacy)+,
ext_contenttype?)>
<!ELEMENT referential (sourcedid | indexid | (sourcedid, indexid))>
<!ELEMENT status (typename?, date?, description?)>
<!ELEMENT partname (typename?, text?)>
<!ELEMENT date (typename?, datetime, description?, ext_date?)>
<!ELEMENT organization (typename?, description?)>
<!ELEMENT level (text, level?)>
<!ELEMENT evaluation (typename?, comment?, contenttype?, evaluationid?, date*,
evalmetadata?, objectives*, status?, noofattempts?, duration*, result*,
description?, evaluation*, ext_evaluation?)>
<!ELEMENT testimonial (typename?, comment?, contenttype?, date*, description?,
ext_testimonial?)>
<!ELEMENT definition (typename?, comment?, contenttype?, definitionfield*,
description?, definition*, ext_definition?)>
<!ELEMENT evalmetadata (typename?, evalmetadatafield+)>
<!ELEMENT objectives (comment?, (media | contentref)+, ext_objectives?)>
<!ATTLIST objectives
    view (All | Administrator | AdminAuthority | Assessor | Author | Candidate |
InvigilatorProctor | Psychometrician | Scorer |Tutor) "All">
<!ELEMENT result (comment?, ((interpretscore | score)* | result*))>
<!ELEMENT product (typename?, comment?, contenttype?, date?, description?,
ext_product?)>
<!ELEMENT formname (typename?, comment?, contenttype?, text?)>
<!ELEMENT learningactivityref (sourcedid | text)+>
<!ELEMENT relationship (typename?, comment?, contenttype?, tuple?, description?,
ext_relationship?)>
<!ELEMENT duration (fieldlabel, fielddata)>
<!ELEMENT tuple (tuplesource, tuplerelation, tupledest+)>
<!ELEMENT media (#PCDATA)>
<!ATTLIST media
    mediamode (Text | Image | Video | Audio | Applet | Application) #REQUIRED
    contentreftype (uri | entityref | Base-64) "Base-64"
    mimetype CDATA #REQUIRED>
<!ELEMENT tysource (#PCDATA)>
<!ATTLIST tysource
    sourcetype (imsdefault | list | proprietary | standard) "imsdefault"
    e-dtype NMTOKEN #FIXED "string">
```

```
<!ELEMENT tuplesource (sourcedid?, indexid)>   <!ATTLIST text
<!ELEMENT tuplerelation (typename, text?)>        uri CDATA #IMPLIED
<!ELEMENT tupledest (sourcedid?, indexid)>        xml:lang CDATA "en"
<!ELEMENT units (unitsfield+)>                    entityref ENTITY #IMPLIED
<!ELEMENT typename (tysource?, tyvalue)>          e-dtype NMTOKEN #FIXED "string">
<!ELEMENT comment (#PCDATA)>                    <!ELEMENT title (#PCDATA)>
<!ATTLIST comment                              <!ATTLIST title
   xml:lang CDATA "en"                            xml:lang CDATA "en"
   e-dtype NMTOKEN #FIXED "string">               e-dtype NMTOKEN #FIXED "string">
<!ELEMENT source (#PCDATA)>                     <!ELEMENT registrationno (#PCDATA)>
<!ATTLIST source                               <!ATTLIST registrationno
   e-dtype NMTOKEN #FIXED "string">               e-dtype NMTOKEN #FIXED "string">
<!ELEMENT id (#PCDATA)>                         <!ELEMENT fieldlabel (typename)>
<!ATTLIST id                                    <!ELEMENT sourcedid (source, id)>
   e-dtype NMTOKEN #FIXED "string">             <!ELEMENT indexid (#PCDATA)>
<!ELEMENT short (#PCDATA)>                      <!ATTLIST indexid
<!ATTLIST short                                    e-dtype NMTOKEN #FIXED "string">
   xml:lang CDATA "en"                         <!ELEMENT evaluationid (#PCDATA)>
   e-dtype NMTOKEN #FIXED "string">             <!ATTLIST evaluationid
<!ELEMENT long (#PCDATA)>                          e-dtype NMTOKEN #FIXED "ID">
<!ATTLIST long                                  <!ELEMENT noofattempts (#PCDATA)>
   xml:lang CDATA "en"                         <!ATTLIST noofattempts
   e-dtype NMTOKEN #FIXED "string">               e-dtype NMTOKEN #FIXED "int">
<!ELEMENT fielddata (#PCDATA)>                  <!ELEMENT evalmetadatafield
<!ATTLIST fielddata                            (fieldlabel, fielddata)>
   e-dtype NMTOKEN #FIXED "string">             <!ATTLIST evalmetadatafield
<!ELEMENT datetime (#PCDATA)>                      xml:lang CDATA "en">
<!ATTLIST datetime                             <!ELEMENT contentref (#PCDATA)>
   e-dtype NMTOKEN #FIXED "dateTime">           <!ATTLIST contentref
<!ELEMENT text (#PCDATA)>                          e-dtype NMTOKEN #FIXED "ID">
<!ELEMENT tyvalue (#PCDATA)>                    <!ELEMENT ext_learnerinfo (#PCDATA)>
<!ATTLIST tyvalue                              <!ELEMENT ext_contenttype (#PCDATA)>
    xml:lang CDATA "en"                        <!ELEMENT ext_activity (#PCDATA)>
    e-dtype NMTOKEN #FIXED "string">           <!ELEMENT ext_date (#PCDATA)>
<!ELEMENT interpretscore (fieldlabel,          <!ELEMENT ext_definition (#PCDATA)>
fielddata)>                                     <!ELEMENT ext_identification (#PCDATA)>
<!ELEMENT score (fieldlabel, fielddata)>        <!ELEMENT ext_objectives (#PCDATA)>
<!ELEMENT definitionfield (fieldlabel,          <!ELEMENT ext_product (#PCDATA)>
fielddata)>                                     <!ELEMENT ext_qcl (#PCDATA)>
<!ELEMENT unitsfield (fieldlabel, fielddata)>  <!ELEMENT ext_relationship (#PCDATA)>
<!ELEMENT ext_goal (#PCDATA)>                   <!ELEMENT ext_testimonial (#PCDATA)>
<!ELEMENT ext_evaluation (#PCDATA)>
```

CP

```
<!ELEMENT manifest (metadata?,                  <!ELEMENT metadata (schema?,
organizations,  resources, manifest*)>          schemaversion?)>
                                                <!ELEMENT schema (#PCDATA)>
<!ATTLIST manifest                              <!ELEMENT schemaversion (#PCDATA)>
   identifier ID #REQUIRED                      <!ELEMENT title (#PCDATA)>
   version CDATA #IMPLIED>                       <!ELEMENT resource (metadata?, file+,
<!ELEMENT organizations (organization*)>        dependency*)>
<!ATTLIST organizations                         <!ATTLIST resource
   default IDREF #IMPLIED>                       identifier ID #REQUIRED
<!ELEMENT organization (title?, item*,             type CDATA #REQUIRED
metadata?)>                                        href CDATA #IMPLIED>
<!ATTLIST organization                          <!ELEMENT resources (resource*)>
   identifier ID #REQUIRED>                     <!ELEMENT file (metadata?)>
<!ELEMENT item (title?, item*, metadata?)>      <!ATTLIST file
<!ATTLIST item                                     href CDATA #REQUIRED>
```

```
identifier ID #REQUIRED              <!ELEMENT dependency EMPTY>
isvisible CDATA #IMPLIED             <!ATTLIST dependency
parameters CDATA #IMPLIED                identifierref CDATA #IMPLIED>
identifierref CDATA #IMPLIED
a-dtype NMTOKENS "isvisible boolean">
```

Appendix B.   X-Device Object Schemata for Educational Metadata

This appendix contains the object schemata in X-DEVICE for the DTDs of the various educational objects that are used throughout this chapter (see Appendix A). Notice that, due to space limitations, only a small part of the object schema for the LIP DTD is actually shown, which is necessary for understanding the XML document of Appendix C and Appendix D.

LOM
```
xml_seq lom
   attributes
      general              (general, single, optional)
      lifecycle            (lifecycle, single, optional)
      metametadata         (metametadata, single, optional)
      technical            (technical, single, optional)
      educational          (educational, single, optional)
      rights               (rights, single, optional)
      relation             (relation, list, optional)
      annotation           (annotation, list, optional)
      classification       (classification, list, optional)
   meta_attributes
      elem_ord             [general, lifecycle, metametadata, technical, educational,
                              rights, relation, annotation, classification]
xml_seq general
   attributes
      identifier           (string, single, optional)
      title                (string, single, optional)
      catalogentry         (catalogentry, list, optional)
      language             (string, list, optional)
      description          (string, list, optional)
      keyword              (string, list, optional)
      coverage             (string, list, optional)
      structure            (structure, single, optional)
      aggregationlevel     (aggregationlevel, single, optional)
      extension            (string, single, optional)
   meta_attributes
      elem_ord             [identifier, title, catalogentry, language, description,
                              keyword, coverage, structure, aggregationlevel,
                              extension]
xml_seq lifecycle
   attributes
      version              (string, single, optional)
      status               (status, single, optional)
      contribute           (contribute, list, optional)
      extension            (string, single, optional)
   meta_attributes
      elem_ord             [version, status, contribute, extension]
xml_seq metametadata
   attributes
      identifier           (string, single, optional)
      catalogentry         (catalogentry, list, optional)
      contribute           (contribute, list, optional)
      metadatascheme       (string, list, optional)
      language             (string, list, optional)
      extension            (string, single, optional)
```

```
      meta_attributes
         elem_ord                    [identifier, catalogentry, contribute,
                                        metadatascheme, language, extension]
xml_seq technical
      attributes
         format                      (string, list, optional)
         size                        (string, single, optional)
         location                    (location, list, optional)
         requirement                 (requirement, list, optional)
         installationremarks         (string, single, optional)
         otherplatformrequirements   (string, single, optional)
         duration                    (duration, single, optional)
         extension                   (string, single, optional)
      meta_attributes
         elem_ord                    [format, size, location, requirement,
                                        installationremarks, otherplatform
                                        requirements, duration, extension]
xml_seq educational
      attributes
         interactivitytype           (interactivitytype, single, optional)
         learningresourcetype        (learningresourcetype, list, optional)
         interactivitylevel          (interactivitylevel, single, optional)
         semanticdensity             (semanticdensity, single, optional)
         intendedenduserrole         (intendedenduserrole, list, optional)
         context                     (context, list, optional)
         typicalagerange             (string, list, optional)
         difficulty                  (difficulty, single, optional)
         typicallearningtime         (typicallearningtime, single, optional)
         description                 (string, single, optional)
         language                    (string, list, optional)
         extension                   (string, single, optional)
      meta_attributes
         elem_ord                    [interactivitytype, learningresourcetype,
                                        interactivitylevel, semanticdensity,
                                        intendedenduserrole, context, typicalagerange,
                                        difficulty, typicallearningtime, description,
                                        language, extension]
xml_seq rights
      attributes
         cost                        (cost, single, optional)
         copyrightandotherrestrictions (copyrightandotherrestrictions, single,
                                        optional)
         description                 (string, single, optional)
         extension                   (string, single, optional)
      meta_attributes
         elem_ord                    [cost, copyrightandotherrestrictions, description,
                                        extension]
xml_seq relation
      attributes
         kind                        (kind, single, optional)
         resource                    (resource, single, optional)
         extension                   (string, single, optional)
      meta_attributes
         elem_ord                    [kind, resource, extension]
xml_seq annotation
      attributes
         person                      (person, single, optional)
         date                        (date, single, optional)
```

```
        description              (string, single, mandatory)
        extension                (string, single, optional)
     meta_attributes
        elem_ord                 [person, date, description, extension]
xml_seq classification
     attributes
        purpose                  (purpose, single, optional)
        taxonpath                (taxonpath, list, optional)
        description              (string, single, optional)
        keyword                  (string, list, optional)
        extension                (string, single, optional)
     meta_attributes
        elem_ord                 [purpose, taxonpath, description, keyword, extension]
xml_seq catalogentry
     attributes
        catalog                  (string, single, mandatory)
        entry                    (string, single, mandatory)
        extension                (string, single, optional)
     meta_attributes
        elem_ord                 [catalog, entry, extension]
xml_seq contribute
     attributes
        role                     (role, single, mandatory)
        centity                  (centity, list, optional)
        date                     (date, single, optional)
        extension                (string, single, optional)
     meta_attributes
        elem_ord                 [role, centity, date, extension]
xml_seq requirement
     attributes
        type                     (type, single, optional)
        name                     (name, single, optional)
        minimumversion           (string, single, optional)
        maximumversion           (string, single, optional)
        extension                (string, single, optional)
     meta_attributes
        elem_ord                 [type, name, minimumversion, maximumversion, extension]
xml_seq rights
     attributes
        cost                     (cost, single, optional)
        copyrightandotherrestrictions (copyrightandotherrestrictions, single,
                                      optional)
        description              (string, single, optional)
        extension                (string, single, optional)
     meta_attributes
        elem_ord                 [cost, copyrightandotherrestrictions, description,
                                   extension]
xml_seq resource
     attributes
        identifier               (string, single, optional)
        description              (string, single, optional)
        catalogentry             (catalogentry, list, optional)
        extension                (string, single, optional)
     meta_attributes
        elem_ord                 [identifier, description, catalogentry, extension]
xml_seq taxonpath
     attributes
        source                   (string, single, optional)
        taxon                    (taxon, single, optional)
```

```
      meta_attributes
         elem_ord                    [source, taxon]
xml_seq taxon
   attributes
         id                          (string, single, optional)
         entry                       (string, single, optional)
         taxon                       (taxon, single, optional)
   meta_attributes
         elem_ord                    [id, entry, taxon]
xml_seq langstring
   attributes
         content                     (string, single, mandatory)
         xml_lang                    (string, single, optional)
   meta_attributes
         elem_ord                    [content]
         att_lst                     [xml_lang]
xml_seq location
   attributes
         content                     (string, single, mandatory)
         type                        (string, single, optional)
   meta_attributes
         elem_ord                    [content]
         att_lst                     [type]
```

Classes structure, aggregationlevel, status, role, type, name, interactivitytype, learning-resourcetype, interactivitylevel, semanticdensity, intendededuserrole, context, difficulty, cost, copyrightandotherrestrictions, kind, purpose have the following structure:

```
xml_seq ClassName
   attributes
         source                      (source, single, mandatory)
         value                       (value, single, mandatory)
   meta_attributes
         elem_ord                    [source, value]
```

Classes date, duration, typicallearningtime have the following structure:

```
xml_seq ClassName
   attributes
         datetime                    (string, single, optional)
         description                 (description, single, optional)
   meta_attributes
         elem_ord                    [datetime, description]
```

Classes centity, person have the following structure:

```
xml_seq ClassName
   attributes
         vcard                       (string, single, mandatory)
   meta_attributes
         elem_ord                    [vcard]
```

LIP

```
xml_seq learnerinformation
   attributes
         comment                     (comment, single, optional)
         contenttype                 (contenttype, single, optional)
         learnerinformation_alt1 (learnerinformation_alt1, list, optional)
         xml_lang                    (string, single, optional)
   meta_attributes
         elem_ord                    [comment, contenttype,
                                        learnerinformation_alt1]
```

```
        att_lst             [xml_lang]
        alias               [identification-learnerinformation_alt1,
                                goal-learnerinformation_alt1,
                                qcl-learnerinformation_alt1,
                                learnerinformation_alt1,
                                activity-learnerinformation_alt1,
                                competency-learnerinformation_alt1,
                                transcript-learnerinformation_alt1,
                                accessibility-learnerinformation_alt1,
                                interest-learnerinformation_alt1,
                                affiliation-learnerinformation_alt1,
                                securitykey-learnerinformation_alt1,
                                relationship-learnerinformation_alt1,
                                ext_learnerinfo-learnerinformation_alt1]
xml_alt learnerinformation_alt1
    attributes
        identification      (identification, single, optional)
        goal                (goal, single, optional)
        qcl                 (qcl, single, optional)
        activity            (activity, single, optional)
        competency          (competency, single, optional)
        transcript          (transcript, single, optional)
        accessibility       (accessibility, single, optional)
        interest            (interest, single, optional)
        affiliation         (affiliation, single, optional)
        securitykey         (securitykey, single, optional)
        relationship        (relationship, single, optional)
        ext_learnerinfo     (string, single, optional)
xml_seq activity
    attributes
        typename            (typename, single, optional)
        comment             (comment, single, optional)
        contenttype         (contenttype, single, optional)
        date                (date, list, optional)
        status              (status, single, optional)
        units               (units, single, optional)
        activity_alt1       (activity_alt1, list, optional)
        description         (description, single, optional)
        activity            (activity, list, optional)
        ext_activity        (string, single, optional)
    meta_attributes
        elem_ord            [typename, comment, contenttype, date, status, units,
                                activity_alt1, description, activity, ext_activity]
        alias               [learningactivityref-activity_alt1,
                                definition-activity_alt1, product-activity_alt1,
                                testimonial-activity_alt1, evaluation-activity_alt1]
xml_alt activity_alt1
    attributes
        learningactivityref (learningactivityref, single, optional)
        definition          (definition, single, optional)
        product             (product, single, optional)
        testimonial         (testimonial, single, optional)
        evaluation          (evaluation, single, optional)
        att_lst             [xml_lang]
xml_seq contenttype
    attributes
        comment             (comment, single, optional)
        contenttype_alt1    (contenttype_alt1, list, mandatory)
        ext_contenttype     (string, single, optional)
```

```
      meta_attributes
         elem_ord                    [comment, contentype_alt1, ext_contentype]
         alias                       [referential-contentype_alt1,
                                        temporal-contentype_alt1, privacy-contentype_alt1]
xml_alt contentype_alt1
      attributes
         referential                 (referential, single, optional)
         temporal                    (temporal, single, optional)
         privacy                     (privacy, single, optional)
xml_seq referential
      attributes
         referential_alt1            (referential_alt1, single, mandatory)
      meta_attributes
         elem_ord                    [referential_alt1]
         alias                       [sourcedid-referential_alt1, indexid-referential_alt1,
                                        sourcedid-referential_alt1_seq1,
                                        indexid-referential_alt1_seq1]
xml_alt referential_alt1
      attributes
         sourcedid                   (sourcedid, single, optional)
         indexid                     (indexid, single, optional)
         referential_alt1_seq1       (referential_alt1_seq1, single, optional)
      meta_attributes
         alias                       [sourcedid-referential_alt1_seq1,
                                        indexid-referential_alt1_seq1]
xml_alt referential_alt1_seq1
      attributes
         sourcedid                   (sourcedid, single, mandatory)
         indexid                     (indexid, single, mandatory)
      meta_attributes
         elem_order                  [sourcedid, indexid]
xml_seq  definition
      attributes
         typename                    (typename, single, optional)
         comment                     (comment, single, optional)
         contentype                  (contentype, single, optional)
         definitionfield             (definitionfield, list, optional)
         description                 (description, single, optional)
         definition                  (definition, list, optional)
         ext_definition              (string, single, optional)
      meta_attributes
         elem_ord                    [typename, comment, contentype, definitionfield,
                                        description, definition, ext_definition]
xml_seq  sourcedid
      attributes
         source                      (source, single, mandatory)
         id                          (id, single, mandatory)
      meta_attributes
         elem_ord                    [source, id]
xml_seq  typename
      attributes
         tysource                    (tysource, single, optional)
         tyvalue                     (tyvalue, single, mandatory)
      meta_attributes
         elem_ord                    [tysource, tyvalue]
xml_seq  evalmetadatafield
      attributes
         fieldlabel                  (fieldlabel, single, mandatory)
         fielddata                   (fielddata, single, mandatory)
```

```
        xml_lang                (string, single, optional)
    meta_attributes
        elem_ord            [fieldlabel, fielddata]
xml_seq definitionfield
    attributes
        fieldlabel          (fieldlabel, single, mandatory)
        fielddata           (fielddata, single, mandatory)
    meta_attributes
        elem_ord            [fieldlabel, fielddata]
xml_seq  fieldlabel
    attributes
        typename            (typename, single, mandatory)
    meta_attributes
        elem_ord            [typename]
xml_seq tysource
    attributes
        content             (string, single, mandatory)
        sourcetype          (string, single, optional)
        e-dtype             (string, single, mandatory)
    meta_attributes
        elem_ord            [content]
        att_lst             [e-dtype, sourcetype]
xml_seq tyvalue
    attributes
        content             (string, single, mandatory)
        xml_lang            (string, single, optional)
        e-dtype             (string, single, mandatory)
    meta_attributes
        elem_ord            [content]
        att_lst             [e-dtype, xml_lang]
```

Classes `comment`, `source`, `id`, `fielddata`, `indexid` have the following structure:

```
xml_seq ClassName
    attributes
        content             (string, single, mandatory)
        e-dtype             (string, single, mandatory)
    meta_attributes
        elem_ord            [content]
        att_lst             [e-dtype]
```

CP
```
xml_seq manifest
    attributes
        metadata            (metadata, single, optional)
        organizations       (organizations, single, mandatory)
        resources           (resources, single, mandatory)
        manifest            (manifest, list, optional)
        identifier          (string, single, mandatory)
        version             (string, single, optional)
    meta_attributes
        elem_ord            [metadata, organizations, resources, manifest]
        att_lst             [identifier, version]
xml_seq metadata
    attributes
        schema              (string, single, optional)
        schemaversion       (string, single, optional)
    meta_attributes
        elem_ord            [schema, schemaversion]
xml_seq organizations
    attributes
```

```
         organization              (organization, list, optional)
         default                   (string, single, optional)
      meta_attributes
         elem_ord                  [organization]
         att_lst                   [default]
xml_seq organization
   attributes
         title                     (string, single, optional)
         item                      (item, list, optional)
         metadata                  (metadata, single, optional)
         identifier                (string, single, mandatory)
      meta_attributes
         elem_ord                  [title, item, metadata]
         att_lst                   [identifier]
xml_seq item
   attributes
         title                     (string, single, optional)
         item                      (item, list, optional)
         metadata                  (metadata, single, optional)
         identifier                (string, single, mandatory)
         isvisible                 (string, single, optional)
         parameters                (string, single, optional)
         identifierref             (string, single, optional)
         a-dtype                   (string, list, optional)
      meta_attributes
         elem_ord                  [title, item, metadata]
         att_lst                   [identifier, isvisible, parameters, identifierref,
                                      a-dtype]
xml_seq resources
   attributes
         resource                  (resource, list, optional)
      meta_attributes
         elem_ord                  [resource]
xml_seq resource
   attributes
         metadata                  (metadata, single, optional)
         file                      (file, list, mandatory)
         dependency                (dependency, list, optional)
         identifier                (string, single, mandatory)
         type                      (string, single, mandatory)
         href                      (string, single, optional)
      meta_attributes
         elem_ord                  [metadata, file, dependency]
         att_lst                   [identifier, type, href]
         empty                     [dependency]
xml_seq dependency
   attributes
         identifierref             (string, single, optional)
      meta_attributes
         elem_ord                  []
         att_lst                   [identifierref]
xml_seq file
   attributes
         metadata                  (metadata, single, optional)
         href                      (string, single, mandatory)
      meta_attributes
         elem_ord                  [metadata]
         att_lst                   [href]
```

Appendix C.   XML Document Example

This appendix presents an example of an XML document that conforms to the LIP DTD (see Appendix A).

```xml
<learnerinformation>
    <comment>An example of LIP Activity information.</comment>
    <contenttype>
        <referential>
            <sourcedid>
                <source>IMS_LIP_V1p0_Example</source>
                <id>2001</id>
            </sourcedid>
        </referential>
    </contenttype>
    <activity>
        <typename>
            <tysource sourcetype="imsdefault"/>
            <tyvalue>Education</tyvalue>
        </typename>
        <contenttype>
            <referential>
                <indexid>activity_1</indexid>
            </referential>
        </contenttype>
        <definition>
            <typename>
                <tysource sourcetype="imsdefault"/>
                <tyvalue>Course</tyvalue>
            </typename>
            <contenttype>
                <referential>
                    <indexid>DegreeCourse</indexid>
                </referential>
            </contenttype>
            <definition>
                <typename>
                    <tysource sourcetype="imsdefault"/>
                    <tyvalue>Curriculum</tyvalue>
                </typename>
                <contenttype>
                    <referential>
                        <indexid>Year1</indexid>
                    </referential>
                </contenttype>
                <definition>
                    <typename>
                        <tysource sourcetype="imsdefault"/>
                        <tyvalue>Module</tyvalue>
                    </typename>
                    <contenttype>
                        <referential>
                            <indexid>Electronics_101</indexid>
                        </referential>
                    </contenttype>
                    <definitionfield>
                        <fieldlabel>
                            <typename>
                                <tyvalue>Lecture1</tyvalue>
                            </typename>
                        </fieldlabel>
```

```
                        <fielddata>BooleanLogic</fielddata>
                     </definitionfield>
                     <definitionfield>
                        <fieldlabel>
                           <typename>
                              <tyvalue>Lecture2</tyvalue>
                           </typename>
                        </fieldlabel>
                        <fielddata>Transistors</fielddata>
                     </definitionfield>
                  </definition>
                  <definition>
                     <typename>
                        <tysource sourcetype="imsdefault"/>
                        <tyvalue>Module</tyvalue>
                     </typename>
                     <contenttype>
                        <referential>
                           <indexid>Maths_101</indexid>
                        </referential>
                     </contenttype>
                     <definitionfield>
                        <fieldlabel>
                           <typename>
                              <tyvalue>Lecture1</tyvalue>
                           </typename>
                        </fieldlabel>
                        <fielddata>BooleanLogic1</fielddata>
                     </definitionfield>
                     <definitionfield>
                        <fieldlabel>
                           <typename>
                              <tyvalue>Lecture2</tyvalue>
                           </typename>
                        </fieldlabel>
                        <fielddata>BooleanLogic2</fielddata>
                     </definitionfield>
                  </definition>
               </definition>
            </definition>
         </activity>
</learnerinformation>
```

Appendix D.   Object-Oriented Representation of an XML Document

This appendix presents the X-DEVICE object-oriented representation for the XML document of Appendix C.

```
object      0#source                        attributes
  instance  source                            content       ''
  attributes                                  e-dtype       'string'
    content         'IMS_LIP_V1p0_Example'    sourcetype    'imsdefault'
    e-dtype         'string'                object        3#tyvalue
object      1#id                              instance    tyvalue
  instance  id                                attributes
  attributes                                    content       'Education'
      content       '2001'                      e-dtype       'string'
      e-dtype       'string'                    xml_lang      'en'
object      2#tysource                      object        4#indexid
  instance  tysource                          instance    indexid
```

```
   attributes                              object       15#indexid
      content      'activity_1'               instance     indexid
      e-dtype      'string'                   attributes
object       5#tyvalue                          content      'Maths_101'
   instance     tyvalue                         e-dtype      'string'
   attributes                              object       16#fielddata
      content      'Course'                   instance     fielddata
      e-dtype      'string'                   attributes
      xml_lang     'en'                          content      'BooleanLogic1'
object       6#indexid                          e-dtype      'string'
   instance     indexid                 object       17#fielddata
   attributes                              instance     fielddata
      content      'DegreeCourse'             attributes
      e-dtype      'string'                      content      'BooleanLogic2'
object       7#tyvalue                          e-dtype      'string'
   instance     tyvalue                 object       18#sourcedid
   attributes                              instance     sourcedid
      content      'Curriculum'               attributes
      e-dtype      'string'                      source   0#source
      xml_lang     'en'                          id       1#id
object       8#indexid                   object       19#typename
   instance     indexid                    instance     typename
   attributes                              attributes
      content      'Year1'                       tysource    2#tysource
      e-dtype      'string'                      tyvalue     3#tyvalue
object       9#tyvalue                   object       20#referential
   instance     tyvalue                    instance     referential
   attributes                              attributes
      content      'Module'                   referential_alt1 21#referential_alt1
      e-dtype      'string'             object       21#referential_alt1
      xml_lang     'en'                    instance     referential_alt1
object       10#indexid                   attributes
   instance     indexid                       indexid     4#indexid
   attributes                                  sourcedid   Ø
      content      'Electronics_101'           referential_alt1_seq1    Ø
      e-dtype      'string'             object       22#typename
object       11#tyvalue                   instance     typename
   instance     tyvalue                    attributes
   attributes                                  tysource    2#tysource
      content      'Lecture1'                  tyvalue     5#tyvalue
      e-dtype      'string'             object       23#referential
      xml_lang     'en'                    instance     referential
object       12#fielddata                 attributes
   instance     fielddata                    referential_alt1
   attributes                                              24#referential_alt1
      content      'BooleanLogic'       object       24#referential_alt1
      e-dtype      'string'                instance     referential_alt1
object       13#tyvalue                   attributes
   instance     tyvalue                       indexid     6#indexid
   attributes                                  sourcedid   Ø
      content      'Lecture2'                  referential_alt1_seq1    Ø
      e-dtype      'string'             object       25#typename
      xml_lang     'en'                    instance     typename
object       14#fielddata                 attributes
   instance     fielddata                       tysource     2#tysource
   attributes                                   tyvalue      7#tyvalue
      content      'Transistors'        object       26#referential
      e-dtype      'string'                instance     referential
```

```
    attributes
        referential_alt1 27#referential_alt1
object          27#referential_alt1
    instance    referential_alt1
    attributes
        indexid     8#indexid
        sourcedid   Ø
        referential_alt1_seq1   Ø
object          28#typename
    instance    typename
    attributes
        tysource    2#tysource
        tyvalue     9#tyvalue
object          29#referential
    instance    referential
    attributes
        referential_alt1  30#referential_alt1
object          30#referential_alt1
    instance    referential_alt1
    attributes
        indexid     10#indexid
        sourcedid   Ø
        referential_alt1_seq1   Ø
object          31#typename
    instance    typename
    attributes
        tysource    Ø
        tyvalue     11#tyvalue
object          32#typename
    instance    typename
    attributes
        tysource    Ø
        tyvalue     13#tyvalue
object          33#referential
    instance    referential
    attributes
        referential_alt1  34#referential_alt1
object          34#referential_alt1
    instance    referential_alt1
    attributes
        indexid     15#indexid
        sourcedid   Ø
        referential_alt1_seq1   Ø
object          35#referential
    instance    referential
    attributes
        referential_alt1
                    36#referential_alt1
object          36#referential_alt1
    instance    referential_alt1
    attributes
        indexid     Ø
        sourcedid   18#sourcedid
        referential_alt1_seq1   Ø
```

```
object          37#contentype
    instance    contentype
    attributes
        comment         Ø
        contentype_alt1
                    [38#contentype_alt1]
        ext_contentype  Ø
object          38#contentype_alt1
    instance    contentype_alt1
    attributes
        referential     20#referential
        temporal        Ø
        privacy         Ø
object          39#contentype
    instance    contentype
    attributes
        comment         Ø
        contentype_alt1
                    [40#contentype_alt1]
        ext_contentype  Ø
object          40#contentype_alt1
    instance    contentype_alt1
    attributes
        referential     23#referential
        temporal        Ø
        privacy         Ø
object          41#contgentype
    instance    contentype
    attributes
        comment         Ø
        contentype_alt1
                    [42#contentype_alt1]
        ext_contentype  Ø
object          42#contentype_alt1
    instance    contentype_alt1
    attributes
        referential     26#referential
        temporal        Ø
        privacy         Ø
object          43#contentype
    instance    contentype
    attributes
        comment         Ø
        contentype_alt1
                    [44#contentype_alt1]
        ext_contentype  Ø
object          44#contentype_alt1
    instance    contentype_alt1
    attributes
        referential     29#referential
        temporal        Ø
        privacy         Ø
object          45#fieldlabel
    instance    fieldlabel
```

```
       attributes
          typename   31#typename
object             46#fieldlabel
   instance      fieldlabel
   attributes
       typename    32#typename
object             47#contentype
   instance      contentype
   attributes
       comment              ∅
       contentype_alt1
                 [48#contentype_alt1]
       ext_contentype       ∅
object             48#contentype_alt1
   instance      contentype_alt1
   attributes
       referential      33#referential
       temporal         ∅
       privacy          ∅
object             49#contentype
   instance      contentype
   attributes
       comment              ∅
       contentype_alt1
                 [50#contentype_alt1]
       ext_contentype       ∅
object             50#contentype_alt1
   instance      contentype_alt1
   attributes
       referential      35#referential
       temporal         ∅
       privacy          ∅
object             51#definitionfield
   instance      definitionfield
   attributes
       fieldlabel    45#fieldlabel
       fielddata     12#fielddata
object             52#definitionfield
   instance      definitionfield
   attributes
       fieldlabel    46#fieldlabel
       fielddata     14#fielddata
object             53#definitionfield
   instance      definitionfield
   attributes
       fieldlabel    45#fieldlabel
       fielddata     16#fielddata
object             54#definitionfield
   instance      definitionfield
   attributes
       fieldlabel    46#fieldlabel
       fielddata     17#fielddata
object             55#definition
   instance      definition
   attributes
       typename      28#typename
       comment          ∅
       contentype    43#contentype
       definitionfield

              [51#definitionfield,
52#definitionfield]
       description          ∅
       definition           []
       ext_definition       ∅
object             56#definition
   instance      definition
   attributes
       typename        28#typename
       comment          ∅
       contentype    47#contentype
       definitionfield
                 [53#definitionfield,
54#definitionfield]
       description          ∅
       definition           []
       ext_definition       ∅
object             57#definition
   instance      definition
   attributes
       typename        25#typename
       comment          ∅
       contentype    41#contentype
       definitionfield  []
       description          ∅
       definition    [55#definition,

56#definition]
       ext_definition       ∅
object             58#definition
   instance      definition
   attributes
       typename        22#typename
       comment          ∅
       contentype    39#contentype
       definitionfield  []
       description          ∅
       definition    [57#definition]

       ext_definition  ∅
object             59#activity
   instance      activity
   attributes
       typename      19#typename
       comment          ∅
       contentype    37#contentype
       date             []
       status           ∅
       units            ∅
       activity_alt1  [60#activity_alt1]

       description          ∅
       activity         []
       ext_activity     ∅
object             60#activity_alt1
   instance      activity_alt1
   attributes
       learningactivityref ∅
       definition          58#definition
```

```
    product         ∅              object        63# learnerinformation_alt1
    testimonial     ∅                instance    learnerinformation_alt1
    evaluation      ∅                attributes
object          61#comment             identification    ∅
  instance      comment                goal              ∅
  attributes                           qcl               ∅
    content       'An example of LIP   activity          59#activity
Activity information.'                 competency        ∅
    e-dtype       'string'             transcript        ∅
    xml_lang      'en'                 accessibility     ∅
object          62#learnerinformation  interest         ∅
  instance      learnerinformation     affiliation       ∅
  attributes                           securitykey       ∅
    comment            61#comment       relationship     ∅
    contenttype        49#contenttype   ext_learnerinfo  ∅
    learnerinformation_alt1    []
    xml_lang           'en'
```

Appendix E.   X-Device Rule Language Syntax

This appendix contains the syntax of the X-DEVICE deductive rule language in BNF notation.

```
<rule> ::= if <condition> then <consequence>
<condition> ::= <inter-object-pattern>
<consequence> ::= {<action> | <conclusion> | <derived_attribute_template>}
<inter-object-pattern> ::= <condition-element> ['and' <inter-object-pattern>]
<inter-object-pattern> ::= <inter-object-pattern> 'and' <prolog_cond>
<condition-element> ::= ['not'] <intra-object-pattern>
<intra-object-pattern> ::= [<inst_expr>'@']<class_expr>['('<attr-patterns>')']
<attr-patterns> ::= <attr-pattern>[','<attr-patterns>]
<attr-pattern> ::= <attr-expr>['.'<path_expr>] {':'<variable> | <predicates>
                                              | ':'<variable> <predicates>
                                              | <list-operator> <variable>}
<path_expr> ::= <nt-attr-expr> ['.'<path_expr>]
<attr-expr> ::= {<nt-attr-expr>|<t-attr>|<normal-attr>'↑'}
<nt-attr-expr> ::= <nt-attr>[{'*'|<integer>}]
<nt-attr-expr> ::= {'*'|'+'}
<nt-attr> ::= {<normal-attr>|<system-attr>}
<t-attr> ::= {<xml-attr>|<empty-attr>}
<normal-attr> ::= <attr>
<system-attr> ::='!'<attr>
<xml-attr> ::= '^'<attr>
<empty-attr> ::= '∅' <attr>
<attr> ::= {<attribute>|<variable>}
<predicates> ::= <rel-operator> <value> [{'&' | ';'} <predicates>]
<predicates> ::= <set-operator> <set>
<rel-operator> ::= '=' | '>' | '>=' | '=<' | '<' | '\=' | '$' | <date-operator>
<date-operator> ::= '$'{'y'|'m'|'d'}
<set-operator> ::='⊂'|'⊄'|'⊆'|'∈'|'∉'|'⊃'|'\⊃'|'⊇'
<list-operator> ::='∋'|'\∋'
<list-operator> ::='∋'<order_expr>
<order_expr> ::= {<abs_order>|<rel_order>|'{'<rel_order>','<abs_order>'}'}
<abs_order> ::= <rel-operator><integer> | <integer>
<rel_order> ::= { 'before' | 'after' }'('<variable>')'
<rel_order> ::= 'between' '('<variable>','<variable>')'
<value> ::= <constant> | <variable> | <arith_expr>
<set> ::= '['<constants>']'
<prolog_cond> ::= 'prolog' '{'<prolog_goal>'}'
```

```
<action> ::= <prolog_goal>
<conclusion> ::= <derived_class_template>
<conclusion> ::= {'xml_result' | 'shallow_result'}'('<elem_expr>')'
<conclusion> ::= {'xml_sorted' | 'shallow_sorted'}'('[<group_list>
                                    ['-'<order_list>]',']<elem_expr>')'
<elem_expr> ::= <derived_class_template>
<elem_expr> ::= <derived_class>'('<derived_class_template>')'
<derived_class_template> ::= <derived_class>'('<templ-patterns>')'
<derived_attribute_template> ::= <variable>'@'{<class>}'('<templ-patterns>')'
<templ-patterns> ::= <templ-pattern> [',' <templ-pattern>]
<templ-pattern> ::= {<normal-attr>|<system-attr>|<xml-attr>}':'{<value> |
<aggregate_expr>}
<templ-pattern> ::= <empty-attr>
<aggregate_expr> ::= <aggregate_function>'('<variable>')'}
<aggregate_expr> ::= 'ord_list('<variable>['-'<group_list>['-'<order_list>]]')'
<aggregate_function> ::= 'count'|'sum'|'avg'|'max'|'min'|'list'|'string'
<group_list> ::= '['<variable>[','<variable>]']'
<order_list> ::= '['<ord_symbol>[','<ord_symbol>]']'
<ord_symbol> ::= {'<' | '>'}
<inst_expr> ::= {<variable>|<class>}
<class_expr> ::= {<variable>|<class>}
<class_expr> ::= <inst_expr>'/'<class>
```

`<class>` ::= *an existing class or meta-class of the OODB schema*
`<derived_class>` ::= *an existing derived class or a non-existing base class of the OODB schema*
`<attribute>` ::= *an existing attribute of the corresponding OODB class*
`<prolog_goal>` ::= *an arbitrary Prolog/ADAM goal*
`<constants>` ::= `<constant>[','<constants>]`
`<constant>` ::= *a valid constant of an OODB simple attribute type*
`<variable>` ::= *a valid Prolog variable*
`<arith_expr>` ::= *a valid Prolog arithmetic expression*

Index